SOIL ORGANIC MATTER AND BIOLOGICAL ACTIVITY

Developments in Plant and Soil Sciences
Volume 16

1. J. Monteith and C. Webb, eds.,
   Soil Water and Nitrogen in Mediterranean-type Environments.
   1981. ISBN 90-247-2406-6
2. J.C. Brogan, ed.,
   Nitrogen Losses and Surface Run-off from Landspreading of Manures.
   1981. ISBN 90-247-2471-6
3. J.D. Bewley, ed.,
   Nitrogen and Carbon Metabolism.
   1981. ISBN 90-247-2472-4
4. R. Brouwer, I. Gašparíková, J. Kolek and B.C. Loughman, eds.,
   Structure and Function of Plant Roots.
   1981. ISBN 90-247-2510-0
5. Y.R. Dommergues and H.G. Diem, eds.,
   Microbiology of Tropical Soils and Plant Productivity.
   1982. ISBN 90-247-2624-7
6. G.P. Robertson, R. Herrera and T. Rosswall, eds.,
   Nitrogen Cycling in Ecosystems of Latin America and the Caribbean.
   1982. ISBN 90-247-2719-7
7. D. Atkinson et al., eds.,
   Tree Root Systems and their Mycorrhizas.
   1983. ISBN 90-247-2821-5
8. M.R. Sarić and B.C. Loughman, eds.,
   Genetic Aspects of Plant Nutrition.
   1983. ISBN 90-247-2822-3
9. J.R. Freney and J.R. Simpson, eds.,
   Gaseous Loss of Nitrogen from Plant-Soil Systems.
   1983. ISBN 90-247-2820-7
10. United Nations Economic Commission for Europe.
    Efficient Use of Fertilizers in Agriculture.
    1983. ISBN 90-247-2866-5
11. J. Tinsley and J.F. Darbyshire, eds.,
    Biological Processes and Soil Fertility.
    1984. ISBN 90-247-2902-5
12. A.D.L. Akkermans, D. Baker, D. Huss-Danell and J.D. Tjepkema, eds.,
    Frankia Symbioses.
    1984. ISBN 90-247-2967-X
13. W.S. Silver and E.C. Schröder, eds.,
    Practical Application of Azolla for Rice Production.
    1984. ISBN 90-247-3068-6
14. P.G.L. Vlek, ed.,
    Micronutrients in Tropical Food Crop Production.
    1985. ISBN 90-247-3085-6
15. T.P. Hignett, ed.,
    Fertilizer Manual.
    1985. ISBN 90-247-3122-4
16. D. Vaughan and R.E. Malcolm, eds.,
    Soil Organic Matter and Biological Activity.
    1985. ISBN 90-247-3154-2

Soil Organic Matter and Biological Activity

Edited by

D. VAUGHAN and R.E. MALCOLM
The Macaulay Institute for Soil Research
Aberdeen
Scotland

1985 **SPRINGER-SCIENCE+BUSINESS MEDIA, B.V.**

Distributors

for the United States and Canada: Kluwer Academic Publishers, 190 Old Derby Street, Hingham, MA 02043, USA
for the UK and Ireland: Kluwer Academic Publishers, MTP Press Limited, Falcon House, Queen Square, Lancaster LA1 1RN, UK
for all other countries: Kluwer Academic Publishers Group, Distribution Center, P.O. Box 322, 3300 AH Dordrecht, The Netherlands

Library of Congress Cataloging in Publication Data

```
Soil organic matter and biological activity.

    (Developments in plant and soil sciences ; v. 16)
    Includes index.
    1. Humus.   2. Plant-soil relationships.   3. Plant
psysiology.   I. Vaughan, D.   II. Malcolm, R. E.
III. Series.
S592.8.S66   1985      631.4'17           85-5050
```

ISBN 978-94-010-8757-5   ISBN 978-94-009-5105-1 (eBook)
DOI 10.1007/ 978-94-009-5105-1

Copyright

© 1985 by Springer Science+Business Media Dordrecht
Originallly published by Martinus Nijhoff/ Dr. W. Junk Publishers, Dordrecht in 1985
Softcover reprint of the hardcover 1st edirion 1985

LIST OF CONTRIBUTORS

H.A. ANDERSON,
   The Macaulay Institute for Soil Research, Aberdeen, Scotland.

M.V. CHESHIRE,
   The Macaulay Institute for Soil Research, Aberdeen, Scotland.

R.D. HARTLEY,
   The Grassland Research Institute, Maidenhead, Berkshire, England.

J.N. LADD,
   C.S.I.R.O. Soils Division, Glen Osmond, South Australia.

D.J. LINEHAN,
   The Macaulay Institute for Soil Research, Aberdeen, Scotland.

J.M. LYNCH,
   Glasshouse Crops Research Institute, Littlehampton, West Sussex, England.

R.E. MALCOLM,
   The Macaulay Institute for Soil Research, Aberdeen, Scotland.

B.G. ORD,
   The Macaulay Institute for Soil Research, Aberdeen, Scotland.

J.W. PARSONS,
   Department of Soil Science, University of Aberdeen, Scotland.

N.M. SCOTT,
   The Macaulay Institute for Soil Research, Aberdeen, Scotland.

G.P. SPARLING,
   D.S.I.R. Lower Hutt, New Zealand. Formerly at The Macaulay Institute for
   Soil Research, Aberdeen, Scotland.

K.R. TATE,
D.S.I.R. Lower Hutt, New Zealand.

D. VAUGHAN,
The Macaulay Institute for Soil Research, Aberdeen, Scotland.

D.C. WHITEHEAD,
The Grassland Research Institute, Maidenhead, Berkshire, England.

FOREWORD

It has long been recognized that soil organic matter is the key to soil fertility. As a nutrient store it gradually provides essential elements which the soil cannot retain for long in inorganic form. It buffers growing plants against sudden changes in their chemical environment and preserves moisture in times of drought. It keeps the soil in a friable, easily penetrated physical condition, well-aerated and free draining, providing young seedlings with an excellent medium for growth. But it has another property, the nature and extent of which have been the subject of argument and controversy ever since scientists began to study the soil, and that is its ability to affect growth directly, other than by providing nutrient elements. Anyone wishing to learn about these effects has been faced with a daunting mass of literature, some confusing, often contradictory, and spread through a multitude of journals. Individual aspects have been covered from time to time in reviews but there has obviously been a need for a modern authoritative text book dealing with the many facets of this subject, so the publication of this volume is timely. The editors and authors are all specialists in their fields, fully familiar with the complex nature of soil organic matter and with the particular difficulties arising in any study of its properties. Where controversies exist they have presented all sides of the argument and have highlighted areas where further work is badly needed. Their treatise should be invaluable to all with an interest in the soil.

G. ANDERSON,
Deputy Director,
The Macaulay Institute for Soil Research, Aberdeen, Scotland.

November, 1984.

ACKNOWLEDGEMENTS

We acknowledge gratefully the help and encouragement that we have received from our colleagues at the Macaulay Institute. We are also greatly indebted to several colleagues further afield who have allowed us to reproduce their published diagrams.

We would also like to thank all the typists, particularly Mrs I. Shand and Mrs C. Smollet and the photographers Mr J. Mitchell and Mr D.J. Riley without whose help the production of this book would have been impossible.

We are grateful to the following organisations for giving permission to reproduce published material:

Elsevier Science Publishers, Amsterdam for Fig. 3, page 18; Fig. 6, page 347; Fig. 3, page 386:

Springer-Verlag, Berlin for Fig. 1, page 331; Fig. 5, page 343:

Marcel Dekker Inc., New York for Fig. 2, page 336:

Pudoc, Wageningen for Fig. 3, page 313:

Cambridge University Press, Cambridge for Fig. 4, page 389:

The British Sulphur Corporation Ltd., London for Fig. 1, page 381:

The Agricultural Institute of Canada for Fig. 3, page 355; Fig. 7, page 355:

The University of New England, Australia for Table 2, page 344.

D. VAUGHAN and R.E. MALCOLM,
Aberdeen U.K.

December, 1984.

LIST OF CONTENTS

INTRODUCTION

SOIL ORGANIC MATTER – A PERSPECTIVE ON ITS NATURE, EXTRAC-
TION, TURNOVER AND ROLE IN SOIL FERTILITY

D. Vaughan and B.G. Ord

Contents (1)

CHAPTER 1

INFLUENCE OF HUMIC SUBSTANCES ON GROWTH AND PHYSIOLOGICAL
PROCESSES

D. Vaughan and R.E. Malcolm

Contents (37)

CHAPTER 2

INFLUENCE OF HUMIC SUBSTANCES ON BIOCHEMICAL PROCESSES IN PLANTS

D. Vaughan, R.E. Malcolm and B.G. Ord

Contents (77)

CHAPTER 3

PHENOLIC ACIDS IN SOILS AND THEIR INFLUENCE ON PLANT GROWTH AND SOIL MICROBIAL PROCESSES

R.D. Hartley and D.C. Whitehead

Contents (109)

CHAPTER 4

ORIGIN, NATURE AND BIOLOGICAL ACTIVITY OF ALIPHATIC SUBSTANCES AND GROWTH HORMONES FOUND IN SOIL

J.M. Lynch

Contents (151)

CHAPTER 5

SOIL ENZYMES

J.N. Ladd

Contents (175)

CHAPTER 6

THE SOIL BIOMASS

G.P. Sparling

Contents (223)

CHAPTER 7

CARBOHYDRATES IN RELATION TO SOIL FERTILITY

M.V. Cheshire

Contents (263)

CHAPTER 8

SOIL NITROGEN: ITS EXTRACTION, DISTRIBUTION AND DYNAMICS

H.A. Anderson and D. Vaughan

Contents (289)

CHAPTER 9

SOIL PHOSPHORUS

K.R. Tate

Contents (329)

CHAPTER 10

SULPHUR IN SOILS AND PLANTS

N.M. Scott

Contents (379)

CHAPTER 11

ORGANIC MATTER AND TRACE ELEMENTS IN SOILS

D.J. Linehan

Contents (403)

CHAPTER 12

ORGANIC FARMING

J.W. Parsons

Contents (423)

INTRODUCTION

SOIL ORGANIC MATTER - A PERSPECTIVE ON ITS NATURE, EXTRACTION, TURNOVER AND ROLE IN SOIL FERTILITY.

D. VAUGHAN and B.G. ORD

The Macaulay Institute for Soil Research, Aberdeen, Scotland.

CONTENTS

INTRODUCTION

"No man is an island, entire of itself; everyman is a piece of the
continent, a part of the main" - John Donne (1571-1631).

   Soil organic matter is one of our most important natural resources
and from antiquity man has recognized that soil fertility may be
maintained or improved by adding organic manures[4]. It has been
calculated that the mass of organic carbon in soil organic matter
(30.1×10^{14}kg) exceeds the total of all other surface carbon reservoirs
(20.8×10^{14}kg) found in the world[13]. Indeed, soil fertility is
dependent on its organic matter content[92] which may vary from less than
1 per cent in young soils and those exhausted by over intensive
cropping, to as much as 95 per cent in some deep peats[120]. Hence it is
hardly surprising that this ubiquitous soil constituent has received
considerable attention from many scientists of different disciplines.
   As with many other areas of research, the progress made in one
discipline may be limited by the current techniques, knowledge, or even
dogma, of another. Thus in the era before the concept of photosynthesis
it was maintained that humus provided the only carbon source for plants,
and carbon was supposedly taken up by plant roots in an elaborated form.
Unfortunately this view continued well after it had been established
beyond all reasonable doubt that plants could fix atmospheric carbon
dioxide[11,12,106]. But misconceptions, although not so great, continue
to this day. This is exemplified in the concept of "organic farming".
The proponents of this concept have quite correctly realized the fact

that soil organic matter plays a vital role in soil fertility. Unfortunately this concept has been taken to the extreme inasmuch as the use of "chemical fertilizers" is considered to be detrimental[120]. This is reminiscent of the now settled arguments of the "Humus" versus "Mineral" theories of the last century (see chapter 1). The fact is that although soil organic matter is essential for the maintenance of soil fertility, crop yields can only be improved dramatically to the levels now required to feed an increasing world population by adding chemical fertilizers (see chapter 12). On a world-wide basis we cannot depend solely on the use of organic fertilizers because insufficient are available for the necessary crop production.

Today, our understanding of the chemistry of soil organic matter, and the manner in which it influences soil fertility, is far from complete. We do, however, know that is contains some 60 per cent of humic substances[45], the importance of which have been recognized since the last century. Fortunately, we are now spared most of the antiquated terminology which persisted until comparatively recent times.

Any student, or research worker, entering the complicated subject of soil organic matter for the first time will be confronted with a vast and often confusing literature. Much of the available data compares like with unlike, or omits adequate controls, so that it is often impossible to make direct comparisons between the results from different laboratories. All too often the results of laboratory experiments are extrapolated incorrectly to field situations. It would be too contentious to site examples, for most of us are not entirely blameless. The aim of this book is to present soil organic matter from the viewpoint of the biologist, a viewpoint which is often sadly neglected, particularly in studies related to humic substances[92]. However, because of the complex nature of the subject, the inclusion of some chemical data and concepts are not only inevitable but highly desirable. The considerable literature available necessitates that only a small, but hopefully representative sample, of references are included in this book. The diversity of the subject leads inevitably to different approaches to research being taken by researchers from different disciplines. This diversity is also reflected in the manner in which the authors have approached the writing of their individual chapters.

This introductory chapter is designed to place into perspective those aspects of soil organic matter which are individually dealt with more thoroughly in later chapters. By necessity, however, it also considers briefly the preparation, nomenclature and nature of humic substances which are referred to frequently throughout the book. For more detailed information on the formation, stability and chemistry of these humic materials, the reviews of Kononova[68], Persson[100], Schnitzer and Khan[112], Bonneau and Souchier[14], and Stevenson[127] are recommended.

1. SOIL ORGANIC MATTER

1.1. Definition of terms

Strictly speaking the term soil organic matter should encompass all the organic components present in the soil. Such a notion presents obvious difficulties when considering living organisms, particularly the macrofauna or roots of higher plants. The difficulty in producing an adequate definition for soil organic matter to satisfy all the authors contributing to this book, let alone all scientists, stems from the heterogeneous nature of organic matter found in soil, its origin and state of decomposition. In fact as McKibbin[87] emphasized over 50 years ago "in an average soil there will be found organic matter existing in an almost infinite number of stages of decomposition". Because soil organic matter is composed of broadly three fractions[92]: (a) the living microflora and fauna that constitute the biomass, (b) fresh debris and readily decomposable material and (c) the humic substances which are fairly resistant to biological attack, most definitions of it are a somewhat arbitrary and subjective compromise.

Stevenson[127] defined soil organic matter as the total of the organic components in soil exclusive of undecayed plant and animal tissues, their "partial decomposition" products (the organic residues) and the soil biomass (living microbial tissue). He used the terms soil organic matter and humus interchangeably as suggested by Waksman[142]. These definitions are perfectly acceptable, but because of difficulties encountered in removing plant remains, particularly fine roots, from the soil prior to analysis we have preferred to use the term soil organic matter in a more practical sense to encompass all the organic components in the soil including the biomass but excluding the macrofauna and macroflora. Like Stevenson, we use the terms soil organic matter and humus interchangeably.

Humus is sometimes classified into two broad types which are characteristic of the environment in which they are formed. Mull humus is characteristic of Cambisols and Rendzinas while mor humus is produced where acidity inhibits humification such as in Podsols. It is now generally agreed that humus can also be conveniently divided into two groups designated as non-humic and humic substances[35,68,108,127]. The non-humic group comprise organic substances which are well defined chemically, are usually uncoloured, and which are not unique to the soil. Most of these substances are simple, low molecular weight compounds which can be utilized as substrates by soil microbes and as such may have a transient existence in the soil. Among these substances may be found carbohydrates, hydrocarbons, alcohols, auxins, aldehydes, resins, amino acids and aliphatic and aromatic acids. Gases such as ethylene and hydrogen sulphide may also be present. In contrast, the humic substances comprise yellow to black coloured, relatively high molecular weight polyphenolic components which are distinctive to, and synthesized in, the soil.

Humic substances can be very stable and may persist in agricultural soils with a mean residence time (MRT) of over a thousand years[22,23,59,107]. There is no certainty,however,that they remain unaltered during chemical extraction, but the method of extraction is used to distinguish three classes of humic materials:- (a) Fulvic acid which is soluble in alkalis and acids, (b) Humic acid which is soluble in alkalis but insoluble in acids and (c) Humin which is insoluble in both alkalis and acids. Alkaline extractions followed by acidification do not of course result in a clear-cut fractionation of humic substances because the solubility of a given fraction can vary with experimental conditions. In addition each humic fraction will contain non-humic compounds having the same solubility because the extraction procedures designed for humic substances will also release non-humic material. The problem then arises as to when an apparently non-humic substance is an integral part of the humic substance or merely an "impurity". In this context Nannipieri and Sequi[92] have emphasized that fresh organic carbon added to soil appears in the humic fraction before humification could possibly have occurred.

To understand some of the difficulties encountered in applying
precise definitions to the organic constituents of the soil it is
essential to consider the manner in which the extraction procedures
have evolved. The word evolved instead of planned is used quite
deliberately because in many ways biologists are still limited in their
thinking by the traditional extraction procedures[92,100].

1.2. Early historical perspectives

The first extraction of humic substances is usually accredited to
Achard (1786)[1]. He extracted peat with an alkaline solution which on
acidification yielded a dark, amorphous precipitate which subsequently
became known as humic acid. Eleven years later, Vauqueline[141] using
alkaline solutions extracted humus-like substances from elm wood
infected with fungi. On acidification these extracts produced a
precipitate which was called ulmin by Thomson[135]. In 1822,
Döbereiner[28] names the dark coloured component of the soil organic
matter humussäure or humus acid, although in 1837, Wiegmann[145]
referred to a similar fraction of peat as humic acid or ulmin. At
this time the terms humus acid, humic acid and ulmin were used
synonymously thus introducing confusion especially between the two
former terms. Indeed it was not until 1936 that Waksman[142] emphasized
that humic acids referred specifically to the precipitate formed after
acidifying alkali extractions of soils, whereas humus acid comprised
the whole alkali soluble material.

During the first part of the nineteenth century considerable
attention was paid to the chemistry of humic substances. A detailed
description of the analysis of humic acid with particular reference
to its acidic properties was given by Sprengel[125,126]. One of his
innovations still in vogue today was the pretreatment of soil with
dilute mineral acid prior to its extraction with alkali. Sprengel also
observed that after removal of all the mineral acid, freshly
precipitated humic acid partially dissolved in cold water, but after
drying, the humic material was converted into a less water-soluble
form for which the term humus coal was proposed.

The first serious attempt at classifying humic substances was that
of Berzelius[10,11,12], who suggested two main groups depending on
solubility in alkaline solutions. Thus humic acid was soluble whereas

mull <u>coal</u>, later called <u>humin</u>, was insoluble. Later, Berzelius added
a third group consisting of <u>crenic</u> and <u>apocrenic</u> acids which were
obtained from mineral waters and slimy mud rich in iron ochre. The
classification by Berzelius was extended by his pupil Mulder[88,89,90,91]
who took into consideration colour in addition to solubility in water or
alkali. This classification may be represented as: (a) brown bodies
comprising either alkali-soluble ulmic acid or alkaline-insoluble ulmin,
(b) black bodies comprising alkaline-soluble humic acid or alkaline-
insoluble humin and (c) water soluble crenic and apocrenic acids.
Although Mulder's contribution was important his adherence to the view
that humic substances were chemically discrete molecules was unhelpful
to the further development of the subject.

But the ultimate in unhelpfulness must surely go to the
classification devised by Hermann[51,52], He, in common with his
contemporaries, divided humic substances into three main groups
depending on solubility but each group was then sub-divided to give
an array of bewildering terms (Table 1). Let us consider how Hermann
arrived at the classification of humins which he based on nitrogen
content. Anitrohumin was obtained artificially from sugar and hence
was nitrogen-free whereas nitrohumin was produced from the soil and so
contained nitrogen. On the other hand, nitrolin was obtained from wood
and contained as much as 12% nitrogen. At least Hermann considered
that humic substances could contain nitrogen, a view which was not
universally accepted at the time. However he persisted in giving
specific formulae for most of the substances in his classification.

Table 1. Hermann's classification of humic substances.

Name of fraction	Solubility status
Humic acids - anitro-humic acid, sugar-humic acid, ligno-humic acid, meta ligno-humic acid. Apocrenic acids - torfic acid, apotorfic acid, arvic acid, apoarvic acid, porla apocrenic acid	Soluble in alkali and insoluble in acids
Humous extract, crenic acid	Soluble in water
Anitrohumin, nitrohumin, nitrolin	Insoluble in alkali and water

Despite the fact that Mulder criticised Hermann's classification, even as late as 1909, Baumann[7] remarked that Hermann's work should be accorded greater emphasis than that of Mulder in the classification of humic substances. As Kononova[68] has emphasized, the misconception that humic substances were chemically individual compounds arose from the absence of detailed investigations on their nature, structure and properties. It is not surprising, therefore, that by the end of the last century there was a prodigious increase in the number of discrete humic substances isolated from the soil including the mudesous acid of Johnson[61], the carbo-ulmic acid of Herz[53] and the hymatomelanic acid of Hoppe-Seyler[54]. Interestingly, this last name (ethanol-soluble humic acid) survives to the present day. Before passing harsh judgements, however, we should remember not only how limited was the knowledge of chemistry at that time but also the relatively primitive equipment available to those working in this field.

This brief historical survey will suffice to highlight the difficulties inherited by twentieth century soil scientists. It is now pertinent to discuss briefly the period leading to our modern day understanding of humic substances because this influences the biological approach to the use of these substances, particularly the widespread use of humic acids.

1.3. Early twentieth century extraction procedures

Despite the conclusions of some researchers in the late nineteenth century, such as those of Van Bemmelen[139], which questioned the relevance of chemical formulae as applied to humic substances, the older views on the discrete nature of these substances persisted into the present century. In 1908, Boudouard[16] recognised four groups of humus bodies and gave them the formulae $C_{18}H_{14}O_6$, $C_{18}H_{14}O_9$, $C_{18}H_{18}O_9$, $C_{18}H_{14}O_{11}$. Implicit in these formulae is also the older idea that humic molecules do not contain nitrogen. Other workers such as Baumann and Gully[8] suggested that the acidic properties associated with humic substances were not the result of functional groups but due to colloidal properties thereby implying that humic substances could not form true salts.

A considerable advance in our understanding of humus was made by Schreiner and Shorey[114]. These workers, using precise chemical analyses,

established that humic acids do not exist as discrete chemical
entities. Furthermore they showed that humus comprised over forty
different components including hydrocarbons, waxes, fats, organic acids
and carbohydrates. Shmuk[115] suggested that it was the diversity of
plant and animal remains which was responsible for the existence of so
many chemical substances in the soil.

Meanwhile Odén[93] in a scheme reminiscent of the earlier work of
Mulder, championed the notion that humic substances could be divided
into four groups depending on solubility and colour (Table 2).
According to this scheme humus coal was similar to the earlier terms
humin and ulmin. Humic acid was thought to be tetrabasic with the
empirical formula $C_{60}H_{52}O_{24}$ $(COOH)_4$, nitrogen being regarded as an
impurity.

Table 2. Odén's classification of humic substances.

Group of substance	Solubility in:			Colour	Other names
	Water	Ethanol	Alkali		
Humus Coal	Insol.	Insol.	Insol.	Black	Humin Ulmin
Humic acid	Sparingly Sol.	Insol.	Sol.	Dark brown, red tinge	Humic acid
Hymatomelanic acid	Difficultly Sol.	Sol.	Sol.	Brown, yellow tinge	Ulmic acid
Fulvic acid	Sol.	Sol.	Sol.	Golden to pale yellow	Crenic acid Apocrenic acid

Odén also considered that hymatomelanic acid, like humic acid, was
a well defined substance. The term fulvic acid was first used by Odén
and under this heading he included Berzelius' crenic and apocrenic acids.
Perhaps surprisingly Odén held the view that whereas humic and
hymatomelanic acids were discrete chemical entities, humus coal and
fulvic acids were groups of complexes. Shmuk[115] emphasised that Odén's
classification for dividing humic substances into groups according to
their solubility was arbitrary. In this connection, like Sprengel, he
demonstrated that when retained in a moist state, humic acid will
partially dissolve in water, but when dried at 100°C this material is no
longer soluble. Although this phenomenon had been reported much

earlier[125,126], Schmuk used it to conclude that researchers were not
dealing with a group of different substances but with only one substance
in a different physical state.

The lasting contribution made by Odén was the introduction of the
term fulvic acid but even this has not been without its critics. In a
scheme for the fractionation of humic substances from soils, Kononova[68]
used the terms crenic and apocrenic acids in preference to fulvic acid.
Odén originally introduced the term fulvic acid to refer to a group of
humic substances in peat waters and under these circumstances it could
be argued that they are indeed equivalent to the crenic and apocrenic
acids isolated by Berzelius from spring waters and water extracts of
soils. Although humic substances remaining in solution after
precipitation of the humic acid by acidification of alkaline extracts
were investigated by workers such as Schreiner and Shorey[114] and Page[94],
it was Tyurin[138] who referred to these substances as fulvic acids.
This application of the term fulvic acid was continued in the 1940's by
workers such as Forsyth[40] and by common usage has become the accepted
definition today. However humic substances present naturally in
drainage waters are predominantly similar to fulvic acids[26,78].

In considering the extraction of humus, it is important to remember
that most of the earlier work concentrated on humic substances. It is
of interest here to consider a different approach taken by Waksman[142].
His method of proximal analysis, based on a method published earlier[143]
accounts for 85 to 98% of the humus constituents in composts and peat in
the form of definite chemical classes. The approach is derived from the
notion that humus is derived mainly from plant remains and as such
contains waxes, oils, fats, resins, carbohydrates and lignoprotein, with
soil microbes providing the proteins. In essence, the humus is
extracted sequentially with ether, ethanol, 2% HCl and finally 80% H_2SO_4
thus removing fats and waxes, resins, hemicellulose and cellulose
respectively. The remaining residue, however, contained about 75% of
the organic matter in a fraction designated by the vague expression
protein plus lignin-humus. Waksman's proximate analysis technique is
usually unsuitable for characterizing humus because it produces only
small differences between widely different soils.

From the brief historical outline presented, it is obvious that
most of the terminology for humic substances was derived when the

chemistry of these substances was in its infancy and poorly understood. The only terms still commonly used are given in Table 3 and this is in accord with the schemes published in recent books by Schnitzer and Khan[112] and Stevenson[127]. This broad classification does not include the use of colour because it is known that colour often depends on the method of extraction and the pH of the preparation. In addition, it is also now known that some humic acids are coloured green[69,70]. It should be remembered, however, that the group name does not imply a specific substance but designates a heterogeneous preparation which has been obtained by a specific extraction procedure.

Table 3. Modern day classification of humic substances according to solubility in alkali and acid solutions.

Group of substance	Solubility in:		
	Water	Alkali	Acid
Humin	Insol.	Insol.	Insol.
Humic acid (HA)	Sparingly Soluble	Soluble	Insol.
Fulvic acid (FA)	Soluble	Soluble	Soluble

1.4. Current extraction procedures

Any extraction method should, as far as possible, take account of criteria such as ensuring that the extractant does not alter the nature of the material being extracted, does not remove inorganic contaminants and is as complete as possible.[127] Such criteria present difficulties because we are unable to judge whether humic acids are altered during extraction by alkali. Nevertheless such criteria must be regarded as useful guidelines.

The most common extraction procedures involve the use of alkaline solutions particularly NaOH, but alternatives such as Na_2CO_3 have been studied.[17,113,137] Of the strong alkalis, NaOH is more efficient than KOH and may remove as much as 80% of the organic matter from some soils[127]. The concentration of the alkali used is important not only in determining the amount of humic substance extracted but also its ash content. For example, 0.1M NaOH is a more efficient extractant than higher concentrations, but 0.5M NaOH isolates humic substances of lower ash content[74]. Because harsh extractants can modify humic substances particularly in the presence of oxygen[130], some alkali extractions have been carried out in a nitrogen only atmosphere[119,123]. Alkali

extractions are usually preceded by treating the soil with a mineral
acid (decalcification) to improve the subsequent yield of alkali
soluble material. This acid pretreatment dissolves a fraction
containing fulvic acid-type material. After extraction with alkali,
the resulting dark brown supernatant is acidified to pH 1.0 to separate
the remaining acid-soluble fulvic acid (FA) from the insoluble humic
acid (HA). The general procedure is outlined in fig. 1. The ratio of
humic to fulvic acid extracted varies with the soil type[79,81] and
usually with the depth of the sample in the soil profile[127]. Humin,
relatively free of ash, can only be obtained by removing the inorganic
component of the soil from the alkaline insoluble humic substances by
treatment with HF[102].

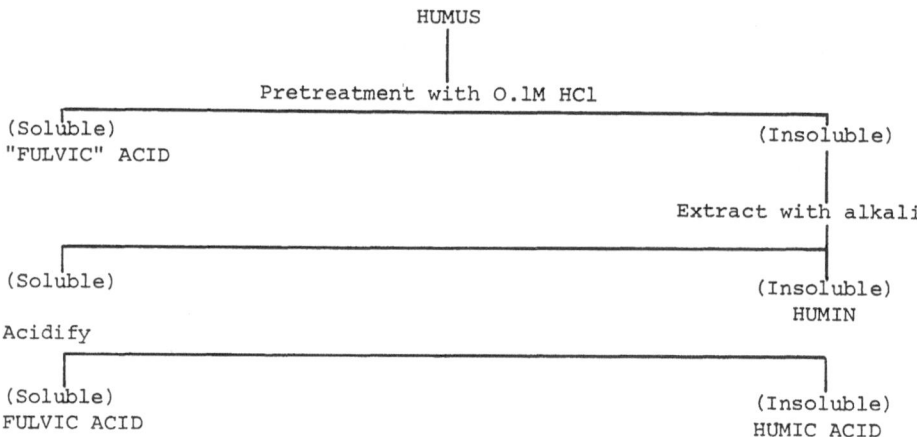

FIGURE 1.   General extraction scheme for humic substances.

The HA and FA may be fractionated further. Forsyth[40] devised a
chromatographic fractionation of FA on a charcoal column and this has
been widely used. The column is eluted successively with 0.1M HCl,
acetone/water (9:1 v/v), water and finally 0.5M NaOH to yield four
fractions designated A, B, C and D by Forsyth[40]. Of these fractions
only B and D are coloured. FA's have also been subjected to other
treatments such as passing through columns of polyvinylpyrrolidone[132] or
Amberlite XAD-8, an uncharged macroreticular resin[127]. HA's have also
been fractionated by subjecting them to treatments such as:
a) refluxing with 6M HCl and/or water to remove proteins, carbohydrates,

phenolic acids and mineral components[48,49,140], b) dissolution in an
electrolyte to yield coagulated grey HA and non-coagulated brown HA[127],
c) partial dissolution in ethanol to yield an ethanol-soluble
hymatomelanic acid[53] or d) gel filtration to produce a mixture of HA's
covering a wide molecular weight range[37,41,72,129,140]. HA's have also
been subjected to various procedures to lower their ash contents such as
shaking with HF followed by treatments with a chelating resin[140] or an
ion exchange resin in the H^+ form[62].

All alkaline extraction procedures remove both humic and non-humic
substances from the soil. Although there is no dubiety regarding the
definition of non-humic substances, some of these components may often
be regarded as part of the humic complex. The distinction between HA
and FA may also be blurred. Thus if allowed to stand at high pH or else
concentrated at moderately high temperatures, FA may acquire some of
the properties associated with HA[43,127].

The exact procedure for the alkaline extraction of humic substances
has in the past varied between different laboratories in a number of
respects: (a) whether or not to use a pretreatment with HCl, (b) the
concentration of alkali, (c) the duration of alkaline extraction and
(d) the subsequent treatment of the HA and FA. With such differences
no exacting comparisons can be made for humic preparations from various
laboratories. But at last the establishment of reference samples of
mineral soils, peats and leonardites from which humic substances can be
extracted by standardized procedures has been under active consideration
at the first international meeting of the International Humic Substances
Society held in 1983 (to be published).

Finally in this section it should be mentioned that in order to
avoid the possible effects of strong alkalis in changing soil
organic matter during extraction, some workers have used milder
extractants. These extractants include sodium fluoride and oxalate[117,118],
sodium pyrophosphate[2,18,137], sodium-EDTA[32,113], and even cupferron and
8-hydroxyquinoline[83,84]. Organic solvents have also been used such as
formic acid[136], a mixture of formic acid and lithium bromide[95],
acetylacetone[84], pyridine[50] and dimethylformamide[146]. Among other
approaches which have been used for extracting humic substances are the
use of chelating resins[29,75], ultrasonic dispersion[5,33] and sequential

extraction with different reagents[30,36,42].

It is thus evident that most work on humus has centred on humic substances. This approach has inevitably operated to the disadvantage of studies investigating the non-humic soil components, although this is now being rectified. The non-humic components are considered in some detail later in the book. It is pertinent now to consider very briefly aspects of the chemistry of humic substances which are of interest to the biologist.

1.5. Nature of humic substances

The term humic substance is used here as a generic expression to include HA and FA and humin (Table 3). It is now generally agreed that humic substances are the products of chemical and biological degradation and modification of plant and animal remains, and the activities of soil microbes[110]. However, little is known of the chemistry of the major reactions which are essential for the production of humic substances. Of the three humic substances present in the soil, HA has received by far the most attention.

HA is a dark-brown, polyphenolic, amorphous substance soluble in alkali and insoluble in mineral acids. Elemental analysis shows HA to comprise predominantly carbon and oxygen but in addition it always contains hydrogen and nitrogen (Table 4). HA's also contain phosphorus and sulphur[130]. The exact extent to which mineral components may be regarded as part of the humic structure is unknown particularly as the ash content depends on the extraction procedure[101] and because it can be lowered by either ultracentrifugation to remove clay materials or treatment with HF and resins[62,140]. Some workers have produced model HA's, by the oxidation of simple phenolic starting materials, which contain essentially no N, P, S or mineral components[110, 140]. These models often have biological properties similar to those of natural HA's[140], and can be used to investigate the active moieties in the natural products. But their use is of limited value in determining the nature of natural humic substances. This is hardly surprising when one considers the complexity of the natural material and the many and varied diverse chemical and biological reactions that are involved in its formation. For this reason only limited reference is made to model substances in this book, although they do appear extensively in the literature.

Table 4.   Elemental composition of humic and fulvic acids and humin
           in a Scottish agricultural soil (from Russell et al [102]).

HUMIC fraction	Ash (%)	Elemental anslysis[A] (%)			
		C	H	N	O[B]
Humic acid	2.2	51.7	5.1	2.9	40.3
Fulvic acid	8.5	43.6	4.9	1.7	49.8
Humin	12.5	55.9	5.9	0.9	37.3

[A] Elemental analyses are expressed on an ash free basis.

[B] Calculated by difference.

In common with the other humic substances, many of the properties of HA are due to the nature of its functional groups. These include carboxyl, aliphatic and aromatic hydroxyl, carbonyl and amide groups which can be readily detected by infrared spectroscopy. The results obtained by chemical methods of functional group analysis are often difficult to interpret and may be contradictory[35]. This in part is due to the presence of similar functional groups in the non-humic components ("impurities") which are extracted along with the humic substances[31]. In addition, HA's also contain free radicals and can participate in electron transfer mechanisms (see chapter 2).

Numerous gel filtration investigations have revealed that HA is composed of a number of coloured components, the molecular weights (MW) of which range from <1000 daltons to over 100,000 daltons[27,72,129]. The lower MW components are probably the youngest[23]. The differences between these components are important in determining their effect on biological activity (chapter 2). High MW material contains more N and P[130] and more aliphatic C-H and acid hydrolysable material[21], than the low MW components. Conversely, the degree of aromaticity and the number of carboxyl groups increase with decreasing MW of the HA fraction[21,131]. (With such chemical differences it is not surprising that the extinction coefficient of HA varies according to its MW[72].) The proportions of carboxylic carbonyl groups and ketonic carbonyl groups can vary according to the method of extracting the HA[133]. Indeed the optical and chemical properties of HA can depend on the extractant[21].

It is likely that HA is a mixture of closely related macromolecules and its composition is determined by the source of the material and by

the method used in its isolation. The extraction method undoubtedly
modifies the structure of the HA and it will probably prove impossible
to write a definitive structural formula of the components, although
further work may lead to a general structure which will enable the
material to be defined in terms of a basic nucleus differing in its
peripheral groups[48,49]. Several structures have been suggested; thus
Thiele and Kettner[134] proposed a structure based on nuclei (benzene,
indole, furan, pyridine etc.) associated with reactive groups (ketones,
hydroxyls, carboxyls, amide, etc.). The nuclei are linked by bridge
units composed of O, N, S etc. (Fig. 2). By selecting the specific
structures involved in this scheme it is possible to formulate a molecule
resembling that of an HA, although no two molecules of a humic substance
may be exactly alike[31].

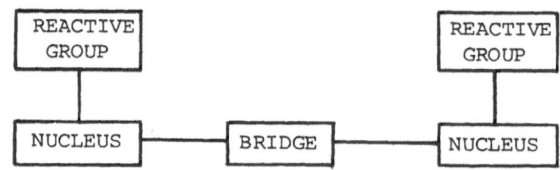

FIGURE 2.   Proposed structure of humic acid as suggested by
Thiele and Kettner[134].

Several other models for the structure of humic substances have
also been postulated[35,48] but it is of interest to give a special
mention to the work of Flaig[39] because this involves the breakdown of
lignins. According to Flaig, HA's are the end products in a sequence
involving the degradation and demethylation of lignin to polyphenol
monomers followed by the oxidation of these polyphenols to quinones
which are subsequently condensed with amino acids to form the humic
substance or its immediate precursors. This hypothesis argues that
although lignin is an essential source of structural units of humic
substances, these substances are the result of the condensation of
phenolic material (The Polyphenol Theory). This differs from the
distinctive view of Waksman[142] in which it was proposed that lignin was
incompletely metabolized by soil microbes with the residues becoming
part of the soil humus (The Lignin Theory). However, it is now well

established that dark coloured products similar to the humic substances can be synthesized by many species of soil fungi in the absence of lignin or externally available phenolic material[3,47,65,85,102]. Indeed it is likely that under natural soil conditions humic substances are predominately of microbial origin[38,65,102].

Kononova[68] has speculated that the alkaline insoluble fraction of the soil, humin, is similar to HA and its insolubility is attributed to the firmness of the binding to the mineral part of the soil. Although there is an extensive literature on the chemical composition and origin of HA, virtually no information exists for humin[68,103]. This is because of the difficulties in isolating humin free from the mineral components of the soil. Recently, humin has been isolated relatively free from ash using large volumes of HF[102]. Its elemental composition is similar to HA although it contains less nitrogen (Table 4). The infrared spectrum of the humin is also similar to that of HA prepared from the same soil and also resembles those of humic-like pigments from several soil fungi[102]. Even when isolated humin is still insoluble in NaOH although about 10% of it is soluble in boiling alkali.

The third group of humic substances, FA, has also been investigated to a lesser extent than HA. Kononova[68] considered FA to possess the same "structural units" or "building blocks" as HA's but the proportion of aromatic units is less and the peripheral aliphatic chain greater than that present in HA. Elemental analyses have established that FA, while containing the same elements as HA, possesses less carbon but more oxygen than the corresponding humic fraction. The major oxygen-containing functional groups in fulvic materials are carboxyls and Linehan[78] considers "purified" FA's to be polycarboxylic acids of MW <1000. An MW of this low order of magnitude suggests that chemical structures for FA molecules should be relatively easy to elucidate but in practice this had not been achieved although a likely molecular model has been suggested by Schnitzer[109]. In his model (Fig. 3), Schnitzer regards the "building blocks" of benzenecarboxylic and phenolic acids as being linked together mainly by hydrogen bonds, Van der Waal's forces and π-bonding. This is said to contrast with high MW HA's where chemical linking by C-O and C-C bonds occur thus increasing the stability of the higher MW material.

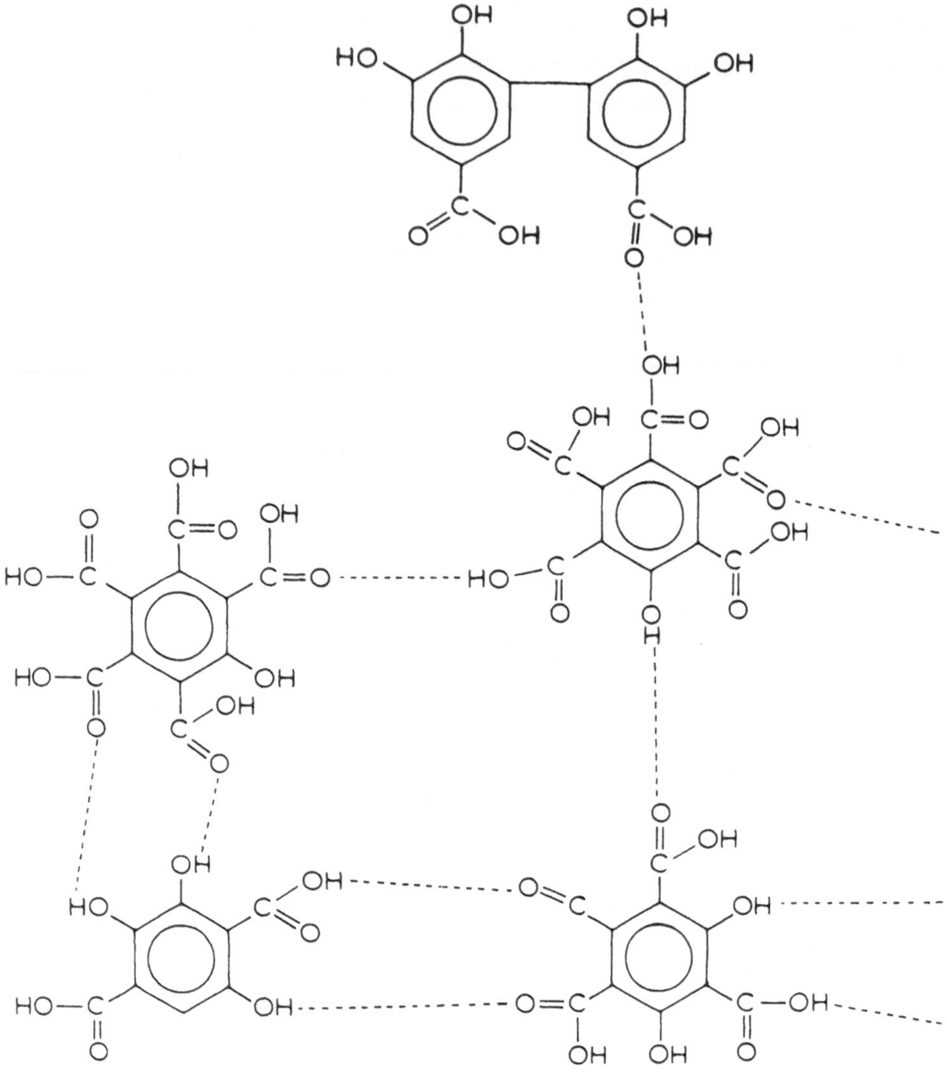

FIGURE 3.   A partial chemical structure for fulvic acid
(from Schnitzer[109]).

This type of macromolecular structure would allow the FA to
react with inorganic and organic constituents in the soil either via
oxygen-containing functional groups on the large external or internal
surfaces, or by trapping them in internal voids. But such a model has
not found universal acceptance. There is some evidence to show that

FA contains aliphatic chains heavily substituted with carboxyl groups[6].
More recently De Haan and his co-workers[26] obtained results which did not
agree with the macromolecular structure of FA as developed by
Schnitzer[109].

From this brief description of HA and FA, it is apparent that
while differing in their solubilities in HCl and in some other minor
respects, these substances are basically similar[112]. Recently,
Schnitzer[110] has postulated that FA is formed by the chemical and/or
enzymatic oxidation of HA. The biologist can best visualize the HA and
FA as being substances at the opposite ends of the same humification
process with appreciable overlaps at the centre, and scope for their
interconversions.

2.   BIOLOGICAL CHANGES IN SOIL ORGANIC MATTER

The preoccupation with investigations using humic substances has
resulted, until comparatively recently, in a dearth of knowledge about
the non-humic soil components. Soil organic matter is very diverse
in its origin and composition so that it is hardly surprising that
there should be a comparable diversity in the chemical and biochemical
reactions that occur in the soil. These reactions, which are the
result of microbial activity and exoenzymes, lead to the decomposition,
utilization and transformation of plant residues to form humic
substances which are more resistant to microbial attack than the original
material. At any given time the biological activity of an agricultural
soil will be in a state of dynamic equilibrium which can be altered by
changes in for example temperature, water content, or by the addition of
more useable organic material.

Organic additions may accelerate decomposition of the native
organic matter. In contrast, some freshly added plant residues may have
a protective effect on indigenous organic matter[121]. This latter
observation is an important consideration in the build up or maintenance
of soil organic matter under certain monocotyledonous crops such as
grass and maize[71,121]. The addition to soils of a simple substrate
such as glucose, which is readily available to microorganisms as an
energy source results in an immediate increase in the metabolism of the
biomass followed by a decrease as the substrate becomes depleted[119,124].
However, much of the glucose is not utilized as an energy source because

substantial amounts of carbon from the glucose enter the microbial tissue[122]. Other carbohydrates and polysaccharides such as cellulose and starch give similar results but differ from glucose in the rates of their decomposition and transformation (see chapter 1). In addition to carbohydrates, soil microorganisms will also utilize substituted aromatic carboxylic acids, phenols and amino acids. Although proteins are utilized, they are generally more stable than amino acids particularly when complexed with other soil components such as metals, clays or humic substances in the soil.

Much of the work carried out on the decomposition and transformation of organic materials in soil has involved ^{14}C-labelled substances and such work has recently been reviewed by Campbell[22] and Smith and Swift[22]. The decomposition in soil of ^{14}C-labelled plant residues has been investigated extensively under aerobic conditions. When residues are added to a soil they are first broken down to their basic organic components by the extracellular enzymes produced by heterotrophs[22]. Other microorganisms then use these components to provide energy sources and build up their body tissues. The result of these processes is that newly added residues change the decomposition rate in the soil as measured in terms of ^{14}CO$_2$ evolution. This change in CO$_2$ output is referred to as the <u>priming action</u>[19,56]. The primary action is usually positive although there are reports of its being negative[19,121]. Sauerbeck[105] has reported both positive and negative priming actions in the same experiment. Care should be taken in the interpretation of priming actions because what appears to be such may be due to other factors such as isotopic exchange[57]. Jenkinson[55] demonstrated that in the temperate climate of England, only about one third of the ^{14}C-labelled carbon from ryegrass residues remained in the soil after 6 months incubation. After this period there was a much slower rate of decomposition with one fifth of the radioactivity remaining in the soil after 4 years. The rate of decomposition of similar residues was considerably faster in the warmer climate of Nigeria[58]. Much of the labelled carbon persisting in the soil for several years was present in microbial tissue[56]. Because sterilization of the soil rendered this ^{14}C-labelled fraction decomposable by fresh microbes, Jenkinson[56] suggested that the size of the biomass could be estimated from the flush of CO$_2$ following soil sterilization with a substance such as chloroform. Jenkinson[57] calculated that of the added carbon, about 10% was present as the biomass after 1 year and this was reduced to 4% after 4 years. This theme is expanded in chapter 6.

Under anaerobic conditions the rate of decomposition of residues is slower than in the presence of oxygen. This results in a greater accumulation of plant residues in poorly drained organic soils. In addition, the final products are different under anaerobic conditions. Thus pyruvic acid, which is produced (as under aerobic conditions) by the degradation of carbohydrates, is not oxidized via the Krebs cycle but accumulates. The pyruvic acid then becomes involved in various reactions resulting in the eventual formation of organic acids of the formic, acetic and butyric series[22]. When proteins are decomposed under anaerobic conditions the products are ammonia, amines, thiols and hydrogen sulphide. These products are considered in more detail in chapter 4.

Different organic components are utilized to very different extents by soil microbes and will therefore remain in the soil for different periods of time. Indeed, humus is in a constant state of decomposition and transformation. It is, however, possible to calculate an average age for the humus as a whole and this is referred to as the mean residence time (MRT). The MRT of the indigenous humus is calculated, not by applying ^{14}C-labelled substances to the soil, but by carbon dating the soil itself. The method, first used by Libby and his co-workers[77], depends on the fact that organic plant material formed as a result of photosynthesis acquires a small amount of ^{14}C present in the atmosphere as a result of cosmic-ray bombardment of atmospheric N_2. When photosynthesis stops there is a slow decay of the ^{14}C and this process provides the basis for estimating the age of any residual carbon in the environment. The long half life of the carbon-14 isotope enables soil carbon as old as 50,000 years to be dated accurately. Several possible sources of error should be taken into account. Firstly, it is assumed that the level of CO_2 in the atmosphere has been constant so that there has been a constant $^{12}CO_2:^{14}CO_2$ ratio for thousands of years. The increased burning of fossil fuels over the last 100 years has altered this ratio and this must be taken into account especially when dating very young soil components. Secondly, allowances should be made for the changes in radioactivity in the atmosphere as a result of nuclear explosions since 1945. Thirdly allowance should be made when fresh debris has recently been added to soils in order to avoid misleadingly low values.

The MRT of a soil varies according to its source and values ranging from 250 years[22,23] to 1310 years[59] or even 1900 years[97,107] have been reported. Campbell and his co-workers[22,23] have also established that HA and humin have a longer MRT than FA. Recent reports[86,102] have demonstrated a similarity in the turnover rates for HA and humin in soil. It would also appear that for HA's the lower MW, material has a lower MRT, and hence a lower stability, than the higher MW components[22]. This latter observation is of interest because FA which is less stable than HA[46] comprises solely low MW components[78].

Several investigators have discovered that the MRT increases with soil depth[82,98]. Thus the MRT of a Canadian chernozem increased from 545 years at the surface, to 4000 years in the lower B horizons to over 8000 years in the deeper horizons[82]. This observation might have been anticipated because the more stable humic acid and humin components increasingly predominate over the other humus fractions with an increase in the depth of the soil profile. The usefulness of using radioactive carbon dating techniques with regard to stability, and therefore turnover, is illustrated by the data of Jenkinson and Ladd[59] in Table 5.

Table 5. The content and turnover of carbon in an unmanured soil under continuous wheat.

Soil or fraction	Result
Weight of soil	2200 Tonnes ha^{-1}
Organic matter in soil	26 Tonnes C. ha^{-1}
Input organic C per year	1.2 Tonnes C. ha^{-1}
Gross turnover of soil organic C	22 years
Turnover of biomass C	2.5 years
Age of soil organic C	1310 years

(From Jenkinson and Ladd[59]).

Using such information, a compartmentalized model can be devised to show the turnover time of different fractions of soil organic matter in top soils. Jenkinson and Rayner[60] suggested that these soil compartments, together with their half lives, comprise: (a) decomposable plant material, 0.165 years, (b) soil biomass, 1.69 years, (c) resistant plant material, 2.31 years, (d) physically stabilized organic matter,

49.5 years and (e) chemically stabilized organic matter, 1980 years. The usefulness of such a model is that it helps us to visualize in relatively simple terms what happens when plant debris is added to the soil. Initially there is rapid decomposition, with a flush of CO_2 as a result of enhanced microbial activity and during this period most of the sugars, starches, hemicelluloses and amino acids, and some of the more resistant materials such as cellulose and lignins, are decomposed. This is followed by a period in which the remainder of the cellulose and lignin are utilized giving rise to the more stable residues, the dark humic substances, which may persist for thousands of years. However, some caution should be applied to carbon dating soil fractions because some humin components may be less stable than the MRT suggests[20,46]. The sequence of events outlined above can influence the soil fertility and in this context it is relevant to mention the carbon:nitrogen ratio of the soil.

In microbial cells the ratio of carbon to nitrogen (C:N) is approximately 6. If the ratio of carbon in the decomposing organic matter greatly exceeds that of nitrogen, growth of the microbial population is restricted and decomposition and transformation are therefore retarded or even stopped[120]. The lowest C:N ratios (12-20) are found in readily decomposable materials derived from young deciduous leaves while materials derived from cereal root stubble and barley straw have higher ratios (40-80) and are less readily decomposable[120]. The only manner in which material with a very high C:N ratio may be induced to decompose is to utilize nitrogen from elsewhere in the system. If the other source is available in organic N, the microbes will compete with plants for that N and N-deficiency may result. This can only be rectified under field conditions by supplying N externally as fertilizers. Bearing this example in mind, the question now arises as to the role of humus in soil fertility.

3. THE ROLE OF SOIL ORGANIC MATTER IN FERTILITY

It has been known since the last century[76] that plants will grow well in the absence of humus provided they are given light and a suitable supply of inorganic nutrients (see chapter 1) and this is now done commercially under conditions using hydroponic or thin-film techniques. Humus is therefore not essential for the growth of plants

commercially on the relatively small scale required for crops such as
lettuce, cucumbers and tomatoes. But most of the essential foods such
as cereals, maize, legumes and potatoes are produced in soils where the
physical environment and the nutrition of the plants cannot be closely
monitored. But fortunately the soils contain humus which plays a
dominant role in controlling these factors. Hence some understanding of
the manner in which humus contributes towards soil fertility is essential
if we are to maximize the use of this vast, natural resource.

The subsequent chapters in this book deal with the various aspects
of soil organic matter in relation to fertility concluding with the
case for and against purely organic farming (chapter 12). Here we
only consider briefly how humus can influence soil fertility. The
inherent complexity of soil organic matter and the many and varied
inputs into it ensure that this soil component influences fertility in
several ways. Firstly there may be a _direct_ effect on plant growth.
In this respect the role of simple organic components such as vitamins,
auxins, aliphatic and aromatic acids (see chapters 3 and 4) on the
inhibition or stimulation of plant growth has been well established.
Many of these substances at phytotoxic concentrations are probably
involved in the phenomenon of "soil sickness"[15,80,144]. These
phytotoxins can arise in the soil under certain conditions during the
decomposition of plant residues[80,96]. However, the role of naturally
occurring, relatively inert, high MW humic substances as _major_ direct
contributors to soil fertility is questionable[147]. It is likely that
the main contribution of soil organic matter towards fertility is
indirect inasmuch as it can influence the physical, chemical and
biological properties of the soil. The important indirect effects may
be summarized:

(i) Humic substances impart a dark brown colour to the soil which
facilitates warming and hence plant growth and yield. This aspect is
particularly important in the temperate regions of the world.

(ii) The maintenance of soil structure and stability is of crucial
important, and the humus helps in processes such as soil aggregation.
The loss of stability, due to a decrease in soil organic matter content
as a consequence of intensive and incorrect agricultural practices, has
resulted all too often in the serious problem of soil erosion. In

assisting in aggregation, soil organic matter also maintains large
pores in the soil thus ensuring adequate drainage and aeration. The
addition of organic matter to a clay soil makes it more porous and
enables better root penetration leading to enhanced root growth[99].
Because of its colloidal properties, soil organic matter (particularly
the humic substances) will retain substantial amounts of water - a very
important property in sandy soils[99]. It should be borne in mind,
however, that some of the water retained by organic matter is
unavailable to plants or microorganisms. Well humified organic matter,
as present in most agricultural soils, can retain up to four times its
own weight of water but only about half of this is available to the
plant[120].

(iii) Soil organic matter acts as a reservoir for plant nutrients
and prevents leaching of elements which are vital to plant growth[120].
This is achieved by the considerable cation exchange capacity (CEC) of
the organic material. Well decomposed humus may have a CEC in excess
of 300 meq/100g humus[120], which is considerably greater than that of the
clay components such as kaolinite (3-15 meq), illite (30-40 meq) or
montmorillonite (80-150 meq). It has been estimated that from 20 to 70%
of the CEC of many soils is attributable directly to the soil organic
matter alone[127] and there is a direct relationship between organic
matter content and the corresponding CEC[64]. The effect of humus in
increasing the CEC of soils is more important in sandy soils than in
clays because in sandy conditions it is difficult to maintain a high
level of humus under the usual cropping practices[63]. The importance of
humus is exemplified by the interest in research for alternative sources
of organic matter such as sewage sludge or the biodegradable fraction of
solid urban waste to maintain or increase existing levels in soils[99].

(iv) The capacity of soil organic matter to buffer soils against
rapid changes in pH is well known. This property arises largely from
the interaction of humic substances and clays producing colloidal
complexes which have negative surface charges and hold cations such as
H^+, Al^{3+}, Fe^{3+} etc. Thus the resistance to pH changes is minimal in
sandy soils but high in soils which have a substantial CEC. As a result
of biological activity, acids and bases are being produced continuously
in the soil so that without a buffering capacity the pH of the soil would
change to the detriment of its biology. In referring to the buffering

capacity of the soil, Allison[4] stated "If this did not exist it would
be difficult to conceive of agriculture as we know it today. We would
have a situation similar to that encountered in hydroponics. Plants
could be grown but at great expense". Sometimes however, the high
buffering capacity of some soils can prove economically disadvantageous
to the farmer. This is encountered in peats which after drainage
require appreciable quantities of lime to bring them to a pH suitable
for growing crops such as cereals. Even under more normal agricultural
conditions, the acidity of root exudates eventually leads to a lowering
of the soil pH and applications of lime are essential to maintain
fertility.

(v) In addition to acting as a reservoir, soil organic matter can
form stable complexes with some metals and thus influence their
availability to plants and micro-organisms. The ability of humic
substances to form complexes is due to their functional groups such
as carboxylic acids, C = O structures and phenolic-, alcoholic-,
and enolic-hydroxyls. Stevenson and Ardakani[128] have emphasized that
metals complexed by HA are largely insoluble while the converse is true
for metals complexed by FA. Thus HA's can complex Cu making it
relatively unavailable to plants[34,73]. In contrast, Al toxicity, which
is a major problem in many acid soils, can be ameliorated by amending
such soils with organic matter thus reducing the availability of the
metal[127].

(vi) During the process of mineralization, soil organic matter
acts as a source of N, P and S, elements which are vital for plant
growth. Much has been written about these elements and according to
Allison[4] N is the most important element in humus when considered from
the economic standpoint. Indeed crop yields are often directly
proportional to the N released from organic matter[63]. As will be shown
in chapters 8, 9 and 10 of this book, considerable amounts of N, P and S
are present in organic forms, many of which (particularly the
S-containing substances) still await identification. Unfortunately
much of the reserves of these elements are unavailable for use by
plants; for example some of the nitrogen becomes "locked up" in stable
humic substances[68,112,127] while some phosphorus may be present as
insoluble calcium, iron or aluminium salts of phosphate esters[44].

During the active growing season, soil becomes depleted of much
of its available N, P and S, although the rate of depletion of N is the
greatest. The levels of these elements may be restored by adding fresh
organic residues but from the viewpoint of enhancing crop yields, the
use of chemical fertilizers containing N, P and S, together with K
and/or some essential micro-nutrients, is rapid in its effect and more
efficient in economic terms. In developed countries, the practice of
adding chemical fertilizers is the most practical way of ensuring the
high crop response demanded of modern agriculture provided that the soil
organic matter content of the soil does not fall below the levels which
would impair other aspects of soil fertility.

(vii) The widespread and commercial use of fungicides, herbicides
and pesticides to control unwanted biological species has implicated
humus in another aspect of soil fertility. These generally low MW
materials can combine with soil organic matter in such a way as to
influence their activity, mobility and biodegradability. Thus the
herbicide paraquat is adsorbed by base exchange and deactivated in
soils[67]. It is now known for example that to obtain an adequate control
of weeds, some herbicides must be applied to the soil at up to 20 times
the normal rate in order to compensate for their adsorption by humus.
Biodegradation in the presence of soil organic matter has been
reported for pesticides such as DDT, heptachlor and endrin[24] and various
herbicides such as 2,4-D[65]. But organic matter may also bring about
non-biological decomposition as reported for 3-amino-triazole and the
slow conversion of aldrin to dieldrin[25,66]. Similarly fungicides such
as dazomet, which are sprayed on crops or used to sterilize soils, are
also decomposed[9]. There is now a considerable literature available on
the interaction of soil organic matter with herbicides, pesticides and
fungicides[63,67,127]. The degrading and detoxifying of these substances
is a most important property of the soil especially in relation to
animal health. But our understanding of the reactions of pesticides in
soil is still limited due to the complexity of humus _per se_[63].

From the foregoing brief resume, it is apparent that soil organic
matter has a vital role in soil fertility. Although its modes of
action are many and varied, it is important to realize that all the
factors outlined above may not be operating to maximum effect at any

given time. A serious loss in efficiency of any one of these
factors, for example the buffering capacity, water retention or
inadequate supplies of micro- or macro-nutrients, can have serious
economic consequences in the loss of crop productivity.

4. CONCLUSIONS

Soil organic matter is in a state of constant decomposition and
transformation due mainly to its active biomass content. Although
extremely complex in both origin and structure, soil organic matter
is usually classified arbitrarily into relatively stable, brown humic
substances and the less stable, and generally lower MW, non-humic
components. For historical reasons, much attention has been given to
the extraction and composition of the humic substances and, until
comparatively recently, this approach has operated against meaningful
studies of the important non-humic components. Indeed, the relevance
of humic substances, particularly HA, as meaningful entities is now
being questioned from the viewpoint of biologists[92,100].

Despite many observations that externally supplied HA and FA can
stimulate the growth of plants[68], it is unlikely that these humic
substances have a major, direct effect on plants under field conditions[147].
Even so, there is a substantial literature on the direct effects of
humic substances, particularly HA (chapters 1 and 2). It is likely
that the major effect of soil organic matter on fertility is the result
of its influencing such factors as for example providing a reservoir
of plant nutrients, supplying N, P and S as a result of microbial
activity, influencing the buffering capacity of the soil and helping
in water retention. These topics are expanded in detail in subsequent
chapters of this book.

5. REFERENCES

1. ACHARD F.K. 1786. Chemische Untersuchung des Torfs. Chemische
   Annalen für die Freunde der Naturlehere von L. CRELL, 2, 391-403.
2. ALEKSANDROVA L.N. 1960. The use of sodium pyrophosphate for
   separating free humic substances and their organomineral compounds
   from soil . Soviet Soil Science, 190-197.
3. ALEKSANDROVA L.N. 1962. Modern ideas on the nature of humus and its
   organomineral derivatives . Problems in soil science. Izd. ANN SSS.
4. ALLISON F.E. 1973. Soil organic matter and its role in crop
   production Developments in Soil Science. Volume 3. Elsevier Press,
   New York.

5. ANDERSON G. 1958. Identification of derivatives of deoxyribonucleic acid in humic acid. Soil Science, 88, 169-174.
6. ANDERSON G. 1982. Recent observations on humus composition and properties. Promocet, 3-10.
7. BAUMANN A. 1909. Untersuchungen über die Humussäuren 1. Geschichte der Humussäuren. Mitteilungen der K bayerischen Moorkulturanstalt, 3, 52-123.
8. BAUMANN A. and GULLY E. 1910. Unterschungen über die Humussäuren des Hochmoores. Mitteilungen der K bayerischen Moorkulturanstalt, 4, 31-156.
9. BEDFORD J.L. 1972. Dazomet. Report of the agricultural division technical department BASF United Kingdom Ltd.
10. BERZELIUS J.J. 1813. Ulmin. Annals of Philosophy, London, 2, 314.
11. BERZELIUS J.J. 1833. Undersökning al vattnet i Porla-kalla. Kunglig Svenska Vetenskapsakademiens handlingar, 18-92.
12. BERZELIUS J.J. 1839. Lehrbuch der Chemie, third edition. Translated by Wohler, 8, 11-16; also 384-431.
13. BOHN H.L. 1976. Estimate of organic carbon in world soils. Journal of the Soil Science Society of America Proceedings, 40, 468-470.
14. BONNEAU M. and SOUCHIER B. 1982. Constituents and properties of soils. Academic Press, London.
15. BORNER H. 1955. Die Ausscheindung organischer Verbindungen aus der Samen von Roggen (*Secale cereale* L) und Gerste (*Hordeum vulgare* L.) wahrend der Quellung. Naturwissenschaften, 42, 583-584.
16. BOUDOUARD O. 1908. Sur les matieres humiques des charbons. Comptes rendus hebdomadaires des Seances de l'Academie des Sciences, 14, 284-286.
17. BREMNER J.M. 1949. Studies on soil organic matter. Part 111. Journal of Agricultural Science, Cambridge, 39, 280-282.
18. BREMNER J.M. and LEES H. 1949. Studies on soil organic matter. Part 11. Journal of Agricultural Science, Cambridge, 39, 275-279.
19. BROADBENT F.E. 1947. Nitrogen release and carbon loss from soil organic matter during decomposition of added plant residues. Proceedings of the Soil Science Society of America, 12, 246-249.
20. BRUCKERT S. 1982. Analysis of the organo-mineral complexes of soils. In Constituents and Properties of Soils. Eds. Bonneau M. and Souchier B. Academic Press, London.
21. BUTLER J.H.A. and LADD J.N. 1968. Effect of extractant and molecular size on the optical and chemical properties of soil humic acids. Australian Journal of Soil Research, 7, 229-239.
22. CAMPBELL C.A. 1978. Soil organic carbon, nitrogen and fertility. In Soil Organic Matter. Developments in Soil Science, Volume 8. Eds. Schnitzer M. and Khan S.U. Elsevier Press, New York.
23. CAMPBELL C.A., PAUL E.A., RENNIE D.A. and McCALLAM K.J. 1967. Applicability of the carbon-dating method of analysis to soil humus studies. Soil Science, 104, 217-224.
24. CASTRO T.F. and YOSHIDA T. 1974. Effect of organic matter on the biodegradation of some organochlorine insecticides in submerged soils. Soil Science and Plant Nutrition, 20, 363-370.
25. CROSBY D.G. 1970. The nonbiological degradation of pesticides in soils. In International Symposium on Pesticides in Soil. pp.86-94. Michigan State University, East Lansing.
26. DEHAAN H., WERLEMARK G. and DEBOER T. 1983. Effect of pH on molecular weight and size of fulvic acids in drainage water from peaty grassland in NW Netherlands. Plant and Soil, 75, 63-73.

27. DELL'AGNOLA G. and FERRARI G. 1969. Gel filtrazione dell'humus. Agrochimica, 13, 327-335.

28. DÖBEREINER J.W. 1822. Zur pneumatischen Chemie 111. Zur pneumatischen Phytochemie, pp 64-74. Jena.

29. DORMAAR J.F. 1972. Chemical properties of organic matter extracted from a number of Ah horizons by a number of methods. Canadian Journal of Soil Science, 52, 67-77.

30. DUCHAUFOUR P. and JACQUIN F. 1964. (Results of recent investigations on the evolution of organic matter in soils). Compte rendu hebdomadaire des seances de l'Academie d'agriculture, 50, 376-387.

31. DUBACH P. and MEHTA N.C. 1963. The chemistry of soil humic substances. Soils and Fertilizers, 26, 293-300.

32. DUBACH P., MEHTA N.C. and DEVEL H. 1961. Extraktion von Huminstoffen aus dem B-Horizant eines Podsols mit Äthylendiamtetraessigssäure. Zeitschrift für Planzenernährung Dungung Bodenkunde, 95, 119-123.

33. EDWARDS A.P. and BREMNER J.M. 1967. Dispersion of soil particles by sonic vibrations. Journal of Soil Science, 18, 47-63.

34. ENNIS M.T. and BROGAN J.C. 1961. The availability of copper-humic complexes. Irish Journal of Agricultural Research, 1, 35-42.

35. FELBECK G.T. 1965. Structural chemistry of soil humic substances. Advances in Agronomy, 17, 327-368.

36. FELBECK T.G. 1971. Chemical and biological characterization of humic matter. In Soil Biochemistry Volume 2. Eds. McLaren A.D. and Skujins J. Marcel Dekker, New York.

37. FERRARI G. and DELL'AGNOLA G. 1963. Fractionation of the organic matter of soil by gel filtration through sephadex. Soil Science, 96, 418-421.

38. FILIP A., HAIDER K., BEUTELSPACHER H. and MARTIN J.P. 1974. Comparison of IR-spectra from melanins of microscopic soil fungi, humic acids and model phenol polymers. Geoderma, 11, 37-52.

39. FLAIG W. 1964. Effects of micro-organisms in the transformation of lignin to humic substances. Geochimica et Cosmochimica Acta, 28, 1523-1535.

40. FORSYTH W.G. 1947. Studies on the soluble complexes of soil organic matter. 1. Biochemical Journal, 41, 176-181.

41. GJESSING E.T. 1965. The use of Sephadex gel for the estimation of molecular weight of humic substances in natural water. Nature, London, 200, 1091-1092.

42. GOH K.M. 1979. Organic matter in New Zealand soils. Part 1. Improved methods for determining humic and fulvic acids with low ash content. New Zealand Journal of Science, 13, 669-686.

43. GOH K.M. and REID M.R. 1975. Molecular weight distribution of soil organic matter as affected by acid pretreatment and fractionation into humic and fulvic acid. Journal of Soil Science, 26, 207-222.

44. GREAVES M.P. and WEBLEY D.M. 1969. The hydrolysis of myoinositol hexaphosphate by soil micro-organisms. Soil Biology and Biochemistry, 1, 37-43.

45. GRIFFITH S.M. and SCHNITZER M. 1975. Oxidative degradation of humic and fulvic acids extracted from tropical volcanic soils. Canadian Journal of Soil Science, 55, 251-267.

46. GUILLET B. 1982. Studies of the turnover of soil organic matter using radio-isotopes (14-C). In Constituents and Properties of Soils. Eds. Bonneau M. and Souchier B. Academic Press, London.

47. HAIDER K. and MARTIN J.P. 1967. Synthesis and transformation of phenolic compounds by *Epiccocum nigrum* in relation to humic acid formation. Soil Science Society of America Proceedings, 31, 766-772.

48. HAWORTH R.D. 1970. The chemical nature of humic acid. Soil Science, 111, 71-79.

49. HAWORTH R.D. and ATHERTON N.M. 1965. Humic acid. Bulletin of the National Institute of Sciences of India, 28, 57-60.

50. HAYES M.H.B., SWIFT R.S., WARDLE R.E. and BROWN J.K. 1975. Humic materials from an organic soil-A comparison of extractants and of properties of extracts. Geoderma, 13, 231-245.

51. HERMANN R. 1837. Chemische Untersuchung des Tschernosems oder der schwarzen Ackerede der sudlichen Gouvernements. Journal für praktisch Chemie, 12, 277-292.

52. HERMANN R. 1842. Untersuchungen über den Moder. Journal für pracktische Chemie, 23, 375-386; also 25, 165-206.

53. HERZ. 1880. (see Waksman. reference number 142).

54. HOPPE-SEYLER F. 1889. Uber Huminsubstanzen, ihre Entstenhung und ihre Elgenschaften. Hoppe-Seyler's Zeitschrift für physiologische Chemie, 13, 66-121.

55. JENKINSON D.S. 1965. Studies on the decomposition of plant material in soils. Journal of Soil Science, 16, 104-115.

56. JENKINSON D.S. 1966. The priming action. In The use of isotopes in soil organic matter studies. Report of the FAO/IAEA meeting, Vienna. Pergamon Press, Oxford.

57. JENKINSON D.S. 1971. Studies on the decomposition of 14-C labelled organic matter in soil. Soil Science, 111, 64-70.

58. JENKINSON D.S. and AYANABA A. 1977. Decomposition of carbon-14 labelled plant material under tropical conditions. Soil Science Society of America Journal, 41, 912-915.

59. JENKINSON D.S. and LADD J.N. 1983. Microbial biomass in soil-measurement and turnover. In Soil Biochemistry, volume 5. Eds. Paul E.A. and Ladd J.N. Marcel Dekker, New York.

60. JENKINSON D.S. and RAYNER J.H. 1977. The turnover of soil organic matter in some of the Rothamsted Classical Experiments. Soil Science, 123, 298-305.

61. JOHNSON J.F.W. 1840. On the constitution of pigolite and on the mudesous and mudesic acids. Philosophical Magazine, 17, 382-383.

62. KAHN S.U. 1971. Distribution and characteristics of organic matter extracted from the black solonetzic and black chernozemic soils of Alberta-The humic acid fraction. Soil Science, 112, 401-409.

63. KAHN S.U. 1978. The interaction of organic matter with pesticides. In Soil Organic Matter. Developments in Soil Science Volume 8. Eds. Schnitzer M. and Kahn S.U. Elsevier Press, Oxford.

64. KAMPRATH E.J. and WELCH C.D. 1962. Retention and cation-exchange properties of organic matter in Coastal Plain soils. Soil Science Society of America Proceedings, 26, 263-265.

65. KANG K.S. and FELBECK G.T. 1965. A comparison of the alkaline extracts of tissues of *Aspergillus niger* with humic acids from the soils. Soil Science, 99, 175-181.

66. KAUFMAN D.D. 1970. Pesticides in soil. In International Symposium on Pesticides in Soil. pp. 73-86. Michigan State University, East Lansing.

67. KEARNEY P.C., HARRIS C.I., KAUFMAN D. and SHEETS T.J. 1965. Behaviour and fate of chlorinated aliphatic acids in soil. Advances in Pest Control Research, 6, 1-30.
68. KONONOVA M.M. 1961. Soil Organic Matter, its nature, its role in soil formation and in soil fertility. Pergamon Press, Oxford.
69. KUMADA K. and HURST H.M. 1967. Green humic acid and its possible origin as a fungal metabolite. Nature, London, 214, 631-632.
70. KUMADA K. and SATO O. 1962. Chromatographic separation of green humic acid from podzol humus. Soil Science and Plant Nutrition, 8, 31-33.
71. KURANOV V.N. 1961. Decomposition of plant residues in the soil. Soviet Soil Science, 3, 300-304.
72. LADD J.N. 1969. The extinction coefficients of soil humic acids fractionated by Sephadex gel filtration. Soil Science, 107, 303-306.
73. LEES H. 1950. A note on the copper-retaining power of a humic acid from a peat soil. Biochemical Journal, 46, 450-451.
74. LEVESQUE M. and SCHNITZER M. 1966. Effects of sodium hydroxide on the extraction of organic matter and of the major inorganic constituents from a soil. Canadian Journal of Soil Science 46, 1-12.
75. LEVESQUE M. and SCHNITZER M. 1967. The extraction of soil organic matter by base and chelating resin. Canadian Journal of Soil Science, 47, 76-78.
76. LIEBIG J. 1841. Organic chemistry in its application to agriculture and physiology. Translated by Weber J.W. and Owen J. Cambridge.
77. LIBBY W.F. 1955. Radiocarbon dating. Second Edition. University of Chicago Press.
78. LINEHAN D.J. 1977. A comparison of the carboxylic acids extracted by water from an agricultural top soil with those extracted by alkali. Journal of Soil Science, 28, 369-378.
79. LINEHAN D.J. 1978. Polycarboxylic acids extracted by water and by alkali from agricultural top soils of different drainage status. Journal of Soil Science, 29, 373-377.
80. LYNCH J.M. 1976. Products of soil micro-organisms in relation to plant growth. Critical Reviews in Microbiology, 67-107.
81. MALCOLM R.E. and VAUGHAN D. 1978. Comparative effects of soil organic matter fractions on phosphatase activities in wheat roots. Plant and Soil, 51, 117-126.
82. MARTEL Y.A. and PAUL E.A. 1974. Effects of cultivation on the organic matter of grassland soils as determined by fractionation and radiocarbon dating. Canadian Journal of Soil Science, 54, 419-426.
83. MARTIN A.E. and REEVE R. 1955. The extraction of organic matter from podzolic B horizons with organic reagents. Chemistry and Industry, 356.
84. MARTIN A.E. and REEVE R. 1957. Chemical studies on podzolic illuvial horizons. 1. The extraction of organic matter by organic chelating agents. Journal of Soil Science, 8, 268-285.
85. MARTIN J.P. and HAIDER K. 1969. Phenolic polymers of *Stachybotrys astra, Stachybotrys chartarum and Epicoccum nigrum* in relation to humic acid formation. Soil Science, 107, 260-270.
86. MARTIN J.P., ZUNINO H., PEIRANO P., CAIOZZI M. and HAIDER K. 1982. Decomposition of 14-C labelled lignins, model humic acid polymers and fungal pigments in allophanic soils. Soil Biology and Biochemistry, 14,,289-293.

87. McKIBBIN R.R. 1933. Soil organic matter. Journal of the American Society of Agronomy, 25, 258-266.
88. MULDER G.J. 1839. Über die Harze des Torfes. Journal für praktische Chemie, 16, 495-497; also 17, 444-453.
89. MULDER G.J. 1840. Untersuchungen über die Humussubstanzen. Journal für praktische Chemie, 21, 203-240; also 321-370.
90. MULDER G.J. 1844. Über die Bestandteile der Ackerede. Journal für praktische Chemie, 32, 321-344.
91. MULDER G.J. 1861. Die Chemie der Ackerkrume, Ubers. J. Muller, Berlin, 1, 311-356.
92. NANNIPIERI P. and SEQUI P. 1982. Soil Fertility - the determination and the turnover of soil organic matter. In Evolution du niveau de fertilite des sols dans differents systemes de culture. Criteres pour mesurer cette fertilite, Bari, Italy. Instituto Sperimentale Agronomico, 141-151.
93. ODÉN S. 1919. Die Huminsäuren. Kolloidchemische Beihelte, 11, 75-260.
94. PAGE H.J. 1932. Studies on the carbon and nitrogen cycles in the soil V. The origin of the humic matter of the soil. Journal of Agricultural Science, 22, 291-296.
95. PARSONS J.W. and TINSLEY J. 1960. Extraction of soil organic matter with anhydrous formic acid. Soil Science Society of America Proceedings, 24, 198-201.
96. PATRICK Z.A. 1971. Phytotoxic substances associated with the decomposition in soil of plant residues. Soil Science, 111, 13-18.
97. PAUL E.A., CAMPBELL C.A., REENIE D.A. and McCALLUM K.J. 1964. Investigations of the dynamics of soil humus, utilizing carbon dating techniques. Transactions of the 8th International Congress of Soil Science, Bucharest, 201-208.
98. PAUL E.A. and VAN VEEN J.A. 1978. The use of tracers to determine the dynamic nature of organic matter. Transactions of the 11th Congress of the International Society of Soil Science, 3, 61-102.
99. PERA A., VALLINI G., SIRENO I., BIANCHIN M.L. and deBERTOLDI M. 1983. Effects of organic matter on rhizosphere organisms and root development of *Sorgham* plants in two different soils. Plant and Soil, 74, 3-18.
100. PERSSON J. 1968. Biological testing of chemical humus analysis. Lantbrukshogskolans Annaler, 34, 81-217.
101. POSNER A.M. 1966. The humic acids extracted by various reagents from a soil. 1. Yield, inorganic components and titration curves. Journal of Soil Science, 17, 65-78.
102. RUSSELL J.D., VAUGHAN D., JONES D. and FRASER A.R. 1983. An IR spectroscopic study of soil humin and its relationship to other soil humic substances and fungal pigments. Geoderma, 29, 1-12.
103. SAIZ-JIMENEZ C., HAIDER K. and MEUZELAAR H.L.C. 1979. Comparisons of soil organic matter and its fractions by pyrolysis-mass spectrometry. Geoderma, 22, 25-37.
104. SATO O. and KUMADA K. 1967. The chemical nature of the green fraction of p-type humic acid. Soil Science and Plant Nutrition, 13, 121.
105. SAUERBECK D. 1966. A critical evaluation of incubation experiments on the priming effect of green manures. Report of the FAO/IAEA meeting, Vienna. Pergamon Press, Oxford.
106. SAUSSURE Th. DE. 1804. Recherches chemiques sur la vegetation. Paris.

34

107. SCHARPENSEEL H.W., TAMERS M.A. and PIETIG F. 1968.
     Altersbestimmung von Boden durch die Radiokohlenstoff
     datierungsmethode. Zeitschrift für Pflanzenernähung und
     Bodenkunde, 119, 44-52.
108. SCHEFFER F. and ULRICH B. 1960. Lehrbuch der Agrikulturchemie
     und Bodenkunde. 111 Teil. Humus und Humus Düngung Bd 1. Stuttgart.
109. SCHNITZER M. 1978. Humic substances: Chemistry and reactions.
     In Soil Organic Matter. Developments in Soil Science, Volume 8.
     Eds. Schnitzer M. and Khan S.U. Elsevier Press, Oxford.
110. SCHNITZER M. 1981. Recent advances in humic acid research.
     Proceedings of the International Peat Symposium, Bemidji,
     Minnesota.
111. SCHNITZER M. and KHAN S.U. 1972. Humic substances in the
     environment. Mercel Dekker, New York.
112. SCHNITZER and KHAN S.U. 1978. Soil Organic Matter. Developments
     in Soil Science, Volume 8. Elsevier Press, Oxford.
113. SCHNITZER M., WRIGHT J.R. and DESJARDINGS J.G. 1958. A comparison
     of the effectiveness of various extractants for organic matter
     from two horizons of a podzol profile. Canadian Journal of Soil
     Science, 38, 49-53.
114. SCHREINER O. and SHOREY E.C. 1910. Chemical nature of soil organic
     matter. Bulletin of the Bureau of Soils. U.S. Department of
     Agriculture, 74.
115. SHMUK A.A. 1924. (The chemistry of soil organic matter). Trudy
     Kubanskogo sel'skokhozyaist-vennogo institute, 1, 2.
116. SIMON K. 1929. Über die Herstellung von Humusextrakten mit
     neutralen Mitteln. Zeitschrift für Pflanzenernahrung Düngung
     und Bodenkunde, A14, 252-257.
117. SIMON K. 1930. Über die Vermeidung alkalischer Wirkung bei der
     Darstellung und Reinigung von Huminsauren. Zeitschrift für
     Pflanzenernahrung Düngung und Bodenkunde, A18, 323-336.
118. SIMON K. and SPEICHERMANN H. 1938. Beitrage zur Humusuntersuch-
     ungsmethodik. Bodenkunde und Pflanzenernahrung, 8, 129-152.
119. SIMONARI P. and MAYAUDON J. 1966. Etude des transformations de
     la matiere organique du sol au moyen du carbone-14. The use of
     isotopes in soil organic matter studies. Report of the FAO/IAEA
     Meeting, Vienna, Pergamon Press, Oxford.
120. SIMPSON K. 1983. Soil. Longman, London.
121. SMITH J.H. 1966. Some inter-relationships between decomposition of
     various plant residues and loss of soil organic matter as measured
     by carbon-14 labelling. Report of the FAO/IAEA Meeting, Vienna,
     Pergamon Press, Oxford.
122. SMITH K.A. and SWIFT R.S. 1983. General radioisotope techniques.
     In Soil Analysis-Instrumental Techniques and Related Procedures.
     Ed. Smith K.A. Marcel Dekker, New York.
123. SORENSEN H. 1963. Studies on the decomposition of ^{14}C-labelled
     barley straw in soil. Soil Science, 95, 45-51.
124. SPARLING G.P., ORD B.G. and VAUGHAN D. 1981. Changes in microbial
     biomass and activity in soils amended with phenolic acids. Soil
     Biology and Biochemistry, 13, 455-460.
125. SPRENGEL C. 1826. Über Pflanzenhumus. Humussäure und humussäure
     Salze. Kastner's Archiv für Gesammte Naturlehre, 8, 145-220.
126. SPRENGEL C. 1837. Die Bodenkunde oder die Lehre von Boden. Leipzig.
127. STEVENSON F.J. 1982. Humus Chemistry-Genesis, Composition,
     Reactions. John Wiley and Sons, New York.

128. STEVENSON F.J. and ARDAKANI M.S. 1972. Organic matter reactions involving micronutrients in soils. In Micronutrients in Agriculture, pp.79-114. Eds. Mortvedt J.J., Giordano P.M. and Lindsay, W.L. Soil Science Society of America.

129. SWIFT R.S. and POSNER A.M. 1971. Gel chromatography of humic acid. Journal of Soil Science, 22, 237-249.

130. SWIFT R.S. and POSNER A.M. 1972. Nitrogen, phosphorus and sulphur contents of humic acids fractionated with respect to molecular weight. Journal of Soil Science, 23, 50-57.

131. SWIFT R.S., THORNTON B.K. and POSNER A.M. 1970. Spectral characteristics of a humic acid fractionated with respect to molecular weight, using an agar gel. Soil Science, 110, 93-99.

132. SWINGER G.D., OADES J.M. and GREENLAND D.J. 1969. The extraction, characterization and significance of soil polysaccharides. Advances in Agronomy, 21, 195-200.

133. THENGE B.K.G. and POSNER A.M. 1967. Nature of the carbonyl groups in soil humic acid. Soil Science, 104, 191-201.

134. THIELE H. and KETTNER H. 1953. Humic acids. Kolloidzeitschrift, 130, 131-160.

135. THOMSON T. 1807. A system of chemistry. Third edition, Edinburgh.

136. TINSLEY J. 1956. The extraction of organic matter from soils with formic acid. Report of the Sixth International Congress of Soil Science Communications 1 and 11B, 541-546.

137. TINSLEY J. and SALAM A. 1961. Extraction of soil organic matter with aqueous solvents. Soils and Fertilizers, 24, 81-84.

138. TYURIN I.V. 1940. (The nature of fulvic acids of soil humus). Trudy Pochvennogo Institute imeni V.V. Dokuchaeva Akademiya nauk SSSR, 23.

139. VAN BEMMELEN J.M. 1888. Die Adsorptionsvermogan der Ackererde Landwirtschaftlichen Versuchsstationen, 35, 69-136.

140. VAUGHAN D. 1969. The stimulation of invertase development in aseptic storage tissue slices by humic acids. Soil Biology and Biochemistry, 1, 15-28.

141. VAUQUELINE C. 1797. Sur une maladie des arbres qui attaque specialeme d'orme et qui est analogue a un ulcere. Annales de Chimie, 21, 39-47.

142. WAKSMAN S.A. 1936. Humus. Origin, chemical composition and importance in nature. Balliere, Tindall and Cox, London.

143. WAKSMAN S.A. and STEVENS K.R. 1928. Contribution to the chemical composition of peat. 1. Chemical nature of organic complexes in peat and methods of analysis. Soil Science, 26, 113-137; also 239-252.

144. WANG T.S.C., YANG T. and CHUANG T. 1967. Soil phenolic acids as plant growth inhibitors. Soil Science, 103, 239-248.

145. WEIGMANN A.F. 1837. Über die Entstehung, Bildung und des Wesen des Torfes. Braunschweig.

146. WHITEHEAD D.C. and TINSLEY J. 1964. Extraction of soil organic matter with dimethylformamide. Soil Science, 97, 34-42.

147. WIERSUM L.K. 1974. The activity of specific growth stimulating substances in the soil in relation to the application of organic matter. Transactions of the Tenth Soil Science Conference, Volume 111, 123-129.

148. WILDING L.P., SMECK N.E. and HALL G.F. 1983. Pedogenesis and Soil Taxonomy. 1. Concepts and interactions. Developments in Soil Science, Volume 11. Elsevier Press, Oxford.

37

CHAPTER 1

INFLUENCE OF HUMIC SUBSTANCES ON GROWTH AND PHYSIOLOGICAL PROCESSES

D. VAUGHAN and R.E. MALCOLM

The Macaulay Institute for Soil Research, Aberdeen, Scotland.

CONTENTS

1. INTRODUCTION

It is now generally accepted that soil organic matter (humus) plays a major role in maintaining or improving soil fertility[2,104]. Because of the complexity of soil organic matter, the precise nature of that role has been the subject of a considerable and long lasting debate. The present, and all too often inadequate, approaches to elucidating the mode of action of soil organic matter on plant growth are still partially influenced by historical concepts. An appraisal of the current ideas can only be undertaken by reference, albeit briefly, to the past. More detailed historical accounts are to be found elsewhere[2,104,171,225].

1.1 Historical perspectives

For over 8000 years man has realized that dark soils are usually productive and that colour and productivity are commonly associated with the organic material derived chiefly from decaying plant remains[2]. Aristotle is often reported as being the first to suggest that plants absorb their food in an elaborated form from soils by processes similar to those found in animals[197,225]. The observation that calcined plants yield ash containing inorganic salts which have a beneficial effect on plant growth has also been known since antiquity. In 1563, Palissy[197] recorded that when a plant is burned, it is reduced to an ash called alcaly which, when added to the soil, returns those substances which have been taken away by plants during growth.

During the early seventeenth century, Francis Bacon[6] suggested that water formed the main nourishment of plants and soil was necessary only to keep them upright and to protect them from excessive changes in temperature. He also believed that plants absorbed a _particular juice_ from the soil for their sustenance. Van Helmont too regarded water as

being the only nutrient of plants. In a famous experiment reported by
his son in Ortus Medicinae (1684)[197] a young willow tree was planted in
dry soil and watered with rain water for five years. After this period
the plant had gained 164 lbs.3 ozs. whereas the soil had lost only
2 ozs. in dry weight. Van Helmont concluded that the increase in the
weight of the willow was due to the water but curiously no comment was
made on the small decrease in the weight of the soil.

Many"plant physiologists"of the seventeenth century considered that
plants could absorb their organic nutrients directly from the soil.
Woodward[231] showed that plants can grow in water but grow better in
river water and best of all in a watery soil extract. He accordingly
rejected Van Helmont's conclusion and postulated that earth not water
is the matter that constitutes vegetables. Although this view was well
received, ideas as to the nature of substances absorbed from the soil
were vague. Boerhaave[16] considered that plants absorbed juices of the
earth and Jethro Tull[210] suggested that very minute particles of soil
were freed by the action of water and taken up into the "lacteal mouths
of the roots". Külbel[108] thought that the magma unguinosum present in
humus was a major source of soil fertility.

Considerable progress was made in our understanding of the role of
soil organic matter in soil fertility from about the beginning of the
last century. Ingen-Housz[89] realized that atmospheric CO_2 was absorbed
by leaves and, in the presence of light, was converted into combustible
plant material. Senebier[181] also considered that plants utilized CO_2 in
the presence of light, but in common with many researchers of the period
he held the view that most of the CO_2 was absorbed from the soil by
plant roots. The concept that CO_2 was absorbed from the soil was
responsible for the widely held view that the soil humus was the only
true source of carbon utilized by plants.

The direct utilization of humus substances by plants (the Humus
Theory) was fully developed by Thaer[203,204] who stated that "humus
comprises a more or less considerable portion of the soil; fertility of
the soil depends largely upon it since besides water, humus is the only
material which supplies nutrients to plants". Such ideas were widely
held and to some extent reflect the limited scientific knowledge of
the period. Davy[45] argued that, in general, plants take in carbon
through their roots and he suggested that chimney soot is valuable

because it contains carbon which is "in a state in which it is capable
of being rendered soluble by the action of oxygen and water". Even as
late as 1872, Grandeau[71] still believed that humus was a major component
of plant food and provided plants with carbon and other nutrients.
The Humus Theory was popular because it combined, in an uncritical way,
the views of chemists and biologists of the period. The chemists had
isolated a group of humus components, the humic acids, which were
soluble in alkalis but insoluble in water. It,therefore,seemed reasonable
to suggest that alkalis and alkali earths present in the soil would
dissolve the humus and make it available to plant roots.

A more critical assessment of the role of humus in soil fertility
was made early in the nineteenth century just prior to Thaer's full
development of the Humus Theory. De Saussure[55] established that plants
could synthesize organic substances using atmospheric CO_2 and water.
Furthermore, he concluded from a series of pot experiments that soil
humus provides a small but essential part of plant nutrients including
nitrogen. Sprengel[191,192] who had carried out considerable work in
the chemical investigation of humic substances (see Introductory
chapter), held views similar to those of De Saussure at a time when it
was more fashionable to be in agreement with the Humus Theory. But
evidence against this theory was also presented by Boussingault[21,22]
and Liebig[118,119,120]. Boussingault deduced that much of the dry
matter produced in crops of beets, turnips, clover, wheat and oats
came from substances in the air and rain,but not from humus. But it
was the contribution made by Liebig which strongly influenced the views
on the role of humus in plant nutrition. He concluded that because
humus was largely insoluble in water, it could not supply all the
carbon requirements of plants and hence much of this carbon must come
from atmospheric CO_2. In addition he considered that ammonia was the
main source of nitrogen for plant growth but surprisingly altered his
views and later regarded the atmosphere as providing the source of
nitrogen. The important role of inorganic constituents in plant
nutrition was also emphasized by Liebig who put forward the Mineral
Theory stating that "The crops on a field diminish or increase in
exact proportion to the dimution or increase in the mineral substances
conveyèd to it in manure".

In the U.K., Lawes and Gilbert performed extensive field trials at

Rothamsted and by 1855 had come to four important conclusions[113,171].
Firstly, crops require phosphates and alkalis, but the composition of
the plant ash does not give a reliable estimate as to the amount of
each constituent required. Hence turnips require considerable
quantities of phosphates but their ash contains only small amounts
(<8.0%). Secondly, non-leguminous crops require a supply of inorganic
nitrogenous substances such as nitrates or ammonium salts. Thirdly,
soil fertility may be maintained for some years at least by applying
only artificial fertilizers. Fourthly, the beneficial effect of
fallowing lies in increasing the available nitrogenous substances
in the soil. Ville[223] was more extreme than Lawes and Gilbert and
maintained that the application of artificial fertilizers was the only
way to preserve soil fertility.

Although the controversies between the Humus and Mineral Theories
continued for many decades, a modification of the latter has become
generally accepted. The observations that plants can grow well in
balanced nutrient solutions without humus[102] pointed the way forward
for more exacting experimentation. By the beginning of the present
century it was established that plants can take up and utilize some
sugars[1,112]. Indeed, Knudson[103] showed that added sugars enhanced root
growth to a much greater extent than they did shoot growth. The
ability of plants to take up and assimilate organic nitrogenous
substances such as urea, acetamide and nucleic acids was also
established[87,88,129,180,228].

In a series of fifteen papers published between 1912 and 1921
Bottomley, for example[17,18,19,20,146], came to the conclusion that
humic substances enhanced the growth of Lemna major, Salvinia natans
and Limnobium stoloniferum in solution culture by providing substances
which he termed "auximones". Similar ideas were expressed by
Hillitzer[86] and Chaminade and Boucher[38]. In contrast, Olsen[146]
concluded that humic substances enhanced plant growth by rendering
iron available to the plant and Burk et al.[27] essentially agreed with
this view. On the other hand, Lieske[121,122] suggested that humic
substances alter the permeability of plant membranes and so promote the
uptake of nutrients.

The results and interpretations of experiments such as those
discussed above form the basis of much of our present day approach to

the influence of soil organic matter, particularly humic substances, on
plant growth. The remainder of this chapter is devoted to the effects
of strictly humic substances on the growth and physiology of,mainly,
higher plants. In discussing physiological processes it should be
emphasized that these and biochemical processes are so inter-related
that chapters 1 and 2 should be considered together.

2. INFLUENCE OF HUMIC SUBSTANCES ON GROWTH

2.1 Intact higher plants

Many reports in the literature reveal that humic substances can,
under certain conditions, stimulate plant growth. These reports also
show that the humic substances are supplied to plants in a variety of
ways. When applied to the leaves of Begonia semperflorens, humic
substances increased the fresh weights of both stems and roots relative
to the untreated controls[185,187]. The response of this plant to fulvic
acid (FA) was greater than its response to humic acid (HA)[185]. When
applied in low concentrations, as foliar sprays under field conditions,
HA enhanced the dry weight production of corn [100] and similar
favourable effects have been claimed for many other crops[99,101].
More recently, Varshney and Gaur[214] claimed that spraying soybeans with
HA solutions enhanced crop yields. Most reports on spraying crops with
HA are to be found in the Eastern European literature. An exception to
this is a report from the United States which showed that foliar
application of humic substances produced a small increase in the dry
matter production of maize[114]. The application of solutions of HA
directly to soil under greenhouse conditions can also enhance the dry
matter yields of crops such as Triticum vulgare, Trifolium alexandrinum
and Sesbania aculeata[68]. Similarly the addition of FA to soil enhanced
the growth of alfalfa and bromegrass[117]. The practical value of large
scale application of humic substances to the soil to enhance plant
growth has been questioned by Wiersum[227]. Indeed it would appear that
externally supplied humic substances only have a substantial effect when
added to soils of low organic matter content[114].

Other techniques have also been used for testing the effects of
humic substances on plants,for example, injecting the tissues with
solutions of HA[95,104], or pelleting the seeds with Ca - humate[90].
However, most experiments have been designed to test their effects on

plant growth in sand or water cultures. Under these conditions there is little doubt that humic substances can enhance plant growth as measured in terms of increases in length and dry and/or fresh weight. A selection of references showing the diversity of plant species (all Angiosperms), sources of humic substances and culture conditions is shown in Table 1. Several review articles are also available[15,33,75,99,101,104]. In some cases it is not clear whether the term "water" or "water culture" refers to water alone or to a nutrient solution. In addition, the description of a humic compound is sometimes misleading and does not

Table 1. A selection of references showing the beneficial effect of humic substances from various sources on the growth of plants.

Plant species	Humic substance and source	Culture conditions	Reference number
Mulberry	HA from coal	Water	5
Ryegrass	HA from compost	Nutrient/soil	31
Wheat	Commercial HA	Water/nutrient	41,42
Sunflower	Water extract of peat	Nutrient	47
Mustard	HA from lignite	Nutrient	48
Bean, grass	HA & FA from soil	Nutrient	54
Maize	HA from peat and FYM	Nutrient	58
Wheat	Synthetic HA	Nutrient	60
Maize	HA/FA from soil	Nutrient	63
Tomato	HA from compost	Nutrient	79
Maize	HA/FA from lignite	Added to soil	92
Wheat	HA from soil, shale, peat	Water/nutrient	94
Maize	HA from soil, peat, straw	Nutrient/soil	114
Tomato	FA from soil	Nutrient	123
Tobacco	HA/FA from soil; "aquahumus"	Nutrient	136
Several spp.	HA from peat; extracted FYM	Water/nutrient	140
Green pepper	HA from lignite	Nutrient	151
Wheat	Commercial HA	Water	154
Cereals/rice	"Oxyhumolite"	Added to soil	163
Flax	HA from peat	Water	164
Cucumber	FA from soil	Nutrient	165
Tomato	HA from soil and coal	Nutrient	168
Cereals	HA from soil, peat coal	Water	169
Tomato	HA/FA from garden soil	Nutrient	184
Several spp.	Commercial HA	Nutrient	188
Maize	HA from coal	Nutrient	200
Soybean, peanut	HA/FA from soil	Sand + nutrients	201
Wheat	HA from soil	Nutrient	219

Abbreviations:- HA - humic acid, FA - fulvic acid, FYM - farmyard manure.

always include its origin. There would be less dubiety in using the
term HA to apply solely to a fraction derived from a soil[44].

The response to the humic substance often depends on the age of
the plant as for example shown for the influence of HA on tomato
plants[168] and FA on tobacco[137]. The nature of the response often
depends on the concentration of the humic substance and high
concentrations are usually inhibitory, for example[56,79,94,104,136,137,
165,168]. The range of optimum concentration varies between different
plant species (see Kononova[104]). There are also reports that the exact
response depends on the source of the humic substance[9,58,85,94,106,136,
148,168,190]. Some workers have even shown that the effects of humic
substances on plant growth can be duplicated using the chelating agent
ethylenediamine tetraacetic acid[47,48,79,82,212].

The precise effect of the humic substance can also depend on the
plant species. Thus, for example, Khristeva[94] reported that HA
stimulated the growth of cereals to a greater extent than it did the
growth of legumes. In a subsequent paper, Khristeva[96] also showed that
potatoes, tomatoes and sugar beet were most responsive to HA whereas
sunflower, pumpkin and cotton gave hardly any response. Similarly
Guminski[74] reported that whereas HA increased the growth of rye and
wheat it had little effect on oats and barley. More recently,
Van der Werff[212] has reported that under certain conditions, FA
enhanced the growth of Senecio sylvaticus to a much greater extent than
that of Holcus lanatus. But even for a single plant species the values
quoted in the literature can vary considerably. Rochus[168], showed
that at its optimal concentration in nutrient solution, HA enhanced the
growth of tomato roots on a dry weight basis by less than 25% whereas
Sladky[184] also using a nutrient solution, reported a stimulation of about
360%.

The relative effect of a humic substance on plant growth ultimately
depends on the parameter which is measured. Vaughan and Linehan[219]
reported that under axenic conditions HA enhanced the increase in fresh
weights of shoots by 29% whereas the corresponding dry weight was
increased by only 13%. In an extreme case, Sladky[184] claimed that FA
increased the lengths of tomato roots by a mere 10% but the fresh and
dry weights were enhanced by 245% and 390% respectively. This latter
report also demonstrates the well documented phenomenon that different

organs of intact plants respond to humic substances to different extents. In this case HA, stimulated root lengths by 54% and stem lengths by 146% when whole plants were grown in nutrient solution. The contrast was even more marked when the tomato plants were grown in FA because whereas stem lengths were stimulated by 170%, the roots were enhanced by 10% only. An excellent example of the effect of a humic substance on the growth of different organs in the intact plant has been reported recently by Rauthan and Schnitzer[165] for the effect of FA on the growth of cucumbers in nutrient culture. In this case, FA at a concentration of 100 mgl^{-1} increased the length of the root by 31%; the height of the shoot by 81%, the freeze-dried weight of the plant by 130%, the number of leaves per plant by 40% and the number of flowers per plant by 145%. Many other investigations have shown differences in responses to humic substances of different organs in intact plants and also differences between responses to HA and FA[66,114,115,136,188,212].

In addition to increasing the lengths and fresh and dry weights, humic substances can exert a favourable effect on the development of adventitious roots in nutrient culture[95,104,106,136,188,190]. The favourable effects of humic substances often occurs when the plants are grown in nutrient solutions under axenic conditions in thin film isolators[128,219]. This shows that the beneficial effects are due to the humic substances _per se_ rather than being mediated via the microbial breakdown products as suggested by Chizhevsky and Dikusar[40].

2.2. Influence of culture conditions

The composition of the culture medium and the subsequent culture conditions are critical for the extent to which humic substances may influence plant growth[62,79]. In some reports the growth media comprise water alone[41,42,154,169,188] and under these conditions humic substances would be expected to enhance plant growth by providing a suitable source of nutrients in addition to organic matter. Thus Kononova and Pankova[106] demonstrated that HA enhanced the length of maize roots compared with the distilled water controls. The observation by Pagel[149] that humic substances in a nutrient solution increased the dry matter content of ryegrass roots by 25% whereas with an insufficient nutrient supply there was an increase of some 50% adds weight to the interpretation that the humic material supplies additional nutrients

under these conditions. Smídová[188] also showed that HA, at high
concentrations of 100 mgl^{-1}, enhanced the increase in fresh weight of
wheat roots to a greater extent in water (41%) than in the presence of
a Knop nutrient solution (24%). However, it should be noted that the
presence of nutrients alone enhanced root growth by 137% when compared
with the water controls. Similar results for the effect of HA on the
growth of wheat plants in the presence and absence of a Knop nutrient
solution have been reported by Činčerová[41,42]. Vaughan and Malcolm
have also shown a similar effect of HA on the growth of winter wheat
in a Hoagland nutrient solution under axenic conditions (Table 2).
These results exemplify several reports in the literature that although
HA stimulates plant growth when compared with the water controls, the
effects of nutrient alone are usually greater than the effect of the
humic material. A recent report[62] has shown that even using a
Hoagland nutrient solution Fe must be present for HA to enhance the
growth of maize plants, indicating the key role of Fe in the HA growth
response.

There have been virtually no attempts made to standardize the
nutrient solutions used in different laboratories. Similarly, there have
been virtually no inter-laboratory comparisons of the effects of humic
substances on a single plant species. Among some of the nutrient media
used have been those of Sachs[184], Knop[42,162,168,174,188], Hampe[79,127,219],
Hoagland[165] and Detmer[175] for whole plants, and the media of Street[123],
Hoagland[62] and White[84] for cultured excised roots. The composition of
the nutrient solution is important because it will ultimately determine
the final effect of the humic substances on plant growth.

Chaminade[31] showed that HA stimulated the growth of ryegrass in
the presence of 20 mgl^{-1} inorganic N in the medium and this was enhanced
further in the presence of 50 mgl^{-1} inorganic N. Mortensen[135] speculated
that HA complexes a number of nutrient cations and thus influences the
growth of plants indirectly rather than by the likely direct effect
proposed by Mylonas and McCants[136,137]. In support of this, Fernandez[58]
demonstrated that Zea mays grown in a nutrient solution composed of high
concentrations of nutrients showed signs of toxicity such as a shortening
of the internodes and yellowish leaves, symptoms which were absent in
the presence of HA. Moreover, Lisiak[127] has reported that HA protected
tomato plants against the toxic effects of excessive amounts of Ca in a

Table 2. The effect of 50 mgl^{-1} concentrations of humic acid on the growth of winter wheat in water or Hoagland's nutrient solution under axenic conditions after 15 days.

Culture medium	Plant organ analysed	Fresh weight (mg/plant)	Stimulation (%)
Water	Root	92.7 ± 5.9*	0
	Shoot	185.0 ± 10.2	0
Water + HA	Root	146.0 ± 13.2	57.5
	Shoot	252.0 ± 14.3	36.2
Hoagland	Root	182.0 ± 10.4	96.3
	Shoot	342.0 ± 17.2	84.9
Hoagland + HA	Root	203.0 ± 9.8	119.0
	Shoot	390 ± 16.9	110.8

* Values are means ± SE of 10 plants.

full nutrient solution. Humic substances have been shown to counteract the toxic effects of several other metals such as cadmium[195], zinc[56,212] copper[208] and iron[212].

The interactions between some nutrients in solution often result in their precipitation, especially after autoclaving,thus reducing their effective concentrations in what would otherwise have been a well balanced growth medium. Thus it has been reported[127] that at low pH relatively insoluble iron phosphates are precipitated from solution thus resulting very often in phosphorus deficiency. Under such circumstances humic substances may complex with such nutrients keeping them in solution and hence more available to the plant[27,47,48,125,146]. Kononova and D'yakonova[105] concluded that humic substances prevented the precipitation of Fe and other micro-nutrients in nutrient solutions even at a high-phosphorus concentrations. Biddulph[11] too reported that HA maintains far more Fe in solution in the presence of inorganic-P than do many other complexing agents.

During growth experiments, root exudations often lower the pH of the nutrient medium with a concomitant precipitation of humic substances, especially HA and its complexes with nutrients such as calcium -, zinc - and even iron - humates. Autoclaving also induces precipitation or results in colloidal solutions of HA rather than a "true" solution.

The extent to which precipitation influences the overall effect of humic substances on plant growth is difficult to assess from the literature, but the ever present risk of precipitation must be appreciable in those cases where the initial pH of the media is low, for example 4.0[80,212]. The results from our laboratories indicate that nutrient solutions containing HA should be changed frequently not only to maintain the optimal pH and replace depleted nutrients, but to avoid precipitation of the humic materials. It should, however, be borne in mind that under natural soil conditions most nutrients will not be at an optimal concentration. It is virtually impossible to reproduce these conditions in the laboratory but the direct effect of humic substances can only be ascertained when no other factors are limiting and even in a well balanced medium the presence of the humic material may itself alter this delicate balance.

Other important areas of nutrient culture which are often neglected and referred to infrequently, include adequate aeration and illumination[114,137,219]. The importance of aeration has been stressed by Gumiński[75] who reiterated that "the purpose of aerating a solution is not only to admit oxygen, but also to mix the precipitate of insoluble salts". Gumiński and his co-workers[73,79] showed that HA enhanced the growth of tomato plants in nutrient culture to the greatest extent under conditions of poor aeration. Additionally, even after stimulation by HA under conditions of inadequate aeration, the tomato plants were still considerably smaller than when grown in nutrient solution alone under adequately aerobic conditions.

From the foregoing discussion it is apparent that the effect of a humic substance on plant growth will depend on many factors including the nature of the humic substance, its source and concentrations, in addition to the composition of the culture medium, its pH and culture conditions and finally the plant species and growth parameter being measured. Taking all these factors into consideration it is hardly surprising that there should be wide discrepancies in the literature ranging from no effect on the one hand to several hundred per cent on the other. Indeed some of the effects are difficult to reproduce. Thus one of us (R.E. Malcolm, unpublished) has been unable to confirm a stimulation in the growth of the duckweed Lemna major using HA reported earlier this century[17,18,19,146].

2.3. Growth of plant organs

As in the case with intact plants, much information is available to show that under certain conditions, humic substances can influence the growth of isolated plant organs. Isolated organs can be used for several purposes:- (a) to test whether an effect of humic substances on growth is due to cell expansion, cell division or both processes[84,123,216], (b) to test humic substances for auxin-like properties[145,153,160] and (c) as a quick and reliable test to distinguish between the biological activities of different humic substances or their fractions[215,221]. Many of the factors discussed for whole plants are equally valid for the isolated organs. Hence, humic substances can enhance the growth of excised roots in nutrient culture, for example[62,123,186,227], and this effect is dependent on the composition of the nutrient solution[62] and on the concentration of the humic material[123,186,215,216]. In experiments designed to investigate the effects of humic substances on cell elongation alone in plant organs, it is not necessary to use a full nutrient solution. In this connection, Poapst et al.[153] showed that FA enhanced elongation in isolated stem segments of Pisum sativum and Vaughan[216] has reported that HA stimulated cell elongation of excised root segments of the same plant species. Generally, the optimal concentration of the humic material required by the isolated root system (5 - 25 mgl^{-1})[123,186,216] is considerably lower than that (60 - 100 mgl^{-1}) required by the intact plant, for example[94,104,165,168].

The exact response depends on the nature of the humic substance. In excised tomato roots grown in nutrient culture, HA was marginally more effective than FA in enhancing growth[84]. However, it would appear that these two humic fractions influence different aspects of growth and whereas HA enhances cell elongation to a greater extent than it does cell division, FA has the opposite effect[186]. Even so, HA can also stimulate cell division and the volume of interphase nuclei in onion root tips[70,147]. The stimulation by HA of root elongation has also been demonstrated in root segments of Pisum sativum[145,216], although in this tissue FA was more effective than HA[145]. In excised tissues such as pea root segments, humic substances evoke their greatest response only in the presence of a suitable energy source such as sucrose[216].

In addition to influencing cell elongation, humic substances can enhance the production of laterals in excised tomato roots[123]. Schnitzer and co-workers[178,179] also demonstrated that FA stimulated root initiation in hypocotyl segments of Phaseolus vulgaris. Interestingly, the optimal concentration of FA required in the hypocotyl system ($3000 - 6000$ mgl^{-1}) is considerably greater than the 25 mgl^{-1} reported by Linehan[123] for excised tomato roots. It is likely that the difference in response results from the different bio-assays used rather than from the intrinsic differences in the FA preparations[123].

Schnitzer and Poapst[179] have postulated that the carboxylic and phenolic and alcoholic hydroxyl groups in FA are responsible for the influence of FA on root hypocotyls as a result of their chelating Fe. Carboxylic acids are almost certainly involved in root initiation in tomato root sections because polymaleic acid, a polycarboxylic acid similar to FA[3], evokes a response similar to that of FA.

Root initiation can also arise as a result of auxin activity so that it is reasonable to ascribe auxin activity to humic substances. This latter interpretation has been applied to the initiation of roots in cuttings of Pelargonium hortorum[145] and to the regeneration of tissues in several plant species[160]. Although this aspect of humic substances is discussed in more detail in the conclusions of chapter 2, it is relevant here to mention that plant organs are often used as a bio-assay for auxins. Using the tips of isolated pea plants,[116] O'Donnell[145] concluded that both HA and FA possess auxin activity, although the pea root test system is not specific for auxins. In a more auxin-specific test system of the wheat coleoptile[116], Paszewski et al.[150] found that some alcohol-soluble fractions of humic substances behaved as auxins, although the unfractionated substances were inactive. In contrast, Řeřábek[166,167] was unable to demonstrate any auxin activity in wheat coleoptiles, and flax hypocotyls and roots. Vaughan and Malcolm (unpublished) have reached a similar conclusion using coleoptiles, in the presence and absence of 2% sucrose. Poapst and Schnitzer[152] considered the possibility that FA may act in a manner relatively independent of endogenous auxin in root initiation in bean hypocotyls.

Although not strictly plant organs, seeds are also influenced by humic substances particularly with regard to their germination. There

are many papers reporting that HA can stimulate the rate of seed
germination, for example[9,60,77,151,207]. It has also been claimed that
HA accelerates water uptake into seeds thereby activating growth enzymes
more quickly[189]. Gumiński and Sulej[77] reported that HA stimulated the
germination of photophilous seeds (from Nicotiana tabacum and
Lactuca sativa) which were not illuminated before germination, and
photophobic seeds (from Amaranthus caudatus and Phacelia tanacetifolia)
which were illuminated before germination. These workers favourably
compared the effects of the HA with similar effects produced by
gibberellin. Interestingly, HA had no effect on germination when seeds
were given the incorrect illumination regime suggesting the involvement
of the phytochrome system in the response[77]. The available evidence
suggests that humic substances increase the rate rather than the
percentage of germination[207]. Some evidence has been presented to show
that treating cereal and legume seeds with model HA leads to an
enhancement of growth immediately after germination[60,207]. Vaughan and
Malcolm (unpublished) have observed that soaking wheat seeds in 50 mgl^{-1}
concentrations of HA at pH 6.0 for 24 h prior to germination had a
beneficial effect on subsequent plant growth in a Hoagland nutrient
medium under axenic conditions. (Fig. 1). This effect was more apparent
at the beginning of the growth period and has disappeared after ten days.
This contrasts with the situation where to obtain the maximum stimulation
of growth, plants must be cultured continuously in solutions of HA. It
would, however, appear that in some cases HA can affect the rate of
germination and the initial growth response.

Surprisingly, virtually no information is available on the
influence of FA on seed germination. This is especially so when one
considers that FA is often found in the soil solution. We have shown
(unpublished results) that presoaking winter wheat or barley seeds in
solutions of FA up to a concentration of 100 mgl^{-1} has no effect either
on rate of germination or the subsequent growth of the plants under both
axenic or non-axenic conditions. In this connection it is interesting
to note that hot water-soluble HA, which often produces metabolic effects
similar to FA (see chapter 2), is also without effect on germination
or subsequent growth of winter wheat or barley.

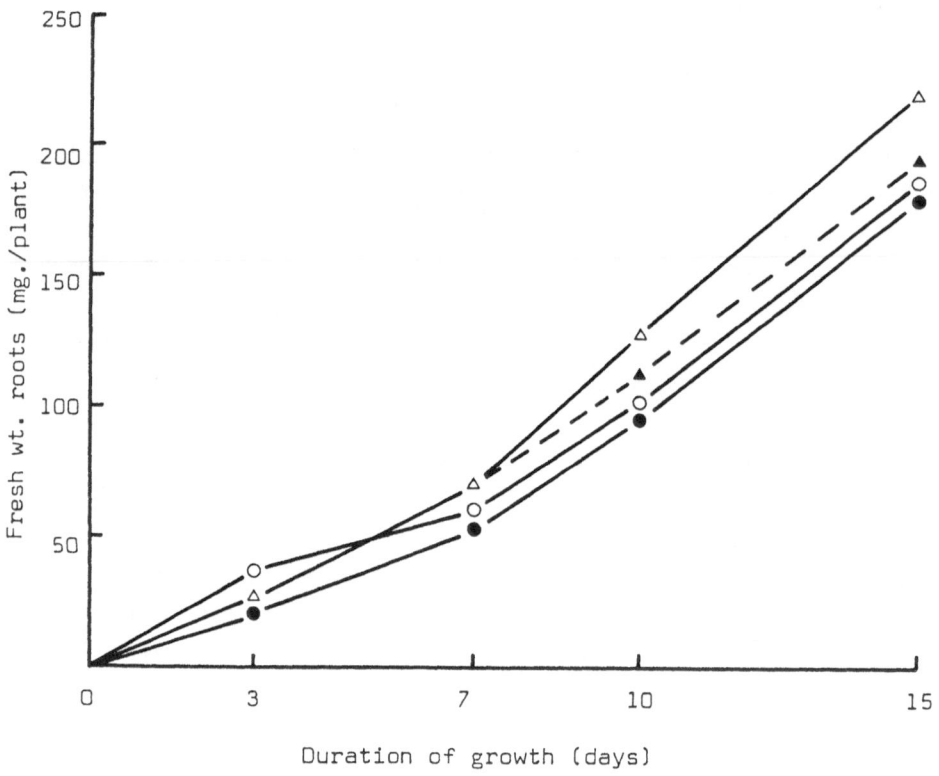

FIGURE 1. Effect of soaking winter wheat seeds in HA on the
subsequent growth of the plants. Treatments were:-
(a) seeds soaked in water for 24 h then grown in Hoagland
solution with (Δ) or without (●) HA,
(b) seeds soaked for 24 h in 50 mgl^{-1} HA then grown in
Hoagland (o),
(c) plants grown in Hoagland plus HA for seven days then
transferred to Hoagland alone (▲).

2.4 Growth of micro-organisms

There are numerous reports of humic substances having a beneficial
effect on the growth of a wide range of microbial species. Thus HA
can enhance the growth of:

(a) algae[23,69,111,114,158,202,208]
(b) bacteria[8,28,49,91,107,133]
(c) fungi[4,26,107,132,199,229,230]
(d) yeasts[76,78,138]
(e) diatoms and dinoflagellates[155,156,157]

Most of these examples refer to the effects on growth measured as increases in weight or cell numbers but other beneficial effects have also been reported. Willoughby and Baker [229,230] demonstrated that HA can induce sporulation in several species of the Actinoplanaceae, while Prakash et al.[156] have shown that HA can increase the chlorophyll contents of some dinoflagellates. Such effects are important because dissolved humic substances occur not only in the soil solution[124] but also in many aquatic environments and by stimulating the growth of, for example, algae can create an environmental problem[69,111,131,182,226]. Both high MW[69] and low MW[157] materials may be involved in the response depending on the microbial species. However, the extent of the growth increase is likely to depend on the source of the humic substance as shown for example for the yeast Torulopsis utilis [138]. But not all micro-organisms respond to humic substances. McLaughlin and Kuster[130] were unable to find any effects of humic substances on the yeast Candida utilis.

Humic substances may produce their beneficial influence on microbes in several ways. Firstly, they can act as a source of food and energy for many micro-organisms[26,159]. Hence Pseudomonas sinosa can use HA as a source of C and N[132] and this is also true for Aspergillus sp.[139] while Bacillus megaterium is capable of using HA as a source of N[4]. Burges and Latter[26] have reported that some fungi such as Polystictus versicolor can use the non-N moiety of HA. It has also been concluded that some bacteria can use FA as a source of C[49]. The breakdown and utilization of humic substances has important connotations in the degradation and transformation of low MW humic material in the soil. The stimulation of growth of some algae such as Scenedesmus obliquus may arise from the direct heterotrophic use of HA or its breakdown products due to microbial decomposition[69]. In some microbial species the exact effect induced by a humic substance may depend on the HA fraction. Willoughby and Baker[229] reported that N-rich HA fractions (obtained by refluxing the HA with water and/or acid) enhance the growth of mycelia of some water fungi whereas the residual HA "core" material[81] induces sporulation in some of these species.

Secondly, humic substances may influence the growth of micro-organisms by virtue of their chelation properties[8,91,157]. Lange[111] has suggested that FA stimulates the growth of blue-green algae because it retains more chelated Fe in solution thus making it available to the micro-organisms. Indeed, there is little doubt that humic substances can increase Fe concentrations in lake waters by chelation[182]. In contrast, Prakash and Rashid[155] have suggested that the chelation of trace metals may reduce their toxicity to dinoflagellates and in this context, Waris[226] has reported that chelation can decrease the toxicity of heavy metals to freshwater algae. Certainly HA can bind metals tightly[123] thus making them relatively unavailable for uptake. Some of these observations may appear contradictory if growth is often enhanced by increasing the solubility of metals by chelation[69].

Thirdly, humic substances may act by modifying cell membranes. This is discussed in more detail for higher plants later in this chapter (page 62). Here it should suffice to state that there are several reports which suggest that the influence of humic substances on micro-organisms may be a membrane phenomenon[155,226]. Gumiński and Sulej[76] working with the yeast Saccharomyces cerevisiae concluded that HA retarded both plasmolysis and deplasmolysis and in this respect behaved like gibberellic acid.

Fourthly, humic substances per se may be taken up by microbes. There is strong evidence that FA is taken up into mycelia of Pisolithus tinctorius[199] and HA by Aspergillus niger[170]. Once taken up, humic materials may affect growth directly via the biochemical mechanisms discussed in chapter 2. Examples include the stimulation of invertase activity and respiration in the yeast Torulopsis utilis[138] and active ion uptake by the mycelia of A. niger[170].

It is likely that in most cases humic substances enhance microbial growth by serving as a food source. In other circumstances, the precise mode of action of the humic material is dependent on the microbial species and factors such as the composition of the growth media, pH and temperature; indeed factors similar to those which can be applied to higher plants.

3. UPTAKE OF HUMIC SUBSTANCES

The assertion that humic substances can have a direct effect on plant growth and metabolism[15,33,59,75,95,100,215] implies that these materials are taken up by plant tissues, but relatively little attention has been devoted to this important subject. Initially, workers relied on observing changes in colour of plant organs to detect the uptake and incorporation into tissues of humic substances. Azo and Sakai[5] immersed seedlings of a mulberry tree in ammonium humate and then homogenized the plant parts. The homogenates were of a brownish colour and so it was claimed that humate had penetrated the tissue. In another report based on visual observation, Prát[161] obtained microscopic evidence that HA was taken into the shoots of several plant species which were allowed to stand with their cut ends in HA solutions. Although this report demonstrated that HA could move along vascular bundles slowly and stain the cell walls of xylem parenchyma, it did not show that the humic material could be taken up by roots.

Most investigations into the uptake and incorporation of humic substances have been carried out using ^{14}C -labelled materials, particularly by Führ and Sauerbeck[64,65,66,67]. These substances are obtained by incubating (^{14}C)-labelled plant residues with soil for several months and then extracting the humic fractions with NaOH in the usual manner[39,66,93,162,176,177,183,220,221]. Unfortunately, up to 70% of the radioactivity can be lost from the soil as $^{14}CO_2$ in the first year[93] so that the resulting humic substances are very low in radioactivity. When ^{14}C -labelled cereal rye straw was incubated with soil for 224 days, the resulting HA had a specific radioactivity of <100 Bq/mg carbon[220,221]. Nevertheless, these humic substances are still sufficiently active to produce reliable results in most experimental systems.

Using ^{14}C -labelled material, Prát and Pospíšil[162] showed that HA accumulated in the roots of <u>Beta saccharifera</u> and <u>Zea mays</u>. Furthermore, only a small fraction of the radioactivity was transported from the root to the shoot. Similar observations have been reported for intact plants by several other workers[64,66,159,212,219]. Some evidence is available to suggest that FA is transported to the shoot to a greater extent than is HA[64,65,67]. Radioactive HA is also taken up from a nutrient solution by wheat roots under axenic conditions with only a limited amount (<5%) of transport to the shoot[219]. Hence HA is taken up directly rather than via its microbial breakdown products, but whether this humic substance is taken up in an unaltered form is unknown.

The available evidence suggests that HA is absorbed at the root surface or accumulates passively in the "apparent free space"[67,220]. But some humic components may also be taken up actively[220]. Führ and Sauerbeck[67] using autoradiography have shown that whereas much of the incorporation of radioactivity from (^{14}C)-HA was into the epidermal cells of sunflower, radish and carrot roots, a substantial amount of activity from the lower MW components also entered the stele. At the cellular level it is mainly those humic substances of low MW which accumulate within the cells[67,221].

In parenchyma storage tissues of Beta vulgaris some of the humic substances are taken up actively[215,221] and in this tissue as much as 15% of the radioactivity becomes associated with mitochondria and 25% with the remaining soluble cell components[217,221]. Using excised pea roots under axenic conditions, it has been shown that initially there is a rapid adsorption of humic substances on to the root surface followed by a period of slower active uptake of the lower MW components[220]. Hence the ratio of adsorbed: unbound HA in the roots decreases with time[219,220]. Because FA comprises only low MW components of <2500 daltons the ratio of uptake for FA:HA increases with incubation time. Similarly the low MW HA fractions are taken up both actively and passively whereas HA components of MW > 50,000 daltons are taken up only passively[220]. It is, therefore, likely that only the low MW humic substances are taken up by a mechanism dependent on metabolism irrespective of whether they are derived from HA or FA. However, both high and low MW substances can be taken up by pinacytosis[59].

The notion that it is mainly the low MW humic substances which are taken up actively is interesting in that these fractions can be more active than their higher MW counterparts in their direct effects on metabolism in many biological tissues (see[215,221] and Chapter 2). Furthermore, Campbell et al.[29] have shown that the lowest MW components are also the youngest of the HA complex so that it may be speculated

that the youngest HA's are the most active in their direct effect on
metabolism and consequently plant growth. It has even been postulated
that it is the amount of low MW humic substances, rather than the total
humic concentration, that gives an indication of soil fertility[57].

4. NUTRIENT CONTENT OF PLANTS

A favourable effect of humic substances on the nutrient uptake and
contents of plants has been reported for the major inorganic elements
added to soil such as nitrogen, phosphorus, potassium and sulphur
(Table 3). In addition, the uptake and contents of nutrients such as
calcium, magnesium, sodium, and copper are also enhanced by humic
substances (Table 4).

The exact nature of the response is dependent on the concentration
of the humic substance[54,218] and high concentrations of both HA and FA
are usually inhibitory[52,137,165,217]. However, for some elements for
example chloride, only inhibitory effects have been reported[80,217], while
for others there are several reports of either no effect or inhibitions
contrasting with the enhancements reported above, for example
potassium[148], magnesium[54,178], iron[54], phosphorus[54], sodium[54,174] and
calcium[54,174].

Most of the reports on the effects of HA on the nutrient contents
of plants are purely descriptive and little attempt was made in the
earlier work to elucidate the mechanisms of the action of the humic
material. The final concentration of a nutrient in a plant can depend
on a number of factors such as the concentration in the nutrient solution,
its interactions with other nutrients and/or humic substances, the pH of
the nutrient medium in addition to the effects of the humic substances
on the ion uptake mechanism and finally translocation within the plant.
Several of these factors are well illustrated in a recent paper by
Gumiński et al.[80]. These workers showed for example that Mg is taken up
to a greater extent at pH 5.0 than at pH 7.0 and that, in general,
humate stimulated Mg uptake at both pH values . Gumiński et al. also
demonstrated that the metabolic inhibitor 2,4-dinitrophenol had an adverse
effect on any stimulation of ion uptake produced by HA. Such a result
demonstrates that humates can influence nutrient concentration in plants
directly through metabolic processes.

58

Table 3.   A selection of references showing the beneficial effect
of humic substances on the uptake or content of N, P, K
and S in plants.

Element	Plant species	Humic substance	Reference numbers
Nitrogen	Cucumber	FA	165
	Maize	HA	58
	Rye	HA	32, 34, 35, 174
	Rye	HA/FA	115
	Tobacco	HA/FA	137
	Tomato	HA	80
Phosphorus	Bean	HA	53
	Cucumber	FA	165
	Maize	HA	114
	Maize	HA/FA	92
	Mustard	HA	141
	Oats, barley	HA	169
	Tobacco	HA/FA	137
	Tomato	HA	80
	Wheat	HA/FA	169, 222
Potassium	Barley	SOM	143
	Dactylis	SOM	13
	Flax	HA	164
	Tobacco	HA/FA	137
	Tomato	HA	80
	Various spp.	SOM	61
	Wheat	HA	14, 34, 35
Sulphur	Barley	HA	52
	Cannabis	HA	141
	Mustard	HA	141

The effects of humic substances on ion uptake may be broadly
divided into metabolic (active) and non-metabolic (passive) processes.
The active uptake can be investigated by interfering with metabolism[80,217]
and is discussed more fully in chapter 2. However, the differences
between the two processes can be illustrated by the uptakes into
storage parenchyma cells of Beta vulgaris of Na, the uptake of which is
active and stimulated by HA, and of Zn which is taken up by passive
forces and inhibited by HA[218]. Different cations are complexed to

different extents by HA, heavy metals more readily than alkaline earth metals[194]. HA will thus complex Zn more readily than Na.

Table 4. A selection of references showing the beneficial effect of humic substances on the uptake or content of some metals in plants.

Element	Plant Species	Humic substance	Reference number
Calcium	Barley	SOM	144
	Cucumber	FA	165
	Maize	HA	148
	Rye	HA	34, 35, 36
	Tobacco	HA/FA	137
	Various	SOM	61
Magnesium	Barley	SOM	144
	Maize	HA	58
	Rye	HA	34, 35, 36
	Tobacco	HA/FA	137
	Tomato	HA	80
Sodium	Barley	SOM	144
	Beetroot	HA	218
Copper	Barley	HA	56
	Cucumber	FA	165
Cadmium	Snapbean/maize	HA	211
	Soybeans	SOM	195

The reversal of the inhibition of Zn uptake by HA in the presence of Cu adds weight to the interpretation that it is simply the complexing of the Zn by HA which inhibits uptake although it should be emphasized that Cu can decrease the uptake of Zn even in the absence of a chelating agent[24]. The effect of humic substances in reducing zinc uptake into whole plants[56,212] may be explained in similar terms. Indeed, Van der Werff[212] has shown for Holcus lanatus that HA, FA and EDTA inhibit the uptake of Zn in proportion to their complex constant values[46,178]. Mortensen[135] has emphasized that HA complexes a number of nutrient cations and in so doing may alter the amounts taken up by plants thus influencing the nutrition of plants indirectly.

There are, however, some reports which claim that humic substances can enhance Zn uptake into some plants[62,165]. In addition to complexation, other factors must therefore operate which are more specific to the plant. Furthermore such factors need not necessarily involve the active uptake mechanism yet may still influence the plant indirectly. One such mechanism for the absorption of Zn by roots could involve differences between the cation exchange capacity of the root tissue and that of the complexing agent. This explanation has been used to account for the differences in Zn uptake by Holcus lanatus and Senecio sylvaticus[212]. Certainly the cation exchange capacity of dicotyledons is higher than that of monocotyledons[43]. Once inside the root, the Zn is translocated to the shoot and there is no evidence that humic substances interfere with this latter process. The amount of Zn in the shoots is dependent on its concentration in the roots[208] and this also applies to several other metals such as Mg and K[80]. Recently Tyler and McBride[211] have reported that HA inhibits the uptake of Cd but has no effect on translocation within the plant. This situation differs appreciably from the uptake and translocation of Fe.

In common with several chelating agents, humic substances inhibit the uptake of Fe into roots[56,80,125,126,206,212] but subsequently enhance the concentration of this cation in the shoots relative to controls without the humic material. The observations for the accumulation of Fe in shoots have been reported for both HA[47,114,206,212] and FA[126,165,212]. In 1955, De Kock[47] demonstrated that the effect of HA was not solely on ion exchange at the root surface, but translocation of Fe into leaves was also promoted. More recently, Guminski et al.[80], using the metabolic inhibitor 2,4-dinitrophenol, have shown that humate facilitates only "passive" transport of Fe to the shoots and this also applies for Ca. Mylonas and McCants[137] have also observed that HA and FA facilitate the translocation of Fe, Ca and Mg in tobacco plants. The manner in which HA influences such apparently "passive" processes is unknown but is nevertheless perplexing because <5% of HA taken up by plants is translocated to the shoot[212,219].

Although humic substances can affect ion uptake, it should be borne in mind that in a nutrient solution the interaction between ions and the chelating agent can make a substantial contribution towards the final result. This can be illustrated by considering Fe and P. In the absence of Fe, humic substances can enhance the uptake of P into plants[80,137,165,217,222]. But with Fe present, the complexing of the P and Fe by FA resulted in a reduction of P availability to corn and alfalfa plants[117]. A similar observation was also documented for tomato plants using HA[80]. Gumiński et al.[80] concluded that whereas Fe binding to HA inhibits P uptake, Ca binding facilitates P accumulation in the plant. The interaction between FA, Mn and P also resulted in a favourable effect on P uptake[117]. Thus the eventual effect on uptake resulting from the interactions between different ions and the humic substance is dependent on the nature of the interacting materials.

The results described above were obtained mainly from laboratory experiments but under natural soil conditions humic substances may influence Zn uptake by preventing the formation of insoluble and immobile hydroxides or inorganic complexes[196,212,224]. Nevertheless, the uptake of the soluble but complexed Zn would probably be less than the uptake of the soluble and free Zn salt. In addition, humic substances may affect the uptake of some cations by influencing their rate of release from the soil mineral components. Pertinent to this, Tan[198] has claimed that both HA and FA are capable of releasing some 24-26% of the K fixed by illite or montmorillonite. Clearly, the liberation of K could have considerable effects on its uptake into plants in addition to the stimulation of the "active" K uptake mechanism produced by humic substances per se[80].

5.   HUMIC SUBSTANCES AND MEMBRANES

It has been established that humic substances can influence the uptake of ions into roots both actively and passively, and at the same time these substances become absorbed on to cell surfaces. Such observations lead many investigators to propose that humic components have an effect on plant nutrition at the cellular membrane level[33]. Indeed there is much evidence available to support this contention[78,80,83,172,209,212,217,218].

In 1936, Prozorovskaya[164] demonstrated that the exosmosis of sugars from bulb scales was increased in the presence of HA. This observation lead Prozorovskaya to conclude that small amounts of HA

62

increase the permeability of the plasma resulting in an increased uptake of nutrients by the plant. More recently, Pagel[149] suggested a similar role for HA and several other workers have concluded that humic substances increase the permeability of plant membranes[36,96,172]. Heinrich[83] has recorded that at pH 7.0, HA can increase the permeability to urea of epidermal cells of Gentiana rochelli and sub-epidermal cells of Majanthenum bifolium. In contrast, HA and FA both decrease the permeability to glucose at root surface cells of Pisum sativum[222]. HA will also retard plasmolysis and deplasmolysis in Nitella gracilis[172] and in leaves of the moss Mnium affine[173].

Such observations indicate that humic substances act at the membrane level but do not reveal the mode of action of these substances in influencing ion uptake or efflux. One suggestion envisages that humic substances increase the rate of mineral absorption by accelerating the surface absorption of mineral ions on the root colloids[14] while another suggests that humic materials can influence active uptake by interfering with specific ion carriers[80,217]. Clearly more work is required in this important area of plant nutrition.

6. CONCLUSIONS

A considerable literature shows that under laboratory conditions, humic substances can have a favourable effect on the growth of higher plants as measured in terms of increases in lengths, and fresh and dry weights of shoots and roots and even on the increase in the numbers of lateral roots and flowers. Humic substances can also influence the growth of micro-organisms as diverse as algae, dinoflagellates, bacteria and yeasts, and sporulation in some water fungi (Actinoplanaceae). These effects are often more pronounced in the presence of inadequate controls such as water or aqueous media which are toxic or deficient in some nutrients. The precise effect usually depends on the nature of the humic substance, its concentration, the composition and pH of the growth medium, the culture conditions, and the plant species being investigated.

Humic substances exert their effects indirectly and/or directly. In the former case, they may, for example, chelate a cation thus lowering its concentration from being supra-optimal to optimal. In the case of a direct effect, the humic material may influence active ion uptake or protein synthesis (see also chapter 2). To produce a direct effect, the humic substance must be taken up by the plant. The available evidence suggests that low MW humic substances are taken up by plant roots but there is little subsequent translocation to the shoot. Whether the humic substances are taken up in an unmodified form is not known.

Although under certain conditions humic substances can influence plant growth directly, the extent of this involvement under natural conditions is uncertain. It is, however, likely to be considerably less than the other ways in which soil organic matter influences growth and fertility such as acting as, for example, a nutrient source for plants, modifying the cation exchange capacity of soils, helping the soil to retain water etc., factors discussed more fully in the Introductory Chapter.

7.   REFERENCES

1.   ACTON E.H. 1889. The assimilation of carbon by green plants from certain organic compounds. Proceedings of the Royal Society, Series B, 150-175.
2.   ALLISON F.E. 1973. Soil organic matter and its role in crop production. Developments in Soil Science, Volume 3, Elsevier Press, New York.
3.   ANDERSON H.A. and RUSSELL J.D. 1976. Possible relationship between soil fulvic acid and polymaleic acid. Nature, London, 260, 597.
4.   ANDREYUK E.I., GORDIENKO S.A., KONOTO I.N. and MARTYNENKO V.A. 1973. Assimilation of humic acid nitrogen by micro-organisms. Mikrobiologichnyi Zhurnal, 35, 139-142.
5.   ASO S. and SAKAI I. 1963. Studies on the physiological effects of humic acid. 1. Uptake of humic acid by crop plants and its physiological effects. Soil Science and Plant Nutrition, 9, 85-91.
6.   BACON F. 1651. Sylva sylvarum, London.
7.   BARTON L.L. and RUOCCO J.J. 1981. Soluble humic complexes and sulfate uptake by Aspergillus niger. Soil Biology and Biochemistry, 1, 435-437.
8.   BHARDWAJ K.K.R. and GAUR A.C. 1971. Studies on the growth stimulating action of humic acid on bacteria. Zentralblatt für Bakteriologie, Parasitenkunde Infektionskrankheiten und Hygiene, Abt. 2. Naturwiss. 126, 694-699.
9.   BIBER V.A. and BOGOLYUBOV N.S. 1951. (Biological activity of soil and peat humic acids). Doklady Akademii nauk SSSR, 76, 313-316.
10.   BIBER V.A. and MAGAZINER K.M. 1951 (The influence of humic and fulvic acids on respiration of isolated plant tissues). Doklady Akademii nauk SSSR 76, 609-612.
11.   BIDDULPH O. and WOODBRIDGE C.G. 1952. The uptake of phosphorus by bean plants with particular reference to the effects of iron. Plant Physiology, 27, 431-440.

64

12. BLAGOWESTSCHENSKI A.W. and PROSOROWSKAJA A.A. 1935. Zur Frage des Einfluss der Huminsaure auf die Assimilation der Pflanzen. Biochemische Zeitschrift, 282, 99-103.

13. BLANC D. 1958. (The effect of humic extracts on the nitrogen metabolism of plants). Comptes rendus de l'Academie des sciences, Paris, 247, 1649-1651.

14. BLANCHET R.M. 1957. Influence des colloides humiques sur differentes phases de l'absorption des elements mineraux par les plantes. Comptes rendus de l'Academie des sciences, Paris, 244, 2418-2420.

15. BLANCHET R.M. 1958. The direct and indirect effect of humified organic matter on the nutrition of vascular plants. Annales agronomiques, 9, 499-532.

16. BOERHAAVE H. 1727. A new method of chemistry. Translated by Shaw P. and Chambers E.

17. BOTTOMLEY W.B. 1914. Some accessory factors in plant growth and nutrition. Proceedings of the Royal Society, Series B, 88, 237-247.

18. BOTTOMLEY W.B. 1914. The significance of certain food substances for plant growth. Annals of Botany, 28, 531-540.

19. BOTTOMLEY W. 1917. Some effects of organic growth-promotion substances (auximones) on the growth of Lemna minor in mineral culture solutions. Proceedings of the Royal Society, Series B, 89, 481-505.

20. BOTTOMLEY W. 1920. The effect of organic matter on the growth of various plants in culture solutions. Annals of Botany, 34, 353-365.

21. BOUSSINGAULT J.B. 1841. De la discussion de la valeur relative des assolements per les resultats de l'analyse elementaire. Annales de Chimie et du Physique (111), 1, 208-246.

22. BOUSSINGAULT J.B. 1851. Economie rurale consideree dans ses rapports avec la chemie, la physique, et la meteorologie. Second Edition. Paris.

23. BOZNIAK E.G. 1969. Laboratory and field studies of phytoplankton communities. PhD Thesis, pp.106. Washington University, St. Louis, Missouri.

24. BRAR M.S. and SEKHON G.S. 1976. Interaction of zinc with other micronutrient cations. Plant and Soil, 45, 137-143.

25. BROWNE C.A. 1944. A Sourcebook of Agricultural Chemistry. Chronica Botanica Company, Massachusetts.

26. BURGES A. and LATTER P. 1960. Decomposition of humic acid by fungi. Nature, London, 186, 404-405.

27. BURK D., LINEWEAVER H. and HORNER C.K. 1932. Iron in relation to the stimulation of growth by humic acid. Soil Science, 33, 413-435.

28. BURK D., LINEWEAVER H. and HORNER C.K. 1932. The physiological nature of humic acid stimulation of Azotobacter growth. Soil Science, 33, 455-487.

29. CAMPBELL C.A., PAUL E.A., RENNIE D.A. and McCALLUM K.J. 1967. Applicability of the carbon-dating method of analysis to soil humus studies. Soil Science, 104, 217-224.

30. CHAMINADE R. 1952. (Research on the role of organic matter in soil fertility). Annales Agronomiques, 3, 95-104.

31. CHAMINADE R. 1956. (The effect of humic acid on the growth of plants and their mineral nutrition). Sixth Congres International de la Science du Sol. Paris Rapports, D, 443-448.

32. CHAMINADE R. 1958. (The effect of humified organic matter on the efficiency of nitrogen). Annales agronomiques, $\underline{9}$, 167-192.

33. CHAMINADE R. 1966. (Physiological effects of soil organic matter constituents on the metabolism of plants). Report of the FAO/IAEA Meeting, Pergamon Press, Oxford.

34. CHAMINADE R. and BLANCHET R. 1951. (Action of colloidal humus on the nutrition of plants). Comptes Rendus hebdomadaire de seances de l'Academie des sciences, $\underline{233}$, 1486-1488.

35. CHAMINADE R. and BLANCHET R. 1952. Influence de l'humus sur la nutrition mineral de la plante. Comptes rendus hebdomadaire de seances de l'Academie des sciences, $\underline{234}$, 878-879.

36. CHAMINADE R. and BLANCHET R. 1953. (Stimulant action of humus on development and mineral nutrition of plants in soil). Comptes rendus hebdomadaire de seances de l'Academie des sciences, $\underline{236}$, 119-121.

37. CHAMINADE R. and BLANCHET R. 1953. (Research on the role of organic matter in soil fertility. Second communication). Annales agronomiques, $\underline{4}$, 399-409.

38. CHAMINADE R. and BOUCHER J. 1940. Recherches sur la presence de substances rhizogenes dans certains milieux naturels. Comptes rendus hebdomadaire des sceances de l'academie d'agriculture de France, $\underline{26}$, 66.

39. CHESHIRE M.V., MUNDIE C.M. and SHEPHERD H. 1973. The origin of soil polysaccharide: Transformation of sugars during the decomposition in soil of plant material labelled with ^{14}C. Journal of Soil Science, $\underline{24}$, 54-68.

40. CHIZHEVSKY M.G. and DIKUSAR M.M. 1955. (The role of humus and micro-organisms in the root nutrition of higher plants in water and sand cultures). Izvestiya Timiryazevskoi se l'sko-khozyaist-vennoi Akademii, $\underline{2}$, 173-192.

41. ČINČEROVÁ A. 1964. The effect of humic acid on transamination in winter wheat plants. Biologia Plantarum, $\underline{6}$, 183-188.

42. ČINČEROVÁ A. 1971. The effect of humic acid on growth and metabolism of wheat seedlings. In Humus et Planta, $\underline{V}$, 609-611.

43. CROOKE W.M. 1964. Measurements of cation-exchange capacity of plant roots. Plant and Soil, $\underline{21}$, 43-49.

44. DAVIES R.I. and COULSON C.B. 1959. Humic acid. Soils and Fertilizers, $\underline{22}$, 159.

45. DAVY H. 1813. Elements of Agricultural Chemistry. Longman, London.

46. DEB D.L., KOHLI C.B.S. and JOSHI O.P. 1976. Stability constants of Zn(II)-humic acid complexes. Fertilizer Technology, $\underline{13}$, 25-29.

47. DEKOCK P.C. 1955. The influence of humic acids on plant growth. Science, $\underline{121}$, 473-474.

48. DEKOCK P.C. and STRMECKI E.L. 1954. An investigation into the growth promoting effect of the lignite. Physiologia Plantarum, $\underline{7}$, 503-512.

49. DE HAAN H. 1974. Effect of a fulvic acid fraction on the growth of a Pseudomonas from Tjeukemeer (The Netherlands). Freshwater Research, $\underline{4}$, 301-310.

50. DE HAAN H. 1976. Evidence for the induction of catechol-1,2-oxygenase by fulvic acid. Plant and Soil, $\underline{45}$, 129-136.

51. DE HAAN H., WERLEMARK G., DEBOER T. 1983. Effect of pH on molecular weight and size of fulvic acids in drainage water from peaty grassland in NW Netherlands. Plant and Soil, $\underline{75}$, 68-73.

52. DELL'AGNOLA G. and FERRARI G. 1971. Effect of humic acids on anion uptake by excised barley roots. In Humus et Planta, $\underline{V}$, 567-569.

53. DOLGOPOLOV N.N. and RUBAN E.L. 1952. (Peat humates and fossil carbonaceous material as stimulators of plant growth). Priroda, 41, 102-104.

54. DORMAAR J.F. 1975. Effects of humic substances from chernozemic Ah horizons on nutrient uptake by Phaseolus vulgaris and Festuca scabrella. Canadian Journal of Soil Science, 55, 111-118.

55. DE SAUSSURE Th. 1804. Recherches chimiques sur la vegetation. Paris.

56. ELGALA A.M., METWALLY A.J. and KHALIL R.A. 1978. The effect of humic acid and Na$_2$ EDDHA on the uptake of Cu, Fe and Zn by barley in sand culture. Plant and Soil, 49, 41-48.

57. ELOFF J.N. and PAULI F.W. 1975. The extraction and electrophoretic fractionation of soil humus substances. Plant and Soil, 42, 413-422.

58. FERNANDEZ V.H. 1968. The action of humic acids of different sources on the development of plants and their effect on increasing concentration of the nutrient solution. Pontificiae Academiae Scientiarum Scripta Varia, 32, 805-850.

59. FLAIG W. 1968. Uptake of organic substances from soil organic matter by plants. Pontificiae Academiae Scientiarum Scripta Varia, 32, 1-48.

60. FLAIG W. and SAALBACH E. 1955. (Humic acids. X. The effect on the root growth of spring wheat of thymohydroquinone as a model substance of pre-humic acids or decomposition products of humic acid). Zeitschrift für Pflanzenernahrung Dungung Bodenkunde, 71, 208-215.

61. FLAIG W. and SOCHTIG H. 1962. Einfluss organischer Stoffe auf die Aufnahme anorganischer Ionen. Agrochimica, 6, 251-264.

62. FORTUN C. and LOPEZ-FANDO C. 1982. (Influence of humic acid on the mineral nutrition and the development of maize roots cultivated in normal nutrient solutions and lacking Fe and Mn). Anales de Edafologiay Agrobiologia, XLI, 335-349.

63. FORTUN C. and POLO A. 1982. Effects of some types of humic compounds on the growth of roots of Zea mays. Agrochemica, 26, 44-54.

64. FÜHR F. 1969. Tracer studies on the uptake of soil organic matter compounds by plant roots. Actes du Symposium International de Radioecologie Centre d'etudes nucleaes de Cadarache, 623-635.

65. FÜHR F. and SAUERBECK D. 1965. Die raumliche und chemische vertei lung durch die wurtzel aufgenommer organiche Rotteproduckte bei Daucus carota. Landwirtschaftliche Forschung, 19, Sonderheft, 153-163.

66. FÜHR F. and SAUERBECK D. 1967. The uptake of straw decomposition products by plant roots. In Report FAO/IAEA Meeting, Vienna, Pergamon Press, Oxford.

67. FÜHR F. and SAUERBECK D. 1967. The uptake of colloidal organic substances by plant roots as shown by experiments with 14C-labelled humus compounds. In Report FAO/IAEA Meeting, Vienna, Pergamon Press, Oxford.

68. GAUR A.C. and BHARDWAJ K.K.R. 1971. Influence of sodium humate on the crop plants inoculated with bacteria of agricultural importance. Plant and Soil, 35, 613-621.

69. GIESY J.P. 1976. Stimulation of growth in Scenedesmus obliquus (Chlorophyceae) by humic acids under iron limited conditions. Journal of Phycology, 12, 172-179.

70. GOROVAYA A.I. and SOLOCHA K.I. 1971. Influence of physiologically active substances of the soil humus and fertilizers on the specific activity of meristematic cells of plant sprouts and the quality of seed grains. In Humus et Planta, V, 557-566.

71. GRANDEAU L. 1872. Recherches sur le rôle des matieres organiques du sol dans les phénômenes de la nutrition des végétaux. Comptes rendus hebdomadaire seances de l'academie des sciences.

72. GRANDEAU L. 1877. Traite d'analyse des matieres agricoles. pp.148. Paris.

73. GUMIŃSKI S. 1950. (Researches on the conditions and mechanism of the action of humic compounds on plant growth). Acta Societatis Botanicorum Poloniae, 22, 589-620.

74. GUMIŃSKI S. 1957. (The mechanism and conditions of the physiological effect of humus substances on plant organs.) Pochvovedenie, 52, 72-78.

75. GUMIŃSKI S. 1968. Present-day views on physiological effects induced in plant organisms by humic compounds. Soviet Soil Science, 1250-1256.

76. GUMIŃSKI S. and SULEJ J. 1975. The stimulating effect of humate of yeasts growing in medium with low acidity. In Humus et Planta, VI, 369-371.

77. GUMIŃSKI S. and SULEJ J. 1979. Effect of Na-humate and tannin on germination of photosensitive seeds. In Humus et Planta, VII, 436-438.

78. GUMIŃSKI S. and SULEJ J. 1979. About the cause of the stimulative effect of humate in yeast cultures. Acta Societatis Botanicorum Poloniae, 43, 279-293.

79. GUMIŃSKI S., GUMIŃSKA A. and SULEJ J. 1965. Effect of humate, agar-agar and EDTA on the development of tomato seedlings in aerated and non-aerated water cultures. Journal of Experimental Botany, 16, 151-162.

80. GUMIŃSKI S., SULEJ J. and GLABISZEWSKI J. 1983. Influence of sodium humate on the uptake of some ions by tomato seedlings. Acta Societatis Botanicorurum Poloniae, 52, 149-164.

81. HAWORTH R.D. 1970. The chemical nature of humic acid. Journal of Soil Science, 111, 71-79.

82. HEATH O.V.S. and CLARK J.E. 1956. Chelating agents as plant growth substances. A possible clue to the mode of action of auxin. Nature, London, 177, 1118-1121.

83. HEINRICH G. 1964. Huminsaure und Permeabilitat. Protoplasma, 58, 402-425.

84. HELANOVA I. and SLADKY Z. 1967. The effect of glycine, humus substances and sucrose on the growth of tomato plants in vitro. Biologia Plantarum, 9, 276-284.

85. HERNANDO V., ORTEGA B.C. and FORTUN C. 1977. (Study of the action of two types of humic acid on the maize plant). In Soil Organic Matter Studies, Volume 2. Report of IAEA Meeting, Vienna. Pergamon Press, Oxford.

86. HILLITZER A. 1932. Uber den Einfluss der Humusstoffe auf das Wurzelwachstum. Beihefte zum Botanischen Zentralblatt, 49, 467-480.

87. HUTCHINSON H.B. and MILLER N.H.J. 1909. Direct assimilation of ammonium salts by plants. Journal of Agricultural Science, 3, 179-194.

88. HUTCHINSON H.B. and MILLER N.H.J. 1912. The direct assimilation of inorganic and organic forms of nitrogen by higher plants. Journal of Agricultural Science, 4, 282-302.

89. INGEN-HOUSZ J. 1796. Essay on the food of plants and the renovation of soils. Agricultural Reports, London. (see also Waksman ref. number 225).

90. ISWARAN V. and CHONKAR P.K. 1971. Action of sodium humate and dry matter accumulation of soybean in saline alkali soil. In Humus et Planta. V, 613-615.

91. JAKUBIEC M., MISZOZAK E. and SZCZERKOWSKA J. 1971. Comparison of the effect of natural and synthetic humates and EDTA on the growth of Escherichia coli. Acta microbiologia Poloniae, Series B, 3, 63-6.

92. JELENIĆ D.B., HAJDUKOVIC M. and ALEKSIC Z. 1966. The influence of humic substances on phosphate utilization from labelled superphosphate. In Report FAO/IAEA Meeting, Pergamon Press, Oxford.

93. JENKINSON D.S. 1965. Studies on the decomposition of plant materials in soil. 1. Journal of Soil Science, 16, 104-115.

94. KHRISTEVA L. 1949. (Nature of the effect of humic acids on the plant). Doklady Vsesoyuznoi Akademii sel'sko-kho-zyaistvennykh nauk im vi Lenina, 7.

95. KHRISTEVA L.A. 1951. (The importance of humic acid for the nutrition of higher plants and humic fertilizers). Transactions of the Dokuchayev Soil Institute of the USSR Academy of Sciences, xxxviii.

96. KHRISTEVA L.A. 1953. (The participation of humic acids and other organic substances in the nutrition of higher plants). Pochvovedenie, 48, 46-59.

97. KHRISTEVA L.A. 1955. (The role of humic acids and other organic substances in the nutrition of higher plants and the agronomic importance of this type of nutrition). Izvestiya Akademii nauk sssr seriya biologicheskoi, No. 4, 58-83.

98. KHRISTEVA L.A. 1968. About the nature of physiologically active substances of the soil humus and of organic fertilizers and their agricultural importance. Pontificiae Academiae Scientiarum Scripta Varia, 32, 701-721.

99. KHRISTEVA L.A. and LUK'YANENKO. 1962. Role of physiologically active substances in the soil-humic acids, bitumens and vitamins B, C, P-P and D in the life of plants and their replenishment. Soviet Soil Science, 1137-1141.

100. KHRISTEVA L.A. and MANOILOVA A.V. 1950. (The nature of the direct effect of humic acid on the growth and development of plants). Doklady Vsesoyuznoi Akademii sel'sko-kho-zyaistvennykh nauk im VI Lenin 11, 10-16.

101. KHRISTEVA L.A., GALUSHKO A.M., GOROVAYA A.I., KOLBASSIN A.A., SHORTSHOI L.P., TKATSHENKO L.K., FOT L.V. and LUK'YAKENKO N.V. 1980. The main aspects of using physiologically active substances of humus nature. Sixth International Peat Congress, Minnesota.

102. KNOP W. 1868. Lehrbuch der Agrikultur Chemie. Leipzig.

103. KNUDSON L. 1916. Influence of certain carbohydrates on green plants. Cornell University Agricultural Experimental Station Memoirs. No. 9.

104. KONONOVA M.M. 1961. Soil organic matter. Its nature, its role in soil formation and in soil fertility. Pergamon Press, Oxford.

105. KONONOVA M.M. and D'YAKONOVA K.V. 1960. Soil organic matter and aspects of plant nutrition. Soviet Soil Science, 229-236.

106. KONONOVA M.M. and PANKOVA N.A. 1950. (The action of humic substances on the growth and development of plants). Doklady Akademii nauk SSSR 73, 1069-1071.

107. KUDRINA E.S. 1951. (The effect of humic acids on certain groups of soil micro-organisms). Trudy Pochvennogo Instituta imeni V.V. Dokuchaeva. Akademiya nauk SSSR.38.

108   KÜLBEL J.A. 1741. Cause de la fertilite des terres. Bordeaux.

109. KUTHY A. and PECZNIK J. 1941. Wirkt die Huminsaure als Hormon oder durch Permeabilitatserhohung auf die Entwicklung der Pflanzenwurzeln. Bodenkunde und Pflanzenernährung, 23, 83-90.

110. KUZIN A.M. and MERENOVA V.L. 1955. (On the assimilation of carbon from manure by plants). Translated from Institute biol. fiziki AN. SSSR, 1,247-255.

111. LANGE W. 1970. Blue-green algae and humic substances. Proceedings of Conference of Great Lakes Research, 13, 58-70.

112. LAURENT J. 1904. Recherches sur la nutrition carbonée des plantes vertes à l'aide de matières organiques. Revue générale de botanique, 16, 14-48; 68-80; 96-119; 120-128; 155-166; 188-202; 231-241.

113. LAWES J.B. and GILBERT J.H. Collected in ten volumes of Rothamsted memoirs and summarised by Hall in The Book of the Rothamsted Experiments, London (1905); also summarized by Russell E.J. (reference number 171).

114. LEE Y.S. and BARTLETT R.J. 1976. Stimulation of plant growth by humic substances. Journal of the Soil Science Society of America, 40, 876-879.

115. LEMAIRE F. 1971. Effect of fractions extracted from the organic matter on the physiology of plants grown with nutrient solutions. In Humus et Planta, V, 511-518.

116. LEOPOLD A.C. 1967. Auxins and Plant Growth. pp.40. University of California Press.

117. LEVESQUE M. 1970. Contribution de l'acide fulvique et des complexes fulvo-metalliques a la nutrition minerale des plantes. Canadian Journal of Soil Science, 50, 385-395.

118. LIEBIG J.v. 1841. Organic chemistry in its applications to agriculture and physiology. Translated by Webster J.W. and Owen J. Cambridge

119. LIEBIG J.v. 1856. On some points of agricultural chemistry. Journal of the Royal Agricultural Society, 17, 284-326.

120. LIEBIG J.v. 1865. Die Chemie in ihrer Anwendung auf Agricultur und Physiologie. Eighth Edition, F. Vieweg. Br aunschweig.

121. LIESKE R. 1931. Untersuchungen über die Verwendbarkeit von Kohlen als Düngemittel. Brennstoffe-Chemie, 12, 81-85.

122. LEISKE R. and WINZER K. 1935. Untersuchungen über die Ursache der Wachstumsförderung durch Braunkohle. Brennstoffe-Chemie, 16, 24-27.

123. LINEHAN D.J. 1976. Some effects of a fulvic acid component of soil organic matter on the growth of cultured excised tomato roots. Soil Biology and Biochemistry, 8, 511-517.

124. LINEHAN D.J. 1977. A comparison of the polycarboxylic acids extracted by water from an agricultural top soil with those extracted by alkali. Journal of Soil Science, 28, 369-378.

125. LINEHAN D.J. 1978. Humic acid and iron uptake by plants. Plant and Soil, 50, 663-670.

126. LINEHAN D.J. and SHEPHERD H. 1979. A comparative study of the effects of natural and synthetic ligands on iron uptake by plants. Plant and Soil, 52, 281-289.

127. LISIAK M.J. 1978. The effect of sodium humate upon phosphorus nutrition of plants with variable doses of iron and calcium in tomato water cultures. Acta Societatis Botanicorum Poloniae, 47, 429-440.

128. LONGLAND J.E. 1973. An investigation into techniques for the growth of plants under axenic conditions using a thin-film isolator. Degree Dissertation, Trent Polytechnic, Nottingham, U.K.

129. LUTZ L. 1898. Recherches sur la nutrition des végétaux. Annales des Sciences naturelles Botanique, 7, 1-103.

130. McLOUGHLIN A.J. and KÜSTER E. 1972. The effect of humic substances on the respiration and growth of micro-organisms. 1. Effects on the yeast Candida utilis. Plant and Soil, 37, 17-25.

131. MARTIN D.F., DOIG M.T. and PIERCE R.H. 1971. Distribution of naturally occurring chelaters (humic acids) and selected trace metals in some west coast Florida streams, 1968-1969. Florida Department for Water Research, St. Petersburg. Technical paper No. 12.

132. MATHUR S.P. and PAUL E.A. 1966. A microbial approach to the problem of soil humic acid structure. Nature, London, 212, 646-647.

133. MATHUR S.P. and PAUL E.A. 1967. Microbial utilization of soil humic acids. Canadian Journal of Microbiology, 13, 573-586.

134. MISTERSKI W. 1957. (Studies on humus. Part 1. Studies of the effect of addition of humate on the utilization of phosphorus). Rocznik nauk rolniczych, 76A, 43-63.

135. MORTENSEN J.L. 1963. Complexing of metals by soil organic matter. Soil Science Society of America Proceedings, 27, 179-186.

136. MYLONAS V.A. and McCANTS C.B. 1980. Effect of humic and fulvic acids on growth of tobacco. 1. Root initiation and elongation. Plant and Soil, 54, 485-490.

137. MYLONAS V.A. and McCANTS C.B. 1980. Effects of humic and fulvic acids on growth of tobacco. 2. Tobacco growth and ion uptake. Journal of Plant Nutrition, 2, 377-393.

138. MÝSKÓW W. 1967. Chemical composition and biological activity of humus obtained from various substances. Bull de l'academie Polonaise des Sciences. CL II, XV, 79-84.

139. NEELAKANTAN S., MISIRA M.M., TEWARI H.K. and VYAS S.R. 1970. Characterization and microbial utilization of humic acid in Hissar soil. Agrochimica, 14, 341-344.

140. NIKLEWSKI B. and WOJCIECHOWSKI J. 1937. Über den Einfluss der Wasserloslichen Humusstoffe auf die Entwicklung einiger Kulturpflanzen. Bodenkunde und Pflanzenernährung, 4, 294-327.

141. NIKLEWSKI B. and WOJCIECHOWSKI J. 1938. (The influence of humic acids on the absorption of ammonium phosphate and sulphate by plants). Acta Societatis Botanicorum Poloniae, 15, 111-151.

142. NIKLEWSKI B. and WOJCIECHOWSKI J. 1947. (The influence of humic acid on the development of plants). Acta Societatis Botanicorum Poloniae, 18, 63-90.

143. NISHITA H., KOWALEWSKY B.W. and LARSON K.H. 1956. Influence of soil organic matter on mineral uptake by barley seedlings. Soil Science, 82, 307-318.

144. NISHITA H., KOWALEWSKY B.W. and LARSON K.H. 1956. Influence of soil organic matter on mineral uptake by tomato plants. Soil Science, 82, 401-407.

145. O'DONNELL R.W. 1973. The auxin-like effects of humic preparations from leonardite. Soil Science, 116, 106-112.

146. OLSEN C. 1930. On the influence of humus substances on the growth of green plants in water culture. Comptes-rendus du Laboratoire Carlsberg, 18, 1-16.

147. OPARTRNA-FISEROVA J. 1964. Der Einfluss von K-Humat auf Teilung und Langenwachstum der Wurzelspitzenzellen von Allium crepa L. Biologia Plantarum, 6, 122-131.

148. ORTEGA C.B., HERNANDO V. and SANCHEZ CONDE M.P. 1968. Different effects on maize of humic acid extracted from manure or peat. In Report IAEA/FAO Meeting Vienna, Pergamon Press, Oxford.

149. PAGEL H. 1960. Über den einfluss von Humusstoffen auf das Pflanzenwachstum 1. Einfluss von Humusstoffen auf Ertrag und Nahrstoffaufnahme. Albrecht-Thaer-Archiv. 4, 492-506.

150. PASZEWSKI A., TROJANOWSKI J. and LOBARZEWSKA W. 1957. (Influence of the humus fraction on the growth of oat coleoptiles). Annales Universitatis Marie Curie-Sklodowska, 12, 1-13.

151. PETROVIC P., VITOROVIC D. and JABLANOVIC M. 1982. Investigations of biological effects of humic acids. Acta Biol. Med. Exp., 7, 21-25.

152. POAPST P.A. and SCHNITZER M. 1971. Fulvic acid and adventitious root formation. Soil Biology and Biochemistry, 3, 215-219.

153. POAPST P.A., GENIER C. and SCHNITZER M. 1970. Effect of soil fulvic acid on stem elongation in peas. Plant and Soil, 32, 367-372.

154. POKORNA V., LUSTINEC J. and PETRU E. 1963. The influence of Na-humate on the respiration of wheat roots and leaves. Biologia Plantarum, 5, 265-270.

155. PRAKASH A. and RASHID M.A. 1968. Influence of humic substances on the growth of marine phytoplankton dinoflagellates (Gonyaulax). Limnology and Oceanology, 13, 598-606.

156. PRAKASH A., JENSEN A. and RASHID M.A. 1972. Humic substances and aquatic productivity. Proceedings of the International Meeting on Humic Substances, Nieuwersluis pp.259-268. Pudoc Wageningen.

157. PRAKASH A., RASHID M.A., JENSEN A. and SUBBA RAO D.V. 1973. Influence of humic substances on the growth of marine phytoplankton: diatoms. Limnology and Oceanography, 18, 516-524.

158. PRÁT S. 1955. The effect of humus substances on algae. Folia Biologica, 1, 321-326.

159. PRÁT S. 1960. Distribution of the humus substances function in plants. Biologia Plantarum, 2, 308-312.

160. PRÁT S. 1962. The effect of humus substances on regeneration of plants. Studies about humus, Praha 223-234.

161. PRÁT S. 1963. Permeability of plant tissues to humic acids. Biologia Plantarum, 5, 279-283.

162. PRÁT S. and POSPÍŠIL F. 1959. Humic acids with 14-C. Biologia Plantarum, 1, 71-80.

163. PRÁT S., CATSKY J. and MELICHAR O. 1957. (The effect of humus substances (oxyhumolites) on plants). Acta Societatis Botanicorum Poloniae, 26, 325-347.

164. PROZOROVSKAYA A.A. 1936. (The effect of humic acid and its
     derivatives on the uptake of nitrogen, phosphorus, potassium and
     iron by plants). Trudy nauch mogo Instituta Udobreniyan
     Insektofunqitisidan, 127.
165. RAUTHAN B.S. and SCHNITZER M. 1981. Effects of soil fulvic acid
     on the growth and nutrient content of cucumber (Cucumis sativus)
     plants. Plant and Soil, 63, 491-495.
166. RERABEK J. 1960. Humic acid interactions in the growth process.
     Biologia Plantarum, 2, 88-97.
167. RERABEK J. 1962. The relation of humic acid to the control of
     straight growth of the plant cell. Studies about Humus, Praha,
     245-254.
168. ROCHUS W. 1967. The effect of humic substances on the growth of
     tomato plants. In Humus et Planta IV, 281-285.
169. ROCHUS W. 1971. Influence of different humic substances on growth
     and phosphate uptake by cereal plants. In Humus et Planta, V,
     535-54.
170. RUOCCO J.J. and BARTON L.L. 1978. Energy-driven uptake of humic
     acids by Aspergillus niger. Canadian Journal of Microbiology, 24,
     533-536.
171. RUSSELL E.J. 1921. Soil conditions and plant growth. pp. 1-29.
     Longmans, Green and Co., London.
172. RYPACEK V. 1962. Der Einfluss isolierter Humusstoffe auf einige
     physiologische Ausserungen der Pflanzenzelle. In Studies about
     Humus, Praha, 234-243.
173. RYPACEK V. 1967. Biological activity of isolated humic acids in
     relation to the time and method of storing. In Humus et Planta,
     IV, 288-290.
174. SAALBACH E. 1956. (The influence of humic substances on the
     metabolism of plants). Sixth Congres International de la Science
     du Sol, Paris. Rapperts, D, 107-111.
175. SAEGER A. 1925. The growth of duckweeds in mineral nutrient
     solutions with and without organic extracts. Journal of General
     Physiology, 7, 517-526.
176. SCHARPENSEEL H.W. 1961. The preparation of plant material and
     humic acid uniformly marked with C-14. Landwirtschaftliche
     Forschung. 14, 42-48.
177. SCHARPENSEEL H.W. 1966. The labelling of soil organic matter.
     In Report of the FAO/IAEA Meeting, Pergamon Press, Oxford.
178. SCHNITZER M. and KAHN S.U. 1972. Humic substances in the
     environment. Marcel Dekker, New York.
179. SCHNITZER M. and POAPST P.A. 1967. Effects of a soil humic
     compound on root initiation. Nature, London, 213, 598-599.
180. SCHREINER O. and SKINNER J.J. 1911. Organic compounds and
     fertilizer action. U.S. Department of Agriculture Bureau for
     Soils, Bulletin 77; also bulletin number 87.
181. SENEBIER J. 1782. Memoirs physico-chymiques. Physiologie
     Végétale, 3, 1800.
182. SHAPIRO J. 1957. Chemical and biological studies on the yellow
     organic acids of lake waters. Limnology and Oceanography, 2,
     161-179.
183. SIMONART P. and MAYAUDON J. 1957. (Studies of the composition of
     organic material in soil by means of radioactive C). Pedologie,
     VII, 282-283.

184. SLADKY Z. 1959. The effect of extracted humus substances on growth of tomato plants. Biologia Plantarum, 1, 142-150.

185. SLADKY Z. 1959. The application of extracted humus substances to overground parts of plants. Biologia Plantarum, 1, 199-204.

186. SLADKY Z. 1967. Effect of humus acids on the growth of isolated roots. In Humus et Planta IV, 286-287.

187. SLADKY Z. and TICHÝ V. 1959. Application of humus substances to overground organs of plants. Biologia Plantarum, 1, 9-15.

188. ŠMÍDOVÁ M. 1960. The influence of humus acid on the respiration of plant roots. Biologia Plantarum, 2, 152-164.

189. ŠMÍDOVÁ M. 1962. Effect of sodium humate on swelling and germination of winter wheat. Biologia Plantarum, 4, 112-118.

190. SOUKUP J. and MATOUS J. 1958. Humic substances in plant nutrition. The physiological effect of humic acids from leaf mould. Sbornik Ceskoslovenské akademie zemedelskych ved. Rostlinna vyroba, XXXl, 11-22.

191. SPRENGEL C. 1832. Ueber Pflanzenhumus,Humussäure und humussäure Salze. Chemie für Landwirthe,Forstmanner und Cameralisten, Göttingen.

192. SPRENGEL C. 1844. Die Bodenkunde oder die Lehre vom Boden. Second Edition. Muller, Leipzig.

193. STEVENSON F.J. and ARDAKANI M.S. 1972. Organic matter reactions involving micronutrients in soils. In Micronutrients in Agriculture, pp.79-114. Eds. Mortvedt J.J., Giordano P.M. and Lindsay W.L. Soil Science Society of America, Madison.

194. STEWART I. 1963. Chelation in the absorption and translocation of mineral elements. Annual Review of Plant Physiology, 14, 295-310.

195. STRICKLAND R.C., CHANEY W.R. and LAMOREAUX R.J. 1979. Organic matter influences phytotoxicity of cadmium to soybeans. Plant and Soil, 52, 393-402.

196. STUMM W. and MORGAN J.J. 1970. Aquatic Chemistry - An introduction emphasizing chemical equilibria in natural waters. Wiley-Interscience, London.

197. SUTCLIFFE J.F. 1962. Mineral Salts Absorption in Plants. Pergamon Press, Oxford.

198. TAN K.H. 1978. Effects of humic and fulvic acids on release of fixed potassium. Geoderma, 21, 67-74.

199. TAN K.H. and NOPAMORNBODI V. 1979. Fulvic acid and the growth of the ectomycorrhizal fungus, Pisolithus tinctorius. Soil Biology and Biochemistry, 11, 651-653.

200. TAN K.H. and NOPAMORNBODI V. 1979. Effect of different levels of humic acids on nutrient content and growth of corn (Zea mays). Plant and Soil, 51, 283-287.

201. TAN K.H. and TANTIWIRAMANOND D. 1983. Effect of humic acids on nodulation and dry matter production of Soybean, Peanut and Clover. Soil Science Society of America Journal, 47, 1121-1124.

202. TATKOWSKA E. and KOBYLANSKA D. 1978. The effects of sodium humate on cultures of Spirodela polyrrhiza BL). Schleiden under aseptic conditions. Ekologia polska, 26, 213-220.

203. THAER A.D. 1808. Grundriss der Chemie für Landwirte. Berlin.

204. THAER A.D. 1846. The principles of agriculture. Eds. Shaw W. and Johnson C.W. New York. pp 166-173.

205. TICHÝ V. 1958. (The influence of some groups of soil humus on growth of wheat). Ceskoslovenská biologie (Praha), 7, 343-346.

206. TICHÝ V. 1979. Maize nutrition with iron under conditions of alternating supply of sodium humate and mineral nutrient solutions. In Humus et Planta, VII, 431-435.

207. TICHÝ V. and CHALUPOVA-JANOVICOVA J. 1958. (The effect of humus on germination and growth of some crops). Folia Biologica (Praha), 4, 281-285.

208. TOLEDO A.P.P., TUNDISI J.G. and D'AQUIRE V.A. 1980. Humic acid influence on the growth and copper tolerance of Chlorella sp. Hydrobiologia, 71, 261-263.

209. TOMCZAK I. 1976. Dissertation for MSc degree, Wroclaw University.

210. TULL J. 1730. The Horse Hoeing Industry, London (The National Reference Library).

211. TYLER L.D. and McBRIDE M.B. 1982. Influence of Ca, pH and humic acid on Cd uptake. Plant and Soil, 64, 259-262.

212. VAN DER WERFF M. 1981. Ecotoxicity of heavy metals in aquatic and terrestrial higher plants. PhD. Thesis, Vrije Universiteit, Amsterdam.

213. VAN HELMONT J.B. 1652. Opera omnia complexionum atque mistionum elementalium figmentum. (see Sutcliffe J.F. reference number 197); also In Ortus Medicinae, Amsterdam.

214. VARSHNEY T.N. and GAUR A.C. 1974. Effect of spraying sodium humate and hydroquinone on Glycine max var. Bragg and Solanum lycoperisicum var. heiz1370. Current Science 43, 95-96.

215. VAUGHAN D. 1969. The stimulation of invertase development in aseptic storage tissue slices by humic acids. Soil Biology and Biochemistry, 1, 15-28.

216. VAUGHAN D. 1974. A possible mechanism for humic acid action on cell elongation in root segments of Pisum sativum under aseptic conditions. Soil Biology and Biochemistry, 6, 241-247.

217. VAUGHAN D. and McDONALD I.R. 1971. Effects of humic acid on protein synthesis and ion uptake in beet discs. Journal of Experimental Botany, 22, 400-410.

218. VAUGHAN D. and McDONALD I.R. 1976. Some effects of humic acid on the cation uptake by parenchyma tissue. Soil Biology and Biochemistry, 8, 415-421.

219. VAUGHAN D. and LINEHAN D.J. 1976. The growth of wheat plants in humic acid solutions under axenic conditions. Plant and Soil, 44, 445-449.

220. VAUGHAN D. and ORD B.G. 1981. Uptake and incorporation of ^{14}C-labelled soil organic matter by roots of Pisum sativum L. Journal of Experimental Botany, 32, 679-687.

221. VAUGHAN D., CHESHIRE M.V. and MUNDIE C.M. 1974. Uptake by beetroot tissue and biological activity of ^{14}C-labelled fractions of soil organic matter. Biochemical Society Transactions, 2, 126-129.

222. VAUGHAN D., ORD B.G. and MALCOLM R.E. 1978. Effect of soil organic matter on some root surface enzymes of and uptakes into winter wheat. Journal of Experimental Botany, 29, 1337-1344.

223. VILLE G. 1879. (On artificial manures, their chemical selection and scientific application to agriculture). Translated by Crookes W, London.

224. WAGEMANN R. and BARICA J. 1979. Speciation and rate of loss of copper from lakewater with implications to toxicity. Water Research, 13, 515-523.

225. WAKSMAN S.A. 1936. Humus origin, chemical composition and importance in nature. Bailliere, Tindall and Cox, London.
226. WARIS H. 1953. The significance for algae of chelating substances in the nutrient solution. Physiologia Plantarum, 6, 538-543.
227. WIERSUM I.K. 1974. The activity of specific growth stimulating substances in the soil in relation to the application of organic matter. Tenth International Congress of Soil Science, 3, 123-129.
228. WILEY H.W. 1897. Über den Einfluss des Humus auf den Stickstoffgehalt des Hafers. Landwirschaftlechen Versuchswesen und die Tatigheit der Landwirtschaftlichen Versuchsstationen Preussens, 49, 193-202.
229. WILLOUGHBY L.G. and BAKER C.D. 1969. Humic and fulvic acids and their derivatives as growth and sporulation media for aquatic Actinomycetes. Verhandlungen der Internationalen Vereinigung fur theoretische und angewandte Limnologie, 17, 795-801.
230. WILLOUGHBY L.G., BAKER C.D. and FOSTER S.E. 1968. Sporangium formation in the Actinoplanaceae induced by humic acids. Experientia, 24, 730-731.
231. WOODWARD J. 1699. Thoughts and experiments on vegetation. Philosophical Transactions, 21, 382-398.

CHAPTER 2

INFLUENCE OF HUMIC SUBSTANCES ON BIOCHEMICAL PROCESSES IN PLANTS

D. VAUGHAN, R.E. MALCOLM and B.G. ORD

The Macaulay Institute for Soil Research, Aberdeen, Scotland.

CONTENTS

1. INTRODUCTION

As the previous chapter testifies, there is a considerable literature to show that under certain conditions, humic substances can influence plant growth. These influences can be conveniently considered as being direct or indirect. An example of the latter would be the complexation by humic substances of a nutrient cation in the growth medium resulting in an enhanced uptake of the nutrient into the plant[31,34,53,87,88,152]. In contrast, direct effects on plant growth are the result of the humic substances directly interfering with metabolic processes such as respiration or nucleic acid and protein synthesis[36,37,42,72,161].

Plant growth is the ultimate expression of a series of inter-related biochemical and physiological processes and thus it is important to investigate the effects of humic substances on all these processes. During the last 40 years particularly, the upsurge of interest in biochemistry generally has been mirrored by the interest shown in metabolic processes by soil scientists. Despite this interest, our current understanding of the effects of humic substances on biochemical processes is still only fragmentary, especially from a mechanistic viewpoint[35,41,52,72,73]. Because of the considerable complexity and diversity of soil organic matter[77,133,145], we shall concentrate in this brief review on the effects on metabolism produced by strictly humic substances as defined in the Introductory Chapter. It is perhaps pertinent to remember that most humic substances are insoluble in water and are not in a physiologically active form[74]. To produce forms which are active in plant metabolism, the humic acids, for example, are usually brought into solution as their Na or K salts and the pH of the resulting media is adjusted with a mineral acid. Accordingly, these humic acids should more accurately be referred to as humates.

2. ENERGY METABOLISM

Aerobic respiration plays a vital role in providing metabolic energy (ATP) in all higher plants, although in the light, plants also produce considerable amounts of ATP directly through photosynthesis. Any substances which interfere with respiration and/or photosynthesis would, therefore, alter the amount of metabolic energy produced by the plant and, as a result, affect plant growth.

2.1. Respiration

There are many reports which show that humic acids (HA), prepared from a wide range of soil types, enhance respiration in higher plants[51,52,54, 69,70,71,72,138,141,142,143,154]. Several model HA's also enhance respiration and these substances, free of mineral components, carbohydrates and proteins, have been used extensively by Flaig and his co-workers[38,41, 42,43,44,45,46,108]. Most results were obtained using water or pot cultures either with[137,141,142] or without[116,142] added nutrient. Other methods of applying HA to plants have also been used. When leaves of *Begonia semper-florens* were sprayed with aqueous solutions of HA, there was a marked increase in O_2 uptake[137,140]; injecting HA into plants effected a similar response[69,70].

Although much of the work has centred on the influence of HA on respiration, some reports indicate that fulvic acid (FA) can sometimes evoke a greater response[6]. Thus Sladky[137] found that when used at a concentration of 50 mg l^{-1} in nutrient solution, FA enhanced O_2 uptake by 38% in roots of tomato plants whereas at the same concentration HA produced an increase of 24%. FA also evoked a greater response than HA in the leaves of *Begonia* after foliar applications[140].

Thus far only the effects on respiration in higher plants have been considered, but some microorganisms also respond. Myśkow[101] has reported that HA enhances O_2 uptake by the yeast *Torulopsis utilis* during culture and a similar observation has been made for the alga *Scenedesmus quadri-cauda*[151]. In contrast humic substances had no effect on respiration in the yeast *Candida utilis*[98], but in this case, unlike *Torulopsis*, the humic substances also had no effect on growth.

In evaluating an effect of humic substances on respiration, both O_2 uptake and CO_2 output should be considered and the Respiratory Quotient ($CO_2:O_2$) calculated. Although the value of the Respiratory Quotient (R.Q.)

has been questioned, it can be a valuable index for the type of substrate being utilized[66]. The theoretical R.Q. value for carbohydrates is unity, for fats *ca.* 0.7 and for carboxylic acids such as malic and succinic acids >1.0. In a detailed investigation, Šmidová[142] found that HA in nutrient solution enhanced both O_2 uptake and CO_2 output in roots of *Triticum vulgare Zea mays* and *Cucurbita maxima*, but in no case was there a significant change in the R.Q. With the exception of the *Cucurbita* incubated in HA, the R.Q. values were always <1.0 and, for example, in *Triticum* ranged from 0.66 to 0.77 in the controls and from 0.70 to 0.81 in the tissue treated with HA. In contrast, the R.Q. of tomato plant roots was increased by HA under conditions of inadequate aeration[54]. This result was accounted for in terms of both a reduced O_2 uptake and enhanced CO_2 output. Under conditions of O_2 deficiency, a substantial amount of the CO_2 produced may well have been attributable to anaerobic respiration[66], the importance of which would have increased relative to aerobic respiration.

The manner in which humic substances influence respiration has been the subject of considerable speculation with as yet no definitive answer. Because humic substances are heterogeneous, comprising, in addition to the brown components, carbohydrates, proteins, phenolic acids and mineral components[77], they may influence respiration indirectly by acting as substrates. This latter interpretation cannot be completely ruled out in experiments where humic substances enhance both growth and respiration in aqueous media without added nutrients. Usually such experiments are performed under non-axenic conditions and the breakdown products from the humic substances arising as a result of microbial activity, would be available for use by plants. The notion that humic substances do not act in a direct manner on the plant but stimulate the development of microflora, the influence of which is the actual cause of improved plant growth, was put forward by Chizhevsk and Dikusar[17]. There is certainly no doubt that humic substances provide a source of food and energy for many micro-organisms[2,12,97,122,147]. To distinguish between direct and indirect participation of humic substances in respiration it is essential to carry out experiments in which they are tested under axenic conditions. It was shown by Vaughan[154] that at optimal concentrations HA enhanced O_2 uptake by 15% in slices of parenchymatous beetroot storage tissue under axenic conditions. More recent work has shown that the residual "core" material remaining after refluxing HA for 12h in 6M HCl, to remove proteins,

carbohydrates, phenolic acids, etc.[57], is as effective as the original
starting material in enhancing O_2 uptake in the beetroot system (Fig. 1).

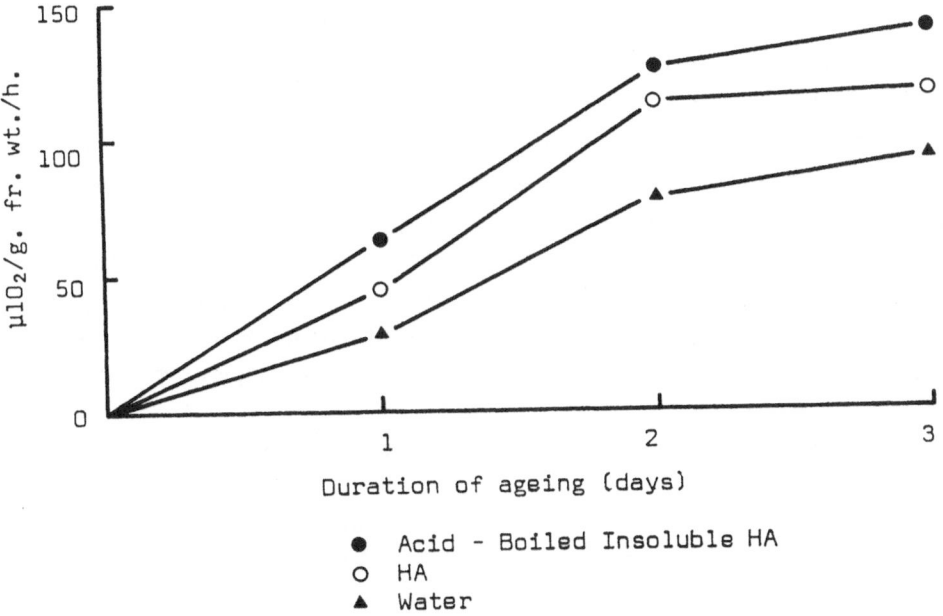

FIGURE 1. Effect of humic acid on the development of respiration in
beetroot slices under axenic conditions.

Clearly HA can have a direct effect on respiration in beetroot storage
tissues and this is also the case for model HA used by Flaig[41].   After
careful and comprehensive work with higher plants, Khristeva[72,76] has
consistently maintained that natural humic substances from many sources
participate directly in respiration and do not act primarily as a food
source.

Several mechanisms have been suggested to explain the effects of humic
substances on respiration.   Under conditions of low O_2 tension HA is
considered to facilitate respiration by a mechanism in which hydroquinones
in the humic material act as hydrogen acceptors[49,50,53,54].   In contrast,
Gumiński et al.[55] attributed the stimulation of tomato seedling growth by
HA under conditions of low O_2 concentration not to a direct effect on
respiration, but to this humic material acting as a complexing agent for
Fe.   Even so, this could still be regarded as a direct effect because Fe
forms part of the enzyme structure of many terminal oxidases, the activity

of which could be rate limiting under such O_2 conditions. The notion that under normal conditions of O_2 tension, humic substances enter the plant and function as hydrogen acceptors has been put forward by several researchers[35,38,41,42,44,45,46,108,143]. Most likely, humic substances enter the plant at early stages of growth and act as respiratory catalysts[73] within the mitochondria. Pertinent to this, experiments using (^{14}C)-labelled HA have revealed that as much as 15% of the radioactivity entering parenchymatous cells becomes associated with the mitochondrial fraction[159]. Khristeva[69,70,77] has suggested that soluble forms of humates are incorporated into the redox chain of respiration in the mitochondria thereby increasing the rate of this process (see Fig. 2). She interprets her results as indicating that HA, because it contains quinone groups, acts as a hydrogen acceptor, and at the same time as an activator of O_2, thereby behaving as a catalyst.

Many of the properties of humic substances are derived from the formation of electron-donor and acceptor complexes between these materials and non-humic substances[174]. Evidence is now accumulating that humic substances, which contain phenolic and quinoid groups[27,106,130], participate in such redox processes, most likely via free radicals which are present mainly as semi-quinones[3,89,128,132,144]. Prát[120], too, has expressed the view that humic substances may enhance biological activity in plants by functioning as free radicals. Support for this view has come from experiments in which the concentration of free radicals in HA was increased using ultraviolet light[150]. HA treated in this way stimulated the growth of lettuce plants to a greater extent than the untreated HA.

$$\text{NADH} \xrightarrow{\text{ATP}} U \text{----} Fp \text{---} Cytb \xrightarrow{\text{ATP}} Cytc \text{---} Cyta \xrightarrow{\text{ATP}} Cyta_3 \longrightarrow O_2$$

    NADH = Nicotinamide Adenine Dinucleotide Phosphate

    ATP = Adenosine Triphosphate Production

    F_p = Flavoprotein   U = Ubiquinone

    Cyt = Cytochrome (for b and c, complexes)

FIGURE 2. Simplified scheme showing the positions on the electron
         transfer chain where ATP is produced from ADP and inorganic-P.

The available evidence suggests that the free radicals of humic substances behave as electron donors or acceptors depending on the situation[136].
Humic substances are, therefore, well suited to participate directly in the
electron-transport system associated with respiration. Furthermore, they
can take part in oxidation-reduction reactions with transition metals[133,136].
In this connection it should be noted that HA can enhance the activity of
cytochrome oxidase[143].

There are however two contrasting views on the nature of its influence
on the formation of high energy phosphate (ATP) by oxidative phosphorylation
(Fig. 2). ATP plays a vital role in all metabolic syntheses. Its
formation from ADP and inorganic-P during aerobic respiration is coupled
to the energy output involved in the oxidation of reduced cytochromes along
the respiratory chain. The efficiency of the conversion of ADP to ATP is
given by the ratio ADP:oxygen. From the data of many experiments using
quinones and other substances, such as 2,4-dinitrophenol, which uncouple
oxidative phosphorylation, Flaig and his co-workers (summarized by Flaig[41])
proposed that HA, too, partially uncouples oxidative phosphorylation.
Experiments using mitochondria obtained from *Brassica olera* indicated that
those substances which uncouple oxidative phosphorylation also increase
the dry matter and sometimes, as in cereals, can also increase yield of
grain. According to Flaig[41,43] therefore the changes in plant constituents
and enzyme activities are only a secondary effect as a result of partial
uncoupling in ATP production. In essence, the increases in dry matter
produced in the presence of humic substances are explained by an increase
in the inorganic-P (Pi) content. This increase in Pi accelerates processes
requiring phosphorylation which result in the accumulation of monomer
building blocks. Although humic substances can produce partial uncoupling
of oxidative phosphorylation, sufficient ATP is still produced to allow
polymerization of some of the monomers resulting in the formation of
components such as cellulose, lignin, nucleic acids and proteins. Indeed,
an increase in the rate of photosynthesis in many plants treated with humic
substances[41] may also be a reason for there being sufficient ATP.

In contrast, Khristeva[70,72,74] has suggested a somewhat more attractive
hypothesis in which it is envisaged that humic substances lead to an
enhanced production of ATP by oxidative phosphorylation as a result of an
increase in respiration. This in turn leads to an intensification in
the synthesis of nucleic acids and proteins. The effects of HA on these

syntheses are considered in more detail later in this review.   In support
of her hypothesis, Khristeva[72] observed that whereas HA can enhance
respiration in wheat, barley and maize, externally supplied ATP is more
effective.   Furthermore, ATP was more effective than HA in increasing
the quantity of free radicals in *Phaseolus aureus* leaves and shoots.   The
observation[18] that there are low levels of free sugars in wheat plants
during intense growth in aqueous solutions containing HA could be interpreted in relation to an enhanced rate of respiration with a concomitant
increase of ATP.

In assessing the relative merits of the different views expressed by
Flaig and Khristeva, it should be borne in mind that whereas Flaig based
his conclusions on data obtained using model HA, Khristeva used only
natural humic substances.   It cannot be assumed that these different
substances always produce identical effects on ATP formation[52].   Clearly
more information is required on the influence of humic substances on the
ADP:oxygen ratio in a wide range of plant species.

2.2. Photosynthesis

Any investigation into the influence of humic substances on energy
production in higher plants would be incomplete without considering photosynthesis.   However, there is a dearth of information available on the
influence of humic substances on the biochemical mechanisms involved in
this important process.   An effect of humic substances on photosynthesis
is always deduced rather than proved.   Thus Sladky and Tichý[140] concluded,
very reasonably, that an increase in dry weight of *Begonia semperflorens*
leaves in the presence of HA was due to an effect on chlorophyll content
and hence a greater photosynthetic assimilation.   But the production of
more chlorophyll does not necessarily mean an increase in plant yield[148].
To our knowledge there are no reports on the influence of humic substances
on photorespiration.   This process is important because photosynthesis
and photorespiration are so inter-related that in the light one does not
proceed without the other.   The influence of photorespiration should not
be ignored in what appears to be a low level of photosynthesis[148].

There are certainly many reports in the literature to show that humic
substances can increase the chlorophyll contents in plant leaves when
supplied either in nutrient solution or as a foliar application[77,102,122,
138,140,143].   Increases in chlorophyll production in leaves of *Zea mays*

after the application of humic substances have also been reported under field conditions[73]. In addition to chlorophylls a and b, humic substances also increase the production of carotene and xanthophyll[139]. HA also results in a stimulation in the growth of plastids containing chlorophyll in the leaves of *Crassula portulacea*[83]. Thus the evidence shows that humic substances can influence chlorophyll production and, in addition, these substances may even prevent chlorosis under conditions where there might be cation deficiency. Relevant to this is the report[35] that HA prevents chlorosis in *Zea mays* presumably by increasing the absorption of either Mg^{2+}, a cation which forms part of the chlorophyll molecule, or Fe^{2+} which is required for chlorophyll formation.

Khristeva[72] has reported that the effect of humic substances on plant growth apparently increases with increasing height above sea level. This result she explains on the basis of the humic material enhancing the absorption of light quanta, by the chloroplasts. Similarly, it was observed that spraying maize plants with HA in the spring and summer effected growth responses which relate more to the composition of light than to the temperature[72]. Clearly more information is now required on the influence of humic substances on the photosynthetic apparatus.

3. NUCLEIC ACIDS

The importance of nucleic acids both in regard to cell replication and growth has been recognized for many years (Fig. 3)[61,62,65,68,158,172]. It is, therefore, surprising to find that apart from the work of a relatively limited number of workers, little information is available on the effects of humic substances on these important cell constituents.

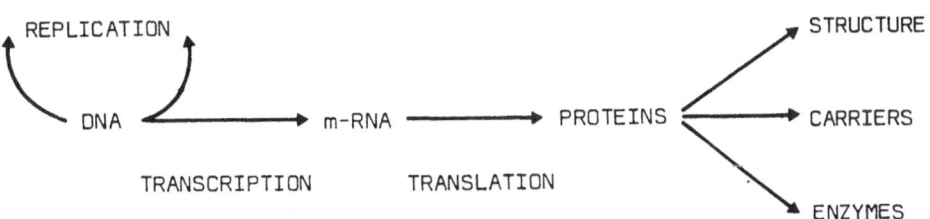

FIGURE 3. Simplified scheme for the role of nucleic acids in protein synthesis.

In eukaryotes gene expression occurs by multiple mechanisms some of which in higher plants involve:-

a. Selective *de novo* messenger RNA (m-RNA) synthesis

b. Processing and translocation of m-RNA

c. Synthesis of new proteins from preformed m-RNA

d. Activation of existing enzymes.

Gorovaya and Solocha[48] noticed that HA increased the volume of interphase nuclei in meristematic tissues of maize, onion and wheat. In a detailed study using maize[72,74] it was established that HA enhanced the mitotic activity of the root meristem and enlarged the size of the interphase nuclei by increasing their DNA content. An enhancement of DNA synthesis induced by HA in meristematic cells was confirmed by autoradiography using ^{3}H -labelled thymidine. A concomitant increase in m-RNA was also observed using ^{3}H -labelled uridine. HA also produced an increase in the rate of RNA synthesis and ^{32}P incorporation into the free nucleotides of root cells of *Phaseolus* and *Helianthus*[75].

In a study on the effect of HA on nucleic acid synthesis in wheat seedlings, Fialová[36,37] showed that when the plants were grown in water alone, the RNA content of their roots was maximal (25 µg/plant) two days after germination, thereafter declining rapidly, whereas in the presence of HA, the RNA reached a maximum (33 µg/plant) after 5-7 days, thereafter declining. However, when cultured in full nutrient solution the RNA content reached 60 µg/plant after 7-9 days before declining more slowly than in those plants cultured in water, with or without HA. These changes in RNA content mirrored changes in plant growth. Fialová also observed that not all aspects of RNA synthesis were affected equally by HA. The greatest effect of HA was on ribosomal-RNA which was enhanced by 60% relative to the water control. Also with regard to total RNA content it was concluded that breakdown prevailed over synthesis in plants which were starved of nutrients. The data of Fialová are important because they can be interpreted as relating RNA synthesis to the regulation of cell elongation by HA. Nevertheless, she observed no effect of HA on net m-RNA synthesis.

Changes in growth and RNA synthesis invoked by HA have also been reported for excised pea root segments under axenic conditions[161]. In the presence of sucrose as an energy source, HA enhanced the increase in the length of these segments. During the 18h duration of the experiment

the total RNA content fell from 15 to 11.5 µg/segment, even in the presence
of HA. Clearly, only certain qualitative changes in RNA metabolism
regulate cell elongation. Pertinent to this is the observation[161] that
actinomycin D, which blocks m-RNA synthesis in elongating tissues[67,135],
produced not only a continuous decrease in the growth rate of excised root
segments, but also annulled the stimulation of cell elongation produced by
HA.

The interaction between HA and actinomycin D has also been studied in
beetroot storage tissues. When slices of this tissue were washed in water
for several days, there was an increase in their RNA content[154]. HA
enhanced the development of RNA synthesis and some other aspects of metabolism such as the development of invertase activity[154]. At suitable
concentrations, actinomycin D completely abolished any stimulation of
metabolism produced by HA[161]. The authors concluded that HA influences
the production m-RNA, the continuous synthesis of which is essential for
the many changes observed in metabolic processes such as invertase activity.
Although continuous RNA and protein synthesis are inter-dependent[26,104,135],
the initial m-RNA synthesis precedes the synthesis of a given protein[26].
As will be shown later in this review, HA appears to have a selective effect
on some aspects of protein synthesis. One likely explanation for this
selective effect is that HA influences the transcription of newly synthesized
m-RNA[41,161], but post-transcriptional control cannot be discounted.

4. PROTEIN METABOLISM

4.1. Protein synthesis

A substantial literature bears witness to the importance of investigating
the effects of humic substances on enzyme changes in plant tissues. In
such studies it is usual to distinguish between an effect on enzyme development (synthesis), and an effect on enzyme activity *per se* which is measured
by adding the humic substances to the enzyme assay media. In this section,
only protein synthesis generally and enzyme synthesis in particular, are
considered because they relate directly to translation and/or transcription
involving m-RNA.

Bukvová and Tichý[11] reported that HA influenced the development of
phosphorylase in wheat plants during growth in sand culture, the nature
of the effect depending on the concentration of the humic substances. At
a concentration of 10mgl^{-1}, HA enhanced the development of phosphorylase

in roots relative to the water control whereas higher concentrations of 100 to 800 mgl^{-1} were inhibitory. Such a result emphasizes the importance in selecting an adequate control in studying the effects of humic substances on growth and metabolism, particularly in situations where the provision of an adequate supply of nutrients is essential. In this connection, Činčerová[19], also using wheat plants, showed convincingly that plants cultured in a Knop nutrient solution with or without HA grew well and had a high level of glutamic-alanine-transaminase activity. Plants cultured in water, however, possessed a low level of the enzyme. Furthermore, a stimulation of the development of glutamic-alanine-transaminase in the presence of HA was only apparent when water was used as the control. In contrast, the development of aspartate-glutamate-transaminase, which was highest in the poorly growing wheat plants in water, was unaffected by HA[20,21].

Humic substances have been shown to influence the development of enzymes in many higher plants, for example catalase, o-diphenoloxidase and cytochrome in tomatoes[143], peroxidase and catalase in *Zea mays*[73] and invertase and peroxidase in beetroot[153,164]. To provide the maximum stimulation of enzyme development, it would appear that the humic substance must be present continuously in the culture medium[154]. The effects of humic substances on enzyme development are not confined to higher plants but can occur in some micro-organisms examples being the stimulations of catechol-1, 2-oxygenase in *Azotobacter*[30], catalase and polyphenoloxidase in *Bacillus mycoides* and *Sorangium cellulosum*[78] and invertase in *Torulopsis utilis*[101]. In contrast, the development of some enzymes such as glucose-6-phosphate dehydrogenase in pea roots[118] and phosphatase in storage tissues[158], are not affected by humic substances. Where a comparison has been made for higher plants, HA has generally a greater effect on enzyme development than has FA[164].

In all cases where humic substances influence the development of an enzyme it is assumed that there is an effect on net enzyme synthesis. There is only a meagre amount of relevant data available to substantiate this assumption. In an investigation into the development of invertase in beetroot storage tissue slices during water washing, it was established that whereas HA stimulated the development of the enzyme it had no effect on the incorporation of ^{14}C -labelled amino acids into proteins in general[158,161]. A similar observation was also reported with the incorporation of radio-actively labelled amino acids into elongating excised pea root segments[161].   Inhibitors of protein synthesis (puromycin, cycloheximide and D-*threo* chloramphenicol) and RNA synthesis (actinomycin D) inhibited both invertase development and the incorporation of ^{14}C -leucine into the proteins of both beetroot slices and excised pea root segments while abolishing the stimulation of invertase development produced by HA. A similar observation using metabolic inhibitors has been reported for the development of peroxidase[159].   Several amino acid analogues such as azetidine-2-carboxylic acid, *cis*-4-hydroxy-L-proline and p-fluorophenyl-alanine, gave responses with HA similar to those described for the "specific" metabolic inhibitors with the difference that their inhibitory effects were ameliorated in the presence of the corresponding natural amino acid[156,161].   These data provide strong evidence for the notion that HA stimulated the *de novo* synthesis of invertase and peroxidase.

The effect of HA on ion uptake was discussed in the previous chapter, but it is now pertinent to consider briefly the development of ion uptake capacity insomuch as it relates to protein synthesis.   HA can have a selective effect on ion uptake[34,159].   The magnitude of the effect is related to the concentration of the humic substance[33,158] and to the pH of the uptake medium[5].   In beetroot slices during washing, HA stimulated the development of uptake capacities of sodium, barium and phosphate, inhibited chloride and had no effect on calcium relative to the water-aged controls[158, 159].   Furthermore, in the presence of HA, cycloheximide and D-*threo* chloramphenicol inhibited the development of protein synthesis and abolished the stimulation of sodium uptake[159].   Although HA enhanced the synthesis of invertase and peroxidase, it had no effect on the synthesis of phos-phatase or on the incorporation or distribution of ^{14}C -labelled amino acids into the proteins of particulate cell fractions[158,159].   Thus HA incluences only some aspects of protein synthesis and this most likely operates via the formation of new m-RNA.   To our knowledge, no information is available to indicate an effect of HA on enzyme induction in higher plants, only on the synthesis of some enzymes already being produced. There is, however, some evidence to suggest that FA can induce enzyme synthesis in bacteria[30].

4.2. Structural wall-proteins

Cell-wall proteins have an important role in cell elongation in higher plants[4,63,82,129]. Despite this, little work has been done on the influence of humic substances on cell-wall proteins during extension growth.

HA enhances wall-protein synthesis in beetroot slices during ageing[154], but no measurements were made on any structural changes in these wall components. However, HA can influence the changes which occur in wall-proteins of excised pea root segments during elongation[155]. The cessation of cell elongation is accompanied by a rapid increase in structural wall-proteins rich in hydroxyproline residues which are linked to the wall-carbohydrates through arabinose[25,82,129,156]. When the rapid increase in wall-bound hydroxyproline is delayed by adding inhibitors to excised pea root segments, cell elongation continues for a longer period than in the non-delayed controls[156]. At concentrations enhancing cell elongation, HA has no effect on protein synthesis generally, but it does inhibit the hydroxylation of proline to hydroxyproline in the cell-wall proteins[155].

The importance of a low rate of hydroxyproline formation in cell-wall proteins in relation to a stimulation of cell elongation has been emphasized[63,156]. Holleman[63] concluded that the stimulation of growth of wheat coleoptiles in the presence of the chelators 2,2'-dipyridyl and 8-hydroxyquinoline[59,60] might be related to a reduced hydroxyproline content of the cell wall. The question now arises as to how HA can influence the hydroxylation of proline. There is a marked similarity between the effects produced by 2,2'-dipyridyl and HA. Because 2,2-dipyridyl specifically binds the ferrous iron[64] which is essential for the hydroxylation of proline[173], it was concluded that HA enhanced cell elongation in the excised pea roots by virtue of its ability to complex ferrous iron within the tissue[155]. Indeed, the stimulation of elongation produced by HA was negated in the presence of sufficient amounts of ferrous iron ion in the culture medium. But this mechanism for the effect of HA on growth in higher plants may not be applicable to all tissues. In a recent report of experiments using beetroot storage tissues[163], it was shown that HA has no effect on either the incorporation of ^{14}C -proline into cell-wall proteins or on its hydroxylation of proline to hydroxyproline in a situation where 2,2'-dipyridyl inhibited hydroxylation.

4.3. Enzyme activities

The possibility that humic substances might behave as growth hormones

has led some workers to investigate the influence of these substances on
the metabolism of indole-3-acetic acid (IAA), particularly its destruction
by IAA-oxidase.   In a series of papers[93,94,95,96], it was shown that both
HA and FA inhibited IAA-oxidase activity to about the same extent in
homogenates of *Lens culinaris*.   Although in all cases the inhibition of
enzyme activity increased with increasing concentration of humic substances,
its magnitude depended on the source of the humic material and the method
of extraction[94,95].   It was postulated that HA could decrease the concentration of IAA free radicals produced in the enzymatic oxidation of IAA.
In this connection it has recently been shown[162] that HA can be involved
directly in the formation and destruction of free radicals in the enzyme
systems of higher plants.

It has now been established that humic substances can affect the
activities of several enzymes present in tissue homogenates, such as
phosphatase[91,92,166], invertase[90,166], choline esterase[29] and peroxidase[160].
Generally, only inhibitions were observed, the effect increasing with
increasing concentration of the humic substance.   Furthermore, where a
comparison was made, HA was usually more effective than FA[92,160].   The
magnitude of the effect depended not only on the enzyme under assay but
also on its source[90,110].

But studies using homogenates may not represent the true nature of
what happens when whole plants are grown in solutions of humic substances.
This aspect has been investigated for invertase using intact wheat plants[166].
In this tissue, the sucrose substrate has access to only 50% of the total
root invertase and this may in practice be regarded as a root surface
enzyme.   Nevertheless, HA inhibits this root surface enzyme in the intact
plant to the same extent as it inhibits the total invertase in the tissue
homogenate.   Whether the tacit assumption can be made that HA influences
enzymes generally in intact tissues awaits further clarification.

Another objection to using tissue homogenates is that crude enzymes may
not respond to humic substances in the same manner as if they were purified.
For this reason many investigators have preferred to use purified enzymes.
In an elegant study using purified proteolytic enzymes, Ladd and Butler[81]
observed that HA inhibited carboxypeptidase A, chymotrypsin A, pronase and
trypsin activities, stimulated papain, ficin and subtilo peptidase and had
no effect on phaseolain and tyrosinase.   HA also inhibited the activity
of aminopeptidase K[111].   Pflug[112], using purified peroxidases from horse-

radish and *Streptococcus faecalis*, showed that HA prepared from many sources inhibited the enzyme, in agreement with observations made using peroxidase in tissue homogenates. For some enzymes the HA-induced inhibition can be partially abolished using a cation such as Zn for amino-peptidase K[111], Mg for phosphatase[91] or Ca for malate dehydrogenase[113].

Kinetic studies have revealed that the inhibition of enzyme activities by humic substances can be competitive, non-competitive or a mixture of both. An example of competitive inhibition was shown for phytase[117,119] using *myo* inositol hexaphosphate as the substrate. In this case the inhibition was caused by the formation of a complex between HA and the substrate thus effectively lowering the availability of the substrate to the enzyme. Peroxidase from *S. faecalis* was inhibited by competitive interactions between the hydrogen donor NADH and HA for the enzyme[112]. Other workers have shown that HA inhibited the activity of pronase in some substrates by lowering the affinity of the enzyme for the substrate, that is, increasing the Km value for the reaction[80].

In many instances, the inhibition produced by HA alters only the maximum velocity of the reaction (non-competitive inhibition) with, for example, phosphatases[91,92], invertase (Fig. 4)[90,166], peroxidase[160] and aminopeptidase K[111].

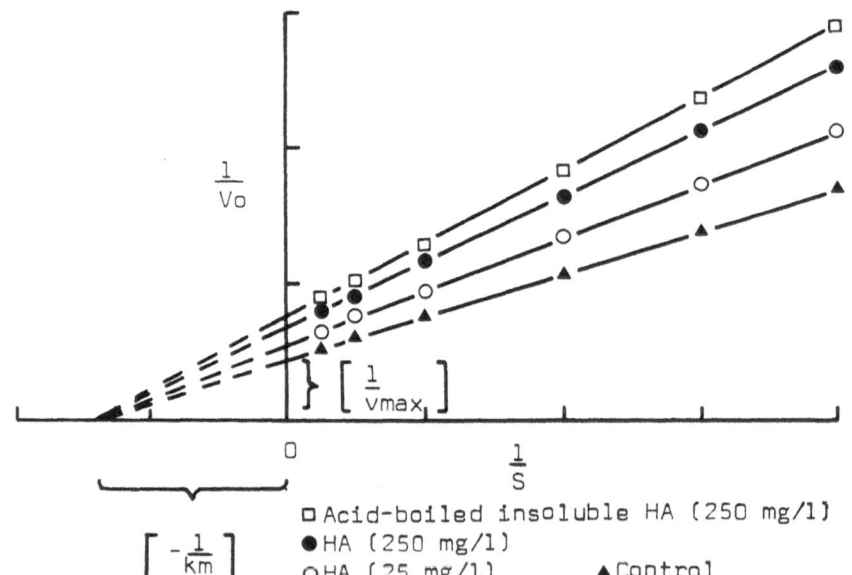

FIGURE 4. Effect of humic acid on the kinetics of wheat root invertase activity (from Malcolm and Vaughan[90]).

The inhibition of malate dehydrogenase is of interest in that for the forward reaction to oxaloacetate it is purely competitive whereas in the reverse direction in the presence of NADH, it is mixed[113].   Of interest too is the use of NADH as the hydrogen donor in the determination of peroxidase because this enzyme was inhibited competitively by competition interactions between NADH and HA for the substrate[112].   No such competitive inhibition was recorded for crude wheat root peroxidase using o-dianisidine as the hydrogen donor[160].

The importance in the choice of substrate in studying the effects of humic substances on enzyme activities was recognised in studies of pronase[80]. In the presence of Z-dipeptides, the inhibition of pronase activity is probably due to the combination of HA with the enzyme.   However, when proteins are used as the substrate, the interaction between HA and the substrate also has to be taken into consideration because such an inter-action can lower the effective concentrations of both substrate and HA. In this context, it has been shown that the inhibitory effect of HA on wheat peroxidase activity depends on the hydrogen donor used.   In the presence of o-dianisidine, HA inhibited the enzyme activity whereas in the presence of guaiacol the effects of HA were variable and less well marked[160].   These effects may be linked to the different responses of the peroxidase isozymes which can differ in their affinities for different hydrogen donors[47].

The question now arises as to the mechanisms involved in the effects of HA on enzyme activities *per se*.   Using Z-peptides as substrates, Ladd and Butler[80,81] have suggested that HA may inhibit pronase activity either by competing with the substrate for the catalytically active sites on the enzyme surface, or by causing conformational changes in the enzyme, resulting in a decreased affinity for the substrate.   In studies involving the enzymes invertase[90,166], phosphatase[91,92] and peroxidase[160], there was no evidence that HA competed with the substrates.   It was considered that HA brought about a conformational change in structure of each of the three enzymes resulting in a lowering of their activities.   A reversible, non-competitive type of inhibition as observed for peroxidase in the presence of o-dianisidine[160], has been considered by Dawes[28] to indicate that the combination of enzyme and inhibitor occurs at a point necessary for enzyme activity but not at the active centre where substrate combination occurs. For peroxidase, it was suggested[160] that HA inhibits enzyme activity by chelation of iron because this enzyme requires haematin to function.

Such an explanation could apply generally to enzymes of the peroxidase-type such as IAA-oxidase[93], or proline hydroxylase[155].

Pflug and Ziechmann[113] concluded that HA could inhibit enzyme activity via functional groups on the surface of the humic material. Certainly the importance of functional groups has been demonstrated by several workers. The influence of HA on some proteolytic enzymes is directly related to the number of carboxyl groups and these groups could combine with basic amino acids on the enzyme molecule[14]. Phenolic hydroxyl groups have been implicated in the inhibitions of IAA-oxidase[95,96] and also of some β-amylases[110,113]. HA is heterogeneous and has many functional groups, so much so that it is doubtful that a single mechanism can account for the effects of this humic material on all enzymes.

5. BIOLOGICAL ACTIVITY AND STRUCTURE OF HUMIC SUBSTANCES

Investigations into the structure and complexity of humic substances are numerous and have been summarized in many excellent reviews[77,133,145]. It is the intention in this current review to indicate only briefly how biological activity can be related to the structure of humic substances. It should, however, be borne in mind that the method of extraction can be one important factor in determining the biological activity of the humic substance[94].

In most investigations, HA has been used more extensively than FA. It would thus be appropriate to have a working model for this humic substance and to evaluate the model in relation to biological activity. For convenience we have selected, without prejudice to other models (see Introductory Chapter) one based on that described by Haworth[57]. It is proposed that HA comprises a polycyclic aromatic "core material" to which are attached carbohydrates, proteins, phenols, mineral components, etc.

The effect of HA on enzyme synthesis, or enzyme activity, is usually unrelated to its ash content[94,95,113,153]. With some exceptions[95], this is also true for the total nitrogen content. Indeed the biological activities of many model humic acids (prepared by the oxidation of simple phenolic acids), which contain essentially no ash or nitrogen, are often similar to those of natural HA[81,90,111,113,153,154].

Where relationships exist between biological activity and HA structure, they usually involve functional carboxyl or hydroxyl groups[14,90,95,113]. In this context Butler and Ladd[14] showed that methylation of carboxyl groups prevented HA from influencing the activities of some proteolytic enzymes. But the study of functional groups in relation to biological activity has not been confined to enzymes. Thus carboxyl and phenolic groups have been implicated in root initiation[131], and the inhibition of sulphate uptake into barley roots is apparently related to the carboxyl content of the humic substance[33].

The influence of specific functional groups on biological activity has been considered in relation to different molecular weight (MW) components of the HA. It was established that only the low MW components (<3500 daltons) are effective in enhancing invertase synthesis in beetroot storage slices during washing in aqueous media (Fig. 5)[153,154,164].

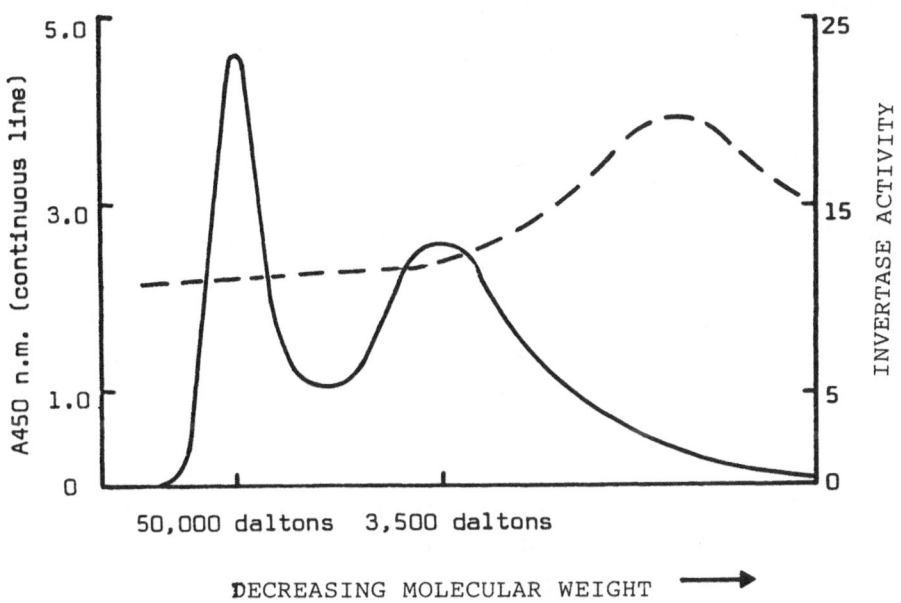

FIGURE 5. Effect of humic acid fractions separated on G-75 Sephadex on invertase synthesis in beetroot discs (from Vaughan[153]).

Similarly IAA-oxidase activity in homogenates of *Lens culinaris* is inhibited to the greatest extent by low MW HA components[95]. Because FA fractions of low MW have the highest carboxyl content[13,119], it might appear that these groups are important in determining the effectiveness of the HA in influencing enzyme activities. However, Mato *et al.*[95]

concluded that phenolic and not carboxyl groups are primarily responsible for the inhibition of IAA-oxidase activity. Similarly, although FA comprises mainly polycarboxylic acid material[1,85] it was the phenolic groups which were primarily involved in IAA-oxidase inhibition[96]. In other studies, for example on the effect of HA on trypsin and papain activities[81], it was demonstrated that the effectiveness of the humic substances increased with increasing MW. In this case it was concluded that HA fractions have properties other than the total carboxyl which are of overriding importance in determining the magnitude of their influence on enzyme activity. A similar conclusion was reached for phosphatase activities[91] indicating that caution should be applied when relating a functional group to the biological activity of the humic substance. In this connexion, similar humic fractions may produce entirely different responses from a similar enzyme such as invertase derived from different species of plants[90]. Some caution should also be expressed in the interpretation of results relating to MW because whereas in a specific test system some low MW components may be biologically active, other apparently similar components in the same HA may be totally inactive[164].

When HA is refluxed in 6M HCl, the residue (equivalent to some 60% of the original HA[160,164]) comprises mainly aromatic "core material". In many instances this "core" is more biologically active than the original HA[90,91,154,160,165] and in such cases the carbohydrates, phenolic acids and proteins cannot be involved in the biological response. Furthermore, the considerable amount of decarboxylation which accompanies the acid refluxing of HA[57] might obviate the necessity for including carboxyl groups as a major factor in the biological response. Thus, whereas HA "core material" is involved in the response, it is virtually impossible to ascribe any particular functional group to its activity.

One approach to this apparently intractable problem has involved preparing "core-type material" by the alkaline oxidation of di-or poly-phenols of known structure, such as catechol and guaiacol[154], and relating the biological responses of these models to the starting phenolic materials. Although these models can be active in biological systems[90,111,113,154,159], there is no evidence that their activities can be correlated with the starting phenolic material. In extreme cases, the response of the model may be entirely different from that of the natural product from which it has been prepared. Thus for wheat root peroxidase, the greatest inhibition is produced by HA "core material" and the least by FA. Polymaleic acid, a model for FA[1], produced an inhibition of peroxidase greater than that of FA and very similar to the HA "core material"[160]. With phosphatase activity in contrast, polymaleic acid evoked a response similar to that of FA[92].

Clearly, functional groups are important in determining the biological activity of humic substances, but the manner in which they act has still to be elucidated. The diversity of responses which humic substances can evoke are most likely the product of separate mechanisms within the plant.

6. CONCLUSIONS

Most studies on the influence of humic substances on biochemical mechanisms have involved the use of HA, either as natural products or synthetic models. For this reason the potential effects only of HA are represented in the diagrammatic model (Fig. 6). The effects are described as potential in that firstly the HA must be in solution to produce the effect and secondly, most of our understanding of the influence of HA on biochemical and physiological processes is fragmentary. The effects of HA on these processes may be summarized as:-

a. An influence on membrane permeability and protein carriers of ions resulting in a more rapid and selective entry of essential elements into the root.

b. Activation of respiration and the Krebs cycle with a concommitant increase in ATP production.

c. An increase in chlorophyll content and photosynthesis giving rise to enhanced formation of ATP, amino acids, carbohydrates and proteins.

d. An effect on nucleic acid synthesis in which not only the amount of RNA but also the transcription of m-RNA is influenced.

e. A selective effect on protein synthesis influencing the relative amounts of enzymes, ion-carriers and structural proteins produced.

f. An effect on enzyme activity, an inhibition or stimulation depending on the enzyme and its source.

This list is not comprehensive and does not for example refer to the possibility of humic substances behaving as growth hormones. In the early part of this century, Bottomley[7,8,9,10] suggested that HA enhanced the growth of *Lemna major* by a mechanism which involved growth substances which he called auximones. The experiments of Mockeridge[99,100] produced data in agreement with Bottomley's views. In contrast, Clark and Roller[23,24]

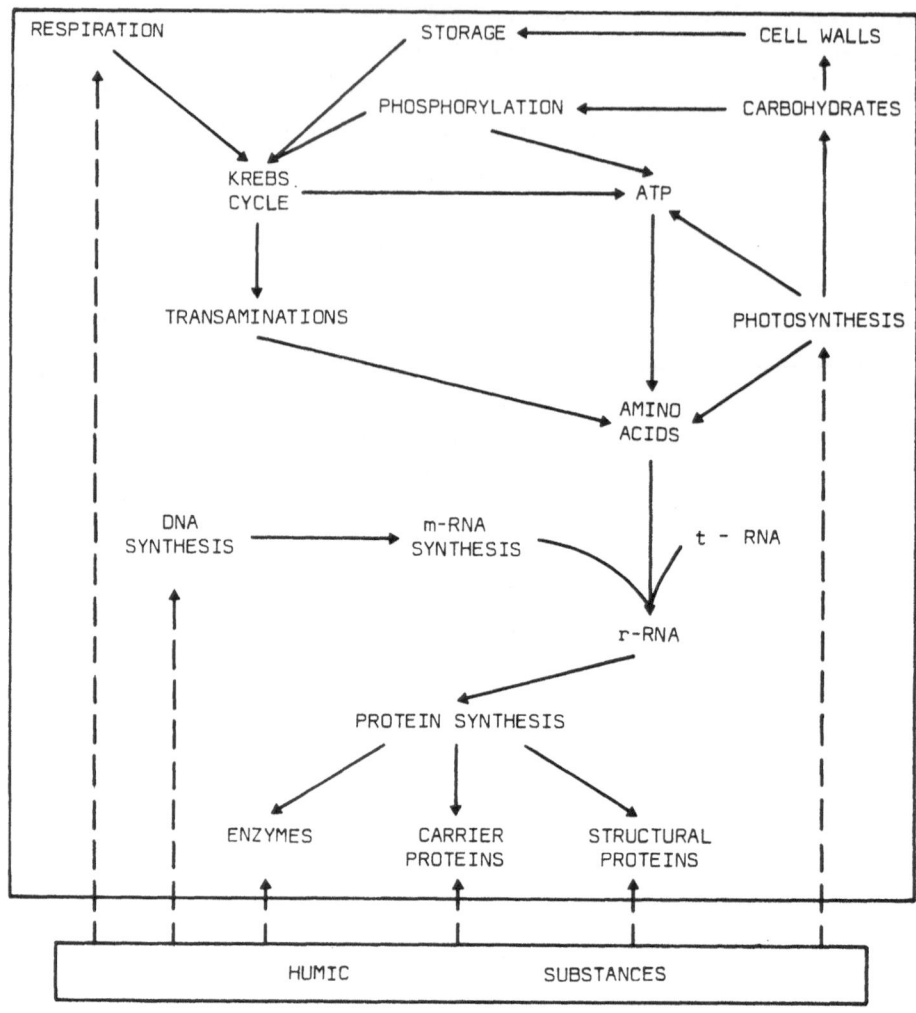

FIGURE 6. Simplified schematic diagram of the likely direct effects of
humic substances on metabolism in plant cells.

found no stimulation on the growth of *Lemna* due to the addition of organic
material and they certainly found no evidence for the presence of auximones.
Wolfe[171] also maintained that Bottomley's auximones did not exist as such
and the influence of humic substances on plant growth was due to the failure
to maintain normal growth in purely inorganic media attributed to the lack

of a proper physiological balance. Indeed, Olsen[107] demonstrated that HA
enhanced the growth of *Lemna* by complexing Fe and making it more readily
available for uptake. Whether humic substances behave as growth hormones
(auxins) has been debated since that time with as yet no conclusive answer.
Since the discovery of growth hormones such as β-indolylacetic acid in the
1930s, several workers have held the view that humic substances can behave
as such[16,79]. Poszewski *et al*[109] used a coleoptile test on the alcoholic
fractions of peat and compost humus and found that although as a whole, the
fractions were inactive, some chromatographically-separated components acted
positively in curving the longitudinal parts of cut coleoptiles. More
recently, O'Donnell[105] using *Pelargonium hortorum* cuttings also considered
that the growth promoting properties of humic substances were caused by
hormone-like material. In contrast, other workers[114,115,126,127,134],
found no evidence for the behaviour of humic substances as auxins. In
recent experiments using *Lemna gibba* we (Vaughan and Malcolm, unpublished)
have confirmed the observations of Clark and Roller[23,24] that humic substances do not influence growth in nutrient culture under sterile conditions
where growth hormones can produce substantial growth enhancement[32,167].

In considering the behaviour of humic substances as auxins it should be
remembered that the elucidation of the mechanisms of auxin action *per se* is
difficult because of the many aspects of physiological/biochemical processes
which an auxin controls. At present we are unable to state whether there
is one basic site and mechanism of auxin action which subsequently leads to
many physiological responses, or whether there are several sites and
mechanisms of action. Certainly, in many systems humic substances can
evoke responses similar to those of auxins, but thus far the case for humic
substances behaving as true auxins, rather than containing some auxin-like
components[56,168,169,170], is unproven. Nevertheless, the variety of
responses which humic substances can induce (for example increasing the
uptake of some ions while decreasing it for others, promoting the activity
of some enzymes without influencing overall protein synthesis) are not
untypical of plant hormones and growth regulators which can bring about a
diversity of apparently unrelated responses. .The net effect of HA on
growth probably involves the interactions of a series of biochemical
stimulations and inhibitions[98].

There is, however, no doubt that once in solution, HA can have a direct
and selective effect on plant metabolism. Whether this can always be

extrapolated to conditions obtaining in the soil remains an open question particularly as the soil solution contains mainly low MW humic substances which closely resemble FA[85,86,166]. FA is physiologically active for example in enhancing plant growth[84,137,140], in influencing enzyme activities[91,96] and having a selective effect on some biochemical processes[166]. When one considers the presence of FA in the soil solution (see Chapter 1), perhaps greater emphasis should now be given to investigating the influence of this relatively neglected humic substance on biochemical processes associated with plant growth.

7. REFERENCES

1. ANDERSON H.A. and RUSSELL J.D. 1976. Possible relationship between soil fulvic acid and polymaleic acid. Nature, London, 260, 597.
2. ANDREYK E.I., GORDIENKO S.A., KONOTO I.N. and MARTYNENKO V.A. 1973. {Assimilation of humic acid nitrogen by micro-organisms}. Mikrobiologichnyi Zhurnal, 35, 139-142.
3. ATHERTON M.M., CRANWELL P.A., FLOYD A.J. and HAWORTH R.D. 1967. Humic acid. I. E.S.R. spectra of humic acids. Tetrahedron, 23, 1653-1667.
4. BARNETT N.M. 1970. Dipyridyl-induced cell elongation and inhibition of cell wall hydroxyproline biosynthesis. Plant Physiology, 45, 188-191.
5. BARTON L.L. and RUOCCO J.J. 1981. Soluble humic complexes and sulphate uptake by Aspergillus niger. Soil Biology and Biochemistry, 13, 435-437.
6. BIBER V.A. and MAGAZINER K.M. 1951. {The influence of humic and fulvic acid on respiration of isolated plant tissues}. Daklady Akademii Nauk SSSR, 76, 609-612.
7. BOTTOMLEY W.B. 1914. The significance of certain food substances for plant growth. Annals of Botany, 28, 531-540.
8. BOTTOMLEY W.B. 1915. Some accessory factors in plant growth and nutrition. Proceedings of the Royal Society, Series B, 88, 237-247.
9. BOTTOMLEY W.B. 1917. Some effects of growth promoting substances (auximones) on the growth of Lemna minor in mineral culture solutions. Proceedings of the Royal Society, Series B, 89, 481-507.
10. BOTTOMLEY W.B. 1920. The growth of Lemna plants in mineral solutions and in their natural medium. The effect of organic matter on the growth of various water plants in cultural solutions. Annals of Botany, 34, 345-365.
11. BUKVOVÁ M. and TICHÝ V. 1967. The effect of humus fractions on the phosphorylase activity of wheat (Triticum aestivum, L.). Biologia Plantarum, 9, 401-406.
12. BURGES A. and LATTER P. 1960. Decomposition of humic acid by fungi. Nature, London, 186, 404-405.
13. BUTLER J.H.A. and LADD J.N. 1969. Effect of extractant and molecular size on the optical and chemical properties of soil humic acids. Australian Journal of Soil Research, 7, 229-239.
14. BUTLER J.H.A. and LADD J.N. 1969. The effect of methylation of humic acids on their influence on proteolytic enzyme activity. Australian Journal of Soil Research, 7, 263-268.
15. CAMPBELL C.A., PAUL E.A., RENNIE D.A. and McCALLUM K.J. 1967. Applicability of carbon dating method of analysis to soil humus studies. Soil Science, 104, 217-224.

16. CHAMINADE R. 1956. The effect of humic acid on the growth of plants. and their nutrition. Sixth International Congress of Soil Science, Paris. _IV_, 443-448.

17. CHIZHEVSKY M.G. and DIKUSAR M.M. 1955. {The role of humus and micro-organisms in the root nutrition of the higher plants in water and sand cultures}. Izvestiya timiryazevskoi sel skokhozyaistvennoi akademii, _2_, 173-192.

18. ČINČEROVÁ A. 1962. Über den Einfluss der Humussaure auf die Veränder-unger der frein Zucker in Weizenpflanzen. In symposium Studies about Humus, Prague, pp. 47-62.

19. ČINČEROVÁ A. 1964. The effect of humic acid on transamination in winter wheat plants. Biologia Plantarum, _6_, 183-188.

20. ČINČEROVÁ A. 1967. Effect of trophic conditions on aspartate trans-amination in wheat plants. Biologia Plantarum, _9_, 64-67.

21. ČINČEROVÁ A. 1969. Effect of trophic conditions on asparagine trans-amination in wheat plants. Biologia Plantarum, _11_, 139-148.

22. ČINČEROVÁ A. 1971. The effect of humic acid on growth and metabolism of wheat seedlings. In Humus et Planta, _V_, 609-611.

23. CLARK N.A. and ROLLER E.M. 1924. "Auximones" and the growth of the green plant. Soil Science, _17_, 193-198.

24. CLARK N.A. and ROLLER E.M. 1931. The stimulation of _Lemna major_ by organic matter under sterile and non-sterile conditions. Soil Science, _31_, 299-309.

25. CLELAND R. and KARLSNES A.M. 1967. A possible role of hydroxyproline containing proteins in the cessation of cell elongation. Plant Physi-ology, _42_, 669-671.

26. CLICK R.E. and HACKETT D.P. 1963. The role of protein and nucleic acid synthesis in the development of respiration in potato tuber slices. Proceedings of the National Academy of Sciences of the United States of America, _50_, 243-250.

27. COMMONER B., TOWNSEND J. and PARK G.E. 1954. Free radicals in bio-logical materials. Nature, London, _174_, 689-691.

28. DAWES E.A. 1962. Quantitative problems in Biochemistry, pp. 124-154, Livingstone, Edinburgh.

29. De ALMEIDA R.M., POSPISIL F., VACKOVA K. and KUŤACEK M. 1980. Effect of humic acids on the inhibition of pea choline esterase and choline a-cyltransferase with malathion. Piologia Plantarum, _22_, 167-175.

30. De HAAN H. 1976. Evidence for the induction of catechol-1, 2-oxygenase by fulvic acid. Plant and Soil, _45_, 129-136.

31. DeKOCK P.C. 1955. Influence of humic acids on plant growth. Science, _121_, 473-474.

32. DeKOCK P.C., VAUGHAN D. and HALL A. 1978. Effect of abscisic and benzyladenine on the inorganic and organic composition of the duckweed _Lemna gibba_ L. New Phytologist, _81_, 505-511.

33. DELL`AGNOLA G. and FERRARI G. 1971. Effects of humic acids on anion uptake by excised barley roots. In Humus et Planta, _V_, 567-569.

34. ELGALA A.M., METWALLY A.J. and KHALIL R.A. 1978. The effect of humic acid and NA_2 EDDHA on the uptake of Cu, Fe and Zn by barley in sand culture. Plant and Soil, _49_, 41-48.

35. FERNANDEZ V.H. 1968. The action of humic acids of different sources on the development of plants and their effect on increasing concentra-tion of the nutrient solution. Pontificiae Academiae Scientiarum Scripta Varia, _32_, 805-850.

36. FIALOVÁ S. 1969. Influence of sodium humate and nutritive conditions on the content of nucleic acids particularly on the ribosomal ribo-nucleic acid in wheat roots. Biologia Plantarum, _11_, 8-22.

37. FIALOVÁ S. 1969. Metabolism of nucleic acids in wheat roots in dependence of nutritive conditions. Biologia Plantarum, 11, 424-431.

38. FLAIG W. 1954. Zur Chemie der Huminsäuren und über die physiologische Wirkung von Modellsubstanzen von Huminsäuren. Arzneimittel-Forchung, 4, 402.

39. FLAIG W. 1961. Die Wirkung von Humusstoffen auf den Staffwechsel der Pflanzen. Maataloustieteellinin Aikakauskiya, 33, 1-16.

40. FLAIG W. 1963. Uber den Einfluss von Humusstoffen auf den Staffwechsel der Pflanzen. International Torf-Kongress, Leningrad.

41. FLAIG W. 1968. Uptake of organic substances from soil organic matter by plants and their influence on metabolism. Pontificiae Academiae Scientiarum Scripta Varia, 32, 1-48.

42. FLAIG W. and OTTO H. 1951. Untersuchungsen über die Einwirkung einiger Quinone als Modellsubstanzen der Auf-und Abbauproduckte von Huminsäure sowie einiger Redoxsubstananzen auf des Wachstum von Pflanzewurzeln. Landwirschaftlicke Forschung, 3, 66-89.

43. FLAIG W. and RIEMEI H. 1971. Contribution to the mechanism of the influence of substances from soil organic matter on plant growth. In Humus et Planta, V, 519-523.

44. FLAIG W. and SAALBACH E. 1955. Zur Kenntnis der Huminsäuren-X. Zeitschrift für Pflanzenernährung und Bodenkunde, 71, 215-224.

45. FLAIG W. and SAALBACH E. 1956. Zur Kenntnis der Huminsäuren-XII. Zeitschrift für Pflanzenernährung und Bodenkunde, 72, 1-7.

46. FLAIG W. and SOCHTIG H. 1962. Einfluss organischer stoffe auf die Aufnahme anorganischer Ionen. Agrochimica, 6, 251-264.

47. GIBSON D.M. and LIV E.H. 1978. Substrate specificities of peroxidase isoenzymes in the developing pea seedling. Annals of Botany, 42, 1075-83.

48. GOROVAYA A.I. and SOLOCHA K.I. 1971. Influence of physiologically active substances of the soil humus and fertilizers on the specific activity meristematic cells of plant sprouts and the quality of seed grains. In Humus et Planta, V, 557-565.

49. GUMIŃSKI S. 1950. {The study of conditions and of the mechanisms of the effect of humus substances on plant organs}. Acta Societatis Botanicorum Polaniae, 20, 589-620.

50. GUMIŃSKI S. 1957. {The mechanism and condition⁻ of the physiological action of humic substances on the plant organism}. Pochvovedenie, 12, 72-78.

51. GUMIŃSKI S. 1967. The effect of humus compounds on some physiological processes and plant nutrition. In Humus et Planta, IV, 255-264.

52. GUMIŃSKI S. 1968. Present day views on physiological effects induced in plant organisms by humic acid. Soviet Soil Science, 1250-1256.

53. GUMIŃSKI S. and GUMINSKA Z. 1953. Studies on the mechanism of the activity of humus on plants. Acta Societatis Botanicorum Polaniae, 22, 45-63.

54. GUMIŃSKI S., CZERWINSKI W., UNGER, E. and BACOWA A. 1955. Studies on the respiration of the root -1. Acta Societatis Botanicorum Polaniae, 24, 723-731.

55. GUMIŃSKI S., GUMIŃSKA Z. and SULEJ J. 1965. Effect of humate agar-agar and EDTA on the development of tomato seedlings in aerated and non-aerated water cultures. Journal of Experimental Botany, 16, 151-162.

56. HAMENCE J.H. 1948. The effects of organic manures on the auxin contents of soils and the "auxin" balance in soils. Journal of the Society of Chemical Industry (London), 67, 277-281.

57. HAWORTH R.D. 1970. The chemical nature of humic acid. Journal of Soil Science, 111, 71-79.

58. HAWORTH R.D. and ATHERTON N.M. 1965. Humic Acid. Bulletin of the National Institute of Sciences of India, 28, 57-60.
59. HEATH O.V.S. and CLARK J.E. 1956. Chelating agents as plant growth substances. A possible clue to the mode of action of auxin. Nature, London, 117, 1118-1121.
60. HEATH O.V.S. and CLARK J.E. 1964. Chelating agents and auxin. Nature, London, 201, 585-587.
61. HEYES J.K. 1960. Nucleic acid changes during cell expansion in the root. Proceedings of the Royal Society, Series B, 152, 218-230.
62. HEYES J.K. and VAUGHAN D. 1967. The effects of 2- thiouracil on growth and metabolism in the root II. The metabolism of isolated root segments. Proceedings of the Royal Society, Series B, 169, 89-105.
63. HOLLEMAN J. 1967. Direct incorporation of hydroxyproline into protein of sycamore cells incubated at growth inhibitory levels of hydroxy- proline. Proceedings of the National Academy of Sciences. United States of America, 57, 50-54.
64. HURYCH J. and CHVAPIL M. 1965. Influence of chelating agents on the biosynthesis of collagen. Biochimica et biophysica acta, 97, 361-363.
65. INGLE J. and HAGEMAN R.H. 1964. Studies on the relationship between ribonucleic acid and rate of growth of corn roots. Plant Physiology, 39, 730-734.
66. JAMES W.O. 1953. In Plant Respiration, Clarendon Press, Oxford, pp. 82-97.
67. KAUFMAN P., GHOSHEN N. and IKUME H. 1968. Promotion of growth and invertase activity by gibberellic acid in developing *Avena* internodes. Plant Physiology, 43, 29-34.
68. KEY J.L. 1964. Ribonucleic acid and protein synthesis as essential processes for cell elongation. Plant Physiology, 39, 365-370.
69. KHRISTEVA L.A. 1951. {The role of humic acid in plant nutrition and humus fertilizers}. Trudy Pochvennogo institute V.V. Dokuchaeva Akademiya nauk SSSR., 39, 108-184.
70. KHRISTEVA L.A. 1953. {The part played by humic acids and other organic substances in the nutrition of higher plants}. Pochvovedenie, 10, 46-59.
71. KHRISTEVA L.A. 1955. {The part played by humic acids and other organic compounds in the nutrition of higher plants and the significance of this nutrition in agronomy}. Izdaniya Anssr Seriya Biologii, 4, 58-83.
72. KHRISTEVA L.A. 1968. About the nature of physiologically active substances of the soil humus and of organic fertilizers and their agricultural importance. Pontificiae Academiae Scientiavum Scripta Varia, 32, 701-721.
73. KHRISTEVA L.A. and LUK'YANENKO N.V. 1962. {The role of physiologically active substances in soil in the life of plants and their replenishment}. Soviet Soil Science, 1137-1141.
74. KHRISTEVA L.A., SOLOCHE, K.I., DYNKINA, R.L., KOVALENKO, V.E. and GOROBAYA, A.I. 1967. {Influence of physiologically active substances of soil humus and fertilizers on nucleic acid metabolism, plant growth and subsequent quality of the seeds}. Humus et Planta, IV, 272-276.
75. KHRISTEVA L.A., REUTOV V.A., GOLIKOVA O.P. and KOZAR D.G. 1971. The influence of NA- humate and vitamin PP on the renewal factor of mole- cules of different forms of nucleic acid. In Humus et Planta, V, 541-546.
76. KHRISTEVA L.A., GALUSHKO A.M., GOROVAYA A.I., KOLBASSIN A.A., SHORTSHAI L.P., TKATSHANKO L.K., FOT L.V. and LUK'YANENKO N.V. 1980. The main aspects of using physiologically active substances of humus nature. Sixth International Peat Congress Proceedings. Minnesota, United States of America, 403-404.

77. KONONOVA M.M. 1966. Soil Organic Matter, its nature, its role in soil formation and soil fertility. Pergamon Press, Oxford.

78. KUNDTINA E.S. 1951. The effect of humic acids in certain groups of soil micro-organisms. Trudy Pochvennogo institua v.v. Dokuchaeva. Akademiya nauk SSSR, 38.

79. KUTHY A. and PECZNIK J. 1941. Wirkt die Huminsäure als Hormon oder durch Permeabilitätserhöhung auf die Entwicklung der Pflanzenwurzeln. Bodenkunde und Pflanzenernahrung, 23, 83-90.

80. LADD J.N. and BUTLER J.H.A. 1969. Inhibitory effect of soil humic compounds on the proteolytic enzyme pronase. Australian Journal of Soil Research, 7, 241-251.

81. LADD J.M. and BUTLER J.H.A. 1971. Inhibition and stimulation of proteolytic enzyme activities by soil humic acids. Australian Journal of Soil Research, 7, 253-261.

82. LAMPORT D.T.A. 1965. The protein component of primary cell walls. Advances in Botanical Research, 2, 151-218.

83. LHOTSKY S. 1975. The effect of simultaneous short term action of streptomycin and humate on plants. Biologia Plantarum, 17, 475-480.

84. LINEHAN D.J. 1976. Some effects of a fulvic acid component of soil organic matter on the growth of cultivated excised tomato roots. Soil Biology and Biochemistry, 8, 511-517.

85. LINEHAN D.J. 1977. A comparison of the polycarboxylic acids extracted by water from an agricultural top soil with those extracted by alkali. Journal of Soil Science, 28, 369-378.

86. LINEHAN D.J. 1978. The uptake of plants of polymaleic acid; a poly-carboxylic acid structurally related to those of soil. Plant and Soil, 50, 625-632.

87. LINEHAN D.J. 1978. Humic acid and ion uptake by plants. Plant and Soil, 50, 663-670.

88. LINEHAN D.J. and SHEPHERD H. 1979. A comparative study of the effects of natural and synthetic ligands on ion uptake by plants. Plant and Soil, 52, 281-289.

89. LISANTI L.E., TESTINI C. and SENESI N. 1976. Research on the para-magnetic properties of humic compounds. - VI. Agrochimica, 21, 47-56.

90. MALCOLM R.E. and VAUGHAN D. 1978. Effects of humic acid fractions on invertase activities in plant tissues. Soil Biology and Biochemistry, 11, 65-72.

91. MALCOLM R.E. and VAUGHAN D. 1979. Humic substances and phosphatase activities in plant tissues. Soil Biology and Biochemistry, 11, 253-259.

92. MALCOLM R.E. and VAUGHAN D. 1979. Comparative effects of soil organic matter fractions on phosphatase activities in wheat roots. Plant and Soil, 51, 117-126.

93. MATO M.C. and MÉNDEZ J. 1970. Inhibition of indoleacetic acid-oxidase by sodium humate. Geoderma, 3, 255-258.

94. MATO M.C. and FÁBREGAS R. and MÉNDEZ J. 1971. Inhibitory effect of soil humic acids on indoleacetic acid oxidase. Soil Biology and Bio-chemistry, 3, 285-288.

95. MATO M.C., OLMEDO M.G. and MÉNDEZ J. 1972. Inhibition of indoleacetic acid oxidase by soil humic acids fractionated in Sephadex. Soil Biology and Biochemistry, 4, 469-473.

96. MATO M.C., GONZÁLEZ-ALONSO L.M. and MÉNDEZ J. 1972. Inhibition of enzymatic indoleacetic acid oxidation by fulvic acids. Soil Biology and Biochemistry, 4, 475-478.

97. MATHUR S.P. and PAUL E.A. 1966. A microbial approach to the problem of soil humic acid structure. Nature, London, 212, 646-647.

98. McLOUGHLIN A.J. and KÜSTER E. 1972. The effect of humic substances on the respiration and growth of micro-organism I. Effects on the yeast *Candida utilis*. Plant and Soil, 37, 17-25.

99. MOCKERIDGE F.A. 1920. The occurrence and nature of the plant growth producing substances in various organic manurial composts. Biochemical Journal, 14, 432-450.

100. MOCKERIDGE F.A. 1924. Auximones I. The formation of plant growth promoting substances by micro-organisms. Annals of Botany, 38, 723-734.

101. MYŚKÓW W. 1967. Chemical composition and biological activity of humus obtained from various substrates. Bulletin de L'Academic Polanaise des Sciences. CL II, XV, 79-84.

102. NIKLEWSKI B. and WOJCIECHOWSKI J. 1937. Über den Einfluss der Wasser-loslichen Humustoffe auf die Entwicklung einiger Kulturpflanzen. Bodenkunde und Pflanzenerndhrung, 4, 294-327.

103. NIKLEWSKI B. and WOJCIECHOWSKI J. 1947. The influence of humic acids on the development of plants. Acta Societatis botanicorum Poloniae, 18, 63-90.

104. NOODEN L.D. and THIMANN K.V. 1966. Action of inhibitors of RNA and protein synthesis on cell enlargement. Plant Physiology, 41, 157-164.

105. O'DONNELL R.W. 1973. The auxin-like effects of humic preparations from leonardite. Soil Science, 116, 106-112.

106. OGNER G. and SCHNITZER M. 1971. Chemistry of fulvic acid, a soil humic fraction and its relation to lignin. Canadian Journal of Chemistry, 49, 1053-1063.

107. OLSEN C. 1930. On the influence of humic substances on the growth of green plants in water culture. Comptesrendus die Laboratoire Carlsberg, 18, 1-16.

108. OTTO H. 1952. Einwirkung von Vorstufen der Synthesehuminsäuren auf das Wurzelwachstum. Zeitschrift für Pflanzenernährung Düngung Bodenkunde, 56, 46.

109. POSZEWSKI A., TROJANOSKI J. and LOBARZEWSKA A. 1957. Wplyw frakciji humusowych na wzrost koleopitle oswa. Annales Universitatus Marie Curie - Sklodowska, 12, 1-13.

110. PFLUG W. 1979. Die Aktivitätsbeeinflussung einiger Dehydrogenasen, Amylasen und Cellulasen durch Huminsäuren. Mittelungen Deutsche Bodenkundlichen Gesellschaft, 29, 443-450.

111. PFLUG W. 1979. Über die Hemmung der Aminopeptidase K (aus *Tritirachium album* Limber) durch Huminstoffe. Zeitschrift für Pflanzenernährung und Bodenkunde, 142, 290-298.

112. PFLUG W. 1980. Effect of humic acids on the activity of two peroxidases. Zeitschrift für Pflanzenernährung und Bodenkunde, 143, 432-449.

113. PFLUG W. and ZIECHMANN W. 1981. Inhibition of malate dehydrogenase by humic acids. Soil Biology and Biochemistry, 13, 293-299.

114. POAPST P.A., GENIER C. and SCHNITZER M. 1970. Effect of soil fulvic acid on stem elongation in peas. Plant and Soil, 32, 367-372.

115. POAPST P.A. and SCHNITZER M. 1971. Fulvic acid and adventitious root formation. Soil Biology and Biochemistry, 3, 215-219.

116. POKORNA V., LUSTINEC J. and PETRU E. 1963. The influence of Na-humate on the respiration of wheat roots and leaves. Biologia Plantarum, 5, 265-270.

117. POSPÍŠIL F. 1971. Humic acids, their properties and effect on phytase activity. In Humus et Planta, V, 343-350.

118. POSPÍŠIL F. 1980. Influence of physiologically active substances of the soil humus on the activity of glucose-6-phosphate - dehydrogenase in Pea (*Pisum sativum*) roots. Biologia Plantarum, 22, 161-166.

119. POSPÍŠIL F. and HRUBCOVÁ M. 1974. The effect of humic acids and of their fractions on phytase activity. Vedecké Práce Výzkumných Ustavu Rostlinné Vyroby Praha-ruzyne, 18, 47-54.

120. PRÁT S. 1955. {Literature and notes on the biological influence of humolites}. Cyklostyl. Praha.

121. PRÁT S. 1955. {Influence of humus substances (capucines) on algae}. Ceskoslovenska Biologie, 4, 535-541.

122. PRÁT S. 1960. {The effect of humus substances on the uptake of mineral salts and on the chlorophyll formation in plants}. Acta agrobotanica, 9, 117-121.

123. PRÁT S. 1960. Distribution of the humus substance function in plants. Biologia Plantarum, 2, 308-312.

124. PRÁT S. 1962. The effect of humus substances in regeneration of plants. Studies about Humus, Prague, 223-234.

125. PRÁT S., CATSKÝ J. and MELICHAR O. 1957. {The effects of humus substances on plants}. Acta Societates botanicorum Poloniae, 26, 325-347.

126. REŘÁBEK J. 1960. Humic acid interactions in the growth process. Biologia Plantarum, 2, 88-97.

127. REŘÁBEK J. 1962. The relation of humic acid to the control of straight growth of the plant cell. Studies about Humus, Prague, 245-254.

128. REX R.W. 1960. Electron paramagnetic resonance studies of stable free radicals in lignins and humic acids. Nature, London, 188, 1185-1186.

129. RIDGE I. and OSBORNE D.J. 1970. Hydroxyproline and peroxidases in cell walls of *Pisum sativum*. Regulation by ethylene. Journal of Experimental Botany, 21, 843-856.

130. RIFFALDI R. and SCHNITZER M. 1972. Effects of diverse experimental conditions on E.S.R. spectra of humic substances. Geoderma, 8, 1-10.

131. SCHNITZER M. and POAPST P.A. 1967. Effects of a soil humic compound on root initiation. Nature, London, 213, 598-599.

132. SCHNITZER M. and SKINNER S.I.M. 1969. Free radicals in soil humic compounds. Soil Science, 108, 383-390.

133. SCHNITZER M. and KHAN S.U. 1978. Developments in Soil Science, Volume 8. Soil Organic Matter. Eds Schnitzer M. and Khan S.U. Elsevier Press, Oxford.

134. SEAGER A. 1925. The growth of duckweeds in mineral nutrient solutions with and without organic extracts. Journal of General Physiology, 7, 517-526.

135. SEITZ K. and LANG A. 1968. Invertase activity and cell growth in lentil epicotyls. Plant Physiology, 43, 1430-1434.

136. SENESI N., CHEN Y. and SCHNITZER M. 1977. The role of free radicals in the oxidation and reduction of fulvic acid. Soil Biology and Biochemistry, 9, 397-403.

137. SLADKÝ Z. 1959. The effect of extracted humus substances on growth of tomato plants. Biologia Plantarum, 1, 142-150.

138. SLADKÝ Z. 1959. The application of extracted humus substances to overground parts of plants. Biologia Plantarum, 1, 199-204.

139. SLADKÝ Z. 1967. Effect of humus acids on the growth of isolated roots. In Humus et Planta, IV, 286-287.

140. SLADKÝ Z. and TICHÝ V. 1959. Applications of humus substances to over ground organs of plants. Biologia Plantarum, 1, 9-15.

141. ŠMÍDOVÁ M. 1957. {The influence of humate on respiration in wheat}. In symposium "Prochnica a roślina", Poznan.

142. ŠMÍDOVÁ M. 1960. The influence of humus acid on the respiration of plant roots. Biologia Plantarum, 2, 152-164.

143. STANCHEV L., TANEV Z. and IVANOV K. 1975. Humus substances as suppressors of biuret phytotoxicity. In Humus et Planta, VI, 373-381.

144. STEELINK C. and TOLLIN G. 1962. Stable free radicals in soil humic acid. Biochimica et biophysica Acta, 59, 25-34.

145. STEVENSON F.J. 1982. Humus Chemistry. John Wiley and Sons, New York.

146. TAN K.H. and NOPAMORNBODI V. 1979. Effect of different levels of humic acids on the nutrient content and growth of corn. (Zea Mays L.). Plant and Soil, 51, 283-287.

147. TAN K.H. and NOPAMORNBODI V. 1979. Fulvic acid and the growth of the ectomycorrhizal fungus, (Pisolithus tinctorus). Soil Biology and Biochemistry, 11, 651-653.

148. THOMAS S.M., THORNE G.N. and PEARMAN I. 1978. Effect of nitrogen on growth, yield and photorespiratory activity in spring wheat. Annals of Botany, 42, 827-837.

149. TICHÝ V. 1956. {The influence of the qualitative composition of humus on the growth and metabolism of young spring wheat plants}. C. Sci. Thesis, Brno. University.

150. TICHÝ V. 1971. Biological activity of ultraviolet irradiated humic acids. In Humus et Planta, V, 553-555.

151. TICHÝ V. and NECHUTOVÁ H. 1967. The influence of humus substances on the calorific value of the plant dry matter and on its production. In Humus et Planta, IV, 291-292.

152. VAN DER WERFF M. 1981. Ecotoxicity of heavy metals to aquatic and terrestrial higher plants. Thesis, Vrije Universiteit, Amsterdam.

153. VAUGHAN D. 1967. Effect of humic acid on the development of invertase activity in slices of beetroot tissues washed under aseptic conditions. In Humus et Planta, IV, 268-271.

154. VAUGHAN D. 1967. The stimulation of invertase development in aseptic storage tissue slices by humic acid. Soil Biology and Biochemistry, 1, 15-28.

155. VAUGHAN D. 1974. A possible mechanism for humic acid action in cell elongation in root segments of Pisum sativum under aseptic conditions. Soil Biology and Biochemistry, 6, 241-247.

156. VAUGHAN D. and CUSENS E. 1973. Effects of hydroxyproline on the growth of excised root segments of Pisum sativum under aseptic conditions. Planta, 112, 243-252.

157. VAUGHAN D. and CUSENS E. 1974. Some effects of analogues of uracil on cell elongation and wall metabolism in excised peat root segments. Planta, 122, 227-238.

158. VAUGHAN D. and MacDONALD I.R. 1971. Effects of humic acid on protein synthesis and ion uptake in beet discs. Journal of Experimental Botany, 22, 400-410.

159. VAUGHAN D. and MacDONALD I.R. 1976. Some effects of humic acid on cation uptake by parenchyma tissue. Soil Biology and Biochemistry, 8, 415-421.

160. VAUGHAN D. and MALCOLM R.E. 1979. Effect of soil organic matter in peroxidase activity of wheat roots. Soil Biology and Biochemistry, 11, 57-63.

161. VAUGHAN D. and MALCOLM R.E. 1979. Effect of humic acid on invertase synthesis in roots of higher plants. Soil Biology and Biochemistry, 11, 247-272.

162. VAUGHAN D. and ORD B.G. 1982. An in vitro effect of soil organic matter and synthetic humic acids on the generation of free radicals. Plant and Soil, 66, 113-116.

108

163. VAUGHAN D. and ORD B.G. 1983. Influence of humic acid on iron required for the formation of hydroxyproline in discs of *Beta vulgaris* L. storage tissue. Plant and Soil, 73, 27-34.
164. VAUGHAN D., CHESHIRE M.V. and MUNDIE C.M. 1974. Uptake by the beetroot tissue and biological activity of ^{14}C-labelled fractions of soil organic matter. Transactions of the Biochemical Society, 2, 126-129.
165. VAUGHAN D., BAKER C.D. and WILLOUGHBY. 1974. Some aspects of humic acid on two different biological systems. Plant and Soil, 40, 429-434.
166. VAUGHAN D., ORD B. and MALCOLM R.E. 1978. Effect of soil organic matter on some root surface enzymes of and the uptake into winter wheat. Journal of Experimental Botany, 29, 1337-1344.
167. VAUGHAN D., DeKOCK P.C. and ORD B.G. 1983. Effect of benzyladenine and abscisic acid on superoxide dismutase in fronds of the duckweed *Lemna gibba* L. Physiologia Plantarum, 58, 239-242.
168. WAINWRIGHT M. and PUGH G.J.F. 1975. Phenol auxins and Erlich reactions in soil. Soil Biology and Biochemistry, 26, 217-223.
169. WHITEHEAD D.C. 1963. Some aspects of the influence of organic matter on soil fertility. Soils and Fertilizers, 26, 217-223.
170. WHITEHEAD D.C., DIBBS H. and HARTLEY R.D. 1982. Phenolic compounds in soil as influenced by the growth of different plant species. Journal of Applied Ecology, 19, 579-588.
171. WOLFE H.S. 1926. The auximone question. Botanical Gazette, 81, 228-231.
172. WOODSTOCK L.W. and SKOOG F. 1960. Relationship between growth rates and nucleic acid content in the roots of hybrid lines of corn. American Journal of Botany, 47, 713-718.
173. YIPP C.C. 1964. The hydroxylation of proline by horseradish peroxidase. Biochemica et biophysica Acta, 92, 395-397.
174. ZIECHMANN W. 1972. Über die E-Donator und Acceptor Eigenschaften von Huminsäuren. Geoderma, 8, 111-131.

CHAPTER 3

PHENOLIC ACIDS IN SOILS AND THEIR INFLUENCE ON PLANT GROWTH AND SOIL
MICROBIAL PROCESSES

R.D. HARTLEY and D.C. WHITEHEAD

The Grassland Research Institute, Hurley, Maidenhead, Berks.

CONTENTS

1. INTRODUCTION

Phenolic compounds of low molecular weight, particularly
p-hydroxybenzoic, vanillic, p-coumaric and ferulic acids, are
of widespread occurrence in soils and occur mainly but not
entirely in chemically-bound forms. They are believed to be
important intermediates in the formation of humic substances and,
at certain concentrations, they may influence the growth of
plants and the activities of soil micro-organisms (see chapter
6). There is evidence that the phenolic acids and related
compounds in soils originate in part from the decomposition
of plant residues and in part from synthesis by soil micro-
organisms.

There are several problems associated with the measurement
of phenolic acids and related compounds in soils. Thus, it
is difficult to distinguish between 'free' and 'bound' phenolics,
and also between those present in the soil and those in
fragments of plant roots. In addition, soils sorb phenolic
acids added to them, and this process undoubtedly occurs
with the phenolic acids present naturally in soils. In recent
years improved methods for the determination of phenolic acids
have become available, but the problems associated with
extraction remain, and consequently there are uncertainties
in deducing possible biological effects from measured
concentrations in soil extracts.

The structural formulae of the important phenolic acids
discussed in this chapter are shown in figure 1.

111

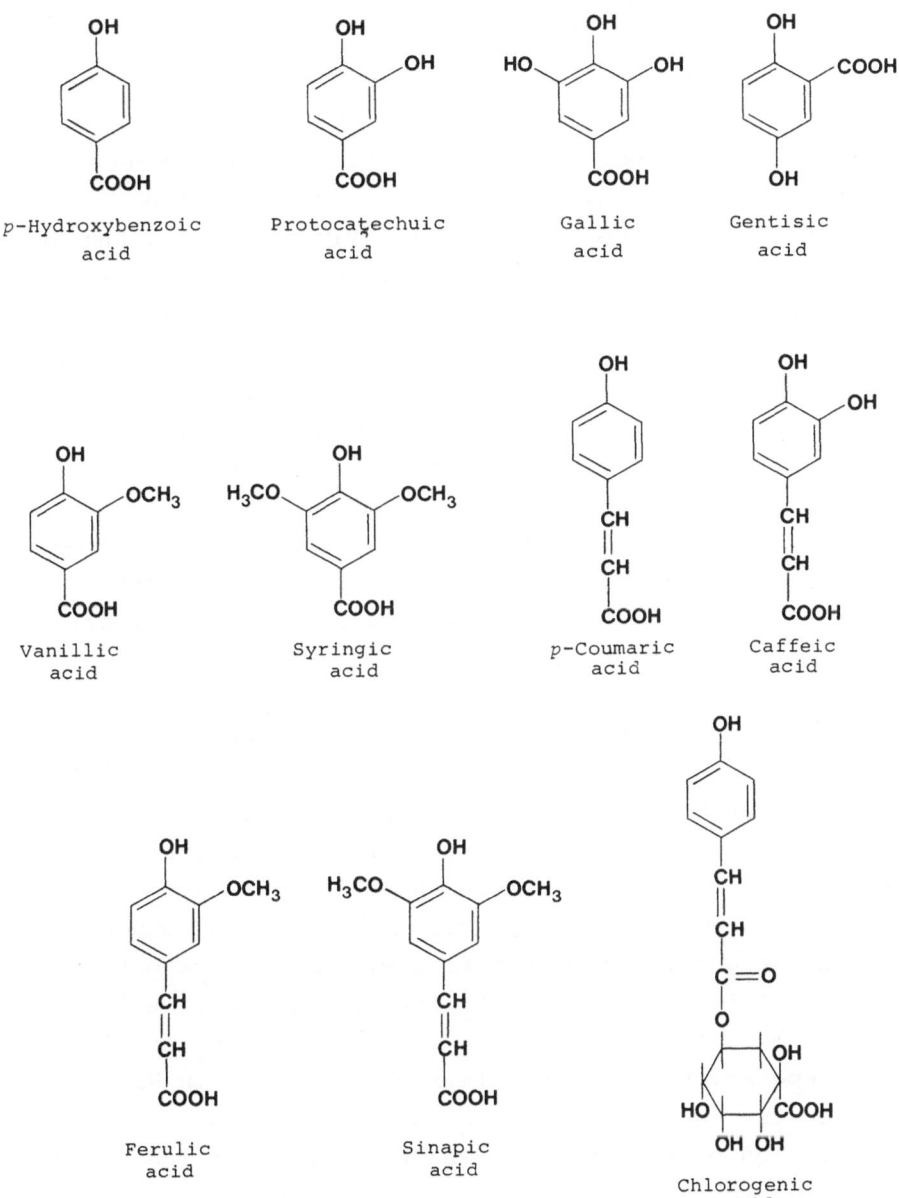

FIGURE 1. Structural formulae of some important phenolic acids.

2. DETECTION AND DETERMINATION

Chromatographic techniques are generally used to determine phenolic acids. In earlier work[14,40,103,131] paper chromatography was often employed, but the technique is slow and is not suitable for simple quantitation. Ion-exchange chromatography has also been used[107]. Thin-layer chromatography (TLC) has been favoured by several workers as it can be performed rapidly, but difficulties with quantitation remain. However, TLC is very useful for qualitative analysis, and using commercial cellulose plates with a toluene-formic acid solvent, several phenolic acids, including *p*-coumaric, ferulic and syringic acids, can be separated readily though without resolution of the *cis* and *trans* isomers[58]. Separation of the *cis* and *trans* forms of the substituted cinnamic acids by TLC can be achieved on cellulose plates with 4% formic acid[57]. Diazotised *p*-nitraniline reagent, and other diazonium salts, applied as a spray to TLC plates give various coloured products for use in qualitative analysis[103]. Stabilised diazonium salts are now available from commercial suppliers and obviate the necessity of preparation *in situ*.

Several workers have employed gas-liquid chromatography (GLC) for the quantitative determination of phenolic acids after their conversion to trimethyl silyl derivatives[57,78,121,126] or ethyl derivatives[108]. Recently, high performance liquid chromatography (HPLC) has been used for quantitative analysis[54,107,137]. It is a more sensitive technique than GLC for the determination of phenolic acids and aldehydes, and does not require the preparation of derivatives before separation. A further important advantage is that while extracts from soils often contain compounds which interfere with GLC separations, such problems rarely occur in HPLC. This is because the UV absorption detectors are set in the aromatic region of the spectrum where the phenolic compounds being determined have high extinction coefficients, and this decreases the possibility of interference. The *cis* and *trans* isomers of the phenolic acids can be separated by both HPLC

and GLC[54,57,107,108]. HPLC has been specifically applied to
phenolic acids and aldehydes derived from plants, and from
the decomposition of organic matter in soil, using reverse-
phase chromatography with C_{18} bonded-phase on silica with
acidic solvents[54]. Very recently new HPLC methods have been
developed with a polyvinylbenzene-based resin which enable
separation in alkaline media. These methods have the advantage
that, when extracting plant or soil materials with alkali,
subsequent extraction with organic solvents is often unnecessary,
and one possible source of loss is therefore avoided[63].

3. PHENOLIC ACIDS IN PLANT TISSUES
3.1. Occurrence of phenolic acids in plants
       Several of the simple phenolic acids, including p-hydroxy-
benzoic, vanillic, syringic and protocatechuic, are widely
distributed amongst the angiosperms. The first three are
also present widely in gymnosperms and ferns. In addition,
many angiosperms contain gentisic and salicylic acids and many
woody species contain gallic acid[45]. The aromatic amino acid
L-phenylalanine is thought to be a major precursor in the
synthesis by plants of both benzoic and cinnamic acid
derivatives[37].
       Benzoic acids and aldehydes have been found in leachates
and exudates from various plants, for example gallic, ellagic,
vanillic and gentisic acids in leaf leachates of *Eucalyptus
globulus*, p-hydroxybenzaldehyde in leaf and rhizome extracts
of Johnson grass and p-hydroxybenzoic and vanillic acids in
leaf leachates of *Camelina alyssum* (see Horsley[67]).
       The cinnamic acids and their derivatives occur almost
universally in higher plants, with one or more of the compounds
p-coumaric, caffeic, ferulic, and sinapic acids, occurring in
'bound' form, often with a sugar as a glycoside or ester[45].
The acids occur mainly in their *trans* forms. Caffeic acid
is often found as the quinic acid ester, chlorogenic acid,
while p-coumaric and ferulic acids, together with other phenolic
acids, have been identified in the 'free' form in water extracts
of several sub-tropical grasses[14].

Cinnamic acids have been found in the leachates or exudates of tissues of a number of plants. For example, ferulic, *p*-coumaric, caffeic, *p*-coumarylquinic and chlorogenic acids have been identified in leaves of *Eucalyptus globulus*, and chlorogenic and *p*-coumaric acids in leaf and rhizome extracts of Johnson grass (see Horsley[67]). There is evidence that the synthesis by plants of some phenolic compounds, including chlorogenic acid, is increased under conditions of poor nutrient supply[1].

Surveys of various classes of phenolic constituents extracted by hydrochloric acid from leaves of monocotyledons and dicotyledons have been published by Bate-Smith[2,3]. In recent years it has been established that cell walls of several families of monocotyledons, including the Gramineae, have phenolic acids (mainly *trans-p*-coumaric and ferulic) attached to their cell walls, apparently by ester links involving the carboxyl group of the acids and hydroxyl groups of sugar units[21,53,66]. Graminaceous cell walls contain 1-3% by weight of phenolic acids: *trans*-ferulic acid, which is often the major constituent, is ester-linked at the 5 position of β-L-arabino-furanose, which in turn is linked to D-xylopyranose[76,115]. The ratio of *p*-coumaric:ferulic acid increases as grasses age[52] probably reflecting a decreasing proportion of primary cell walls, which have low ratios[51]. Ferulic acid in graminaceous cell walls can be located by fluorescence microscopy of plant sections as the acid and its esters fluoresce under ultra-violet radiation[24,30,49,50]. The dimer of ferulic acid, *trans,trans-* diferulic acid, is similarly linked to sugars in graminaceous cell walls[58,84].

A survey of the cell walls of dicotyledons showed that only species in the Caryophyllales, which includes the economically important sugar and fodder beets, contained detectable amounts of bound phenolic acids[56]. Both 'free' and 'bound' forms of several phenolic acids were found in the whole tissue of lucerne shoots by Newby *et al.*[97].

The coumarins (lactones of *o*-hydroxycinnamic acid) also occur fairly widely in higher plants, and have been reviewed by Brown[6].

3.2. Release from water-soluble phenolic glycosides and esters

Aqueous extraction of plant tissues often releases phenolic acids as glycosides or esters[44,103] and the phenolic acids can be separated from these compounds by hydrolysis with alkali (under nitrogen to limit oxidation). Enzymic methods involving treatment with glycosidases or esterases may also be used[62].

3.3. Release from plant cell walls

Cell walls can be rapidly isolated from plant tissues by a modification[61] of the method of Van Soest and Wine[124] which involves extraction with neutral detergent. Most investigators have treated isolated cell walls with alkali to release phenolic acids as their sodium salts[59,66]. The acids can also be released as their carbohydrate esters by degrading the walls with a commercial 'cellulase' preparation from fungal sources (Basidiomycete, *Oxyporous* spp.) having both cellulase and hemicellulase activity[61]. Some commercial 'cellulase' preparations from *Aspergillus niger* and *Trichoderma viride* also contain an esterase which leads to the liberation of free phenolic acids from graminaceous walls[62]. Several workers have used commercial cellulase preparations to predict the biodegradability of plant cell walls in the animal. Cell walls are treated with the enzyme preparation and biodegradability determined by measuring the weight of wall digested, or by measuring the phenolics released which, for graminaceous walls, are closely correlated with the weight of wall digested[61]. Using similar techniques, it has also been shown[55] that the rate of degradation of plant cell walls can be expressed mathematically as two first order reactions occurring simultaneously.

The relationship between the amount of phenolics released by NaOH and cell wall biodegradability measured by a cellulase technique has been examined with maize stem[60]. The walls were treated for varying periods of time to release different

amounts of phenolics. There was a highly significant
correlation (r = 0.98) between biodegradability, measured by
the weight of cell wall degraded, and the amount of phenolic
compounds released by treating with 0.1M NaOH. Similarly,
Lau and Van Soest[79] and Van Soest et al.[123] have used the
measurement of phenolics, released from the walls of cereal
straws and maize stover during treatment with alkali, to
predict the biodegradability of the treated straws by ruminants.

3.4. Role in resistance to pathogens

Comprehensive reviews have recently appeared on the
relationship between phenolics in plants and their resistance
to disease[28,29,46]. Although the plant-pathogen relationship
is often complicated, there is considerable evidence that
phenolics, including p-coumaric and ferulic acids, are involved
in limiting the establishment of pathogens, particularly of
fungi. However, in view of the reviews mentioned above, this
topic will not be discussed further.

4. PHENOLIC ACIDS IN SOILS

4.1. Identity of phenolic acids occurring in soils

The phenolic acids that have been widely identified in
soils, after extraction with mild aqueous or organic reagents,
or after hydrolysis with alkali, are p-hydroxybenzoic, vanillic,
p-coumaric and ferulic acids. Syringic and protocatechuic
acids have been found less frequently and generally in smaller
quantities. Gallic and gentisic acids were reported, in addition
to others, as occurring in Canadian podzols[23]. In addition,
the following acids have been identified in decomposing plant
materials, sometimes on incubation with soil: benzoic,
caffeic, cinnamic, o-coumaric, o-hydroxyphenylacetic,
phenylacetic, 4-phenylbutyric, salicylic[13], gallic and
chlorogenic[12] and p-hydroxyphenylpropionic and 3,4-dihydroxy-
phenylpropionic[83]. Yet others have been found in peat[93,94].
The aldehyde derivatives of some of these acids, particularly
p-hydroxybenzaldehyde and vanillin, have also been identified.

When the organic matter of soil is subjected to more
drastic chemical degradation by, for example, oxidation with
alkaline permanganate, a wider range of phenolic acids,
together with other phenolic compounds, is often produced
(for a review see Stevenson[118]). However these techniques
are designed to degrade the 'core' structure of humic molecules
(see Introductory chapter), and as many of the small molecular
products do not represent compounds that arise from the
decomposition of the organic matter in soils, they are not
considered further in this chapter.

Several of the phenolic acids that have been identified
following relatively mild extractions of soil occur, as
indicated above, in plant cell walls. It is likely that, with
these compounds, a substantial proportion of the amount
apparently present in the soil occurs in the cell walls of
partially decomposed plant residues. A secondary route for
the transfer of phenolic compounds from plants to soil is
through animal excreta, including the application of slurries
from housed livestock[31,85,86,136]. In soils, the synthesis
of phenolics by micro-organisms is also probable. However,
the relative importance of plant residues and microbial
synthesis as sources of phenolic acids in soils is unknown,
and it is likely to vary with environmental and soil
conditions[118].

Once in the soil, phenolic compounds are subject to a
variety of processes which may result either in decomposition,
yielding ultimately carbon dioxide and water, or in polymerisation yielding relatively stable humic substances.

4.2. The extraction of phenolic acids from soils

Most of the reagents that have been employed to extract
'free' or 'bound' phenolic acids from soils have been aqueous
solutions but, in a few studies, organic or mixed aqueous-
organic solvents have been used. In work carried out during
the period 1960-1975, solutions of $Ca(OH)_2$[131] or ethanolic
NaOH at pH 11[92,129] were often used. These reagents were
considered to extract the free phenolic acids, including

fractions that were sorbed or present in soil lipids, but were considered not to cause appreciable hydrolysis. However it is likely that some bound phenolics (for example those ester linked by carboxyl groups to hydroxyl groups of carbohydrates) would be saponified at pH 11 or above, leading to the release of free from previously bound forms.

More recently, the importance of pH in influencing the amounts of phenolic acids released from soils has been clearly demonstrated in a study involving the examination of four phenolic acids and two aldehydes in four soils[132]. A range of extractant pH was achieved by the use of graded amounts of Ca(OH)$_2$ and, for the highest pH, 2M NaOH. The amount of each compound extracted from the four soils was extremely low at pH < 7.5, as measured in the soil suspension after shaking for 20 hours, but the amounts increased sharply above a threshold pH which varied from about 7.5 to 10.5 depending mainly on the compound. Ferulic acid had the highest threshold pH, 10.5, and this was consistent for all four soils. With p-hydroxybenzoic acid, vanillic acid, p-coumaric acid, p-hydroxy-benzaldehyde and vanillin the threshold pH values differed somewhat amongst the soils, but all were between pH 7.5 and 10.5. The amounts of the four major phenolic acids extracted from the soil under permanent grassland at a range of pH values between 6 and 14 are shown in figure 2. With this soil, and also with the others, the amounts extracted increased continuously with increasing pH. The amounts of the phenolic compounds extracted by water alone from these soils were equivalent to concentrations in the soil solution ranging from 1.4 µM for p-hydroxybenzoic acid to < 10 nM for ferulic acid. Amounts up to 2000 times greater than these were extracted by 2M NaOH.

At the concentrations occurring in soils, free and non-sorbed phenolics would be soluble in water. Some glycosidic and ester-bound phenolics would also be soluble to an extent depending mainly on molecular size: release of free phenolics from these soluble yet bound forms can be achieved by treatment with alkali. In water extracts of four soils, varying proportions,

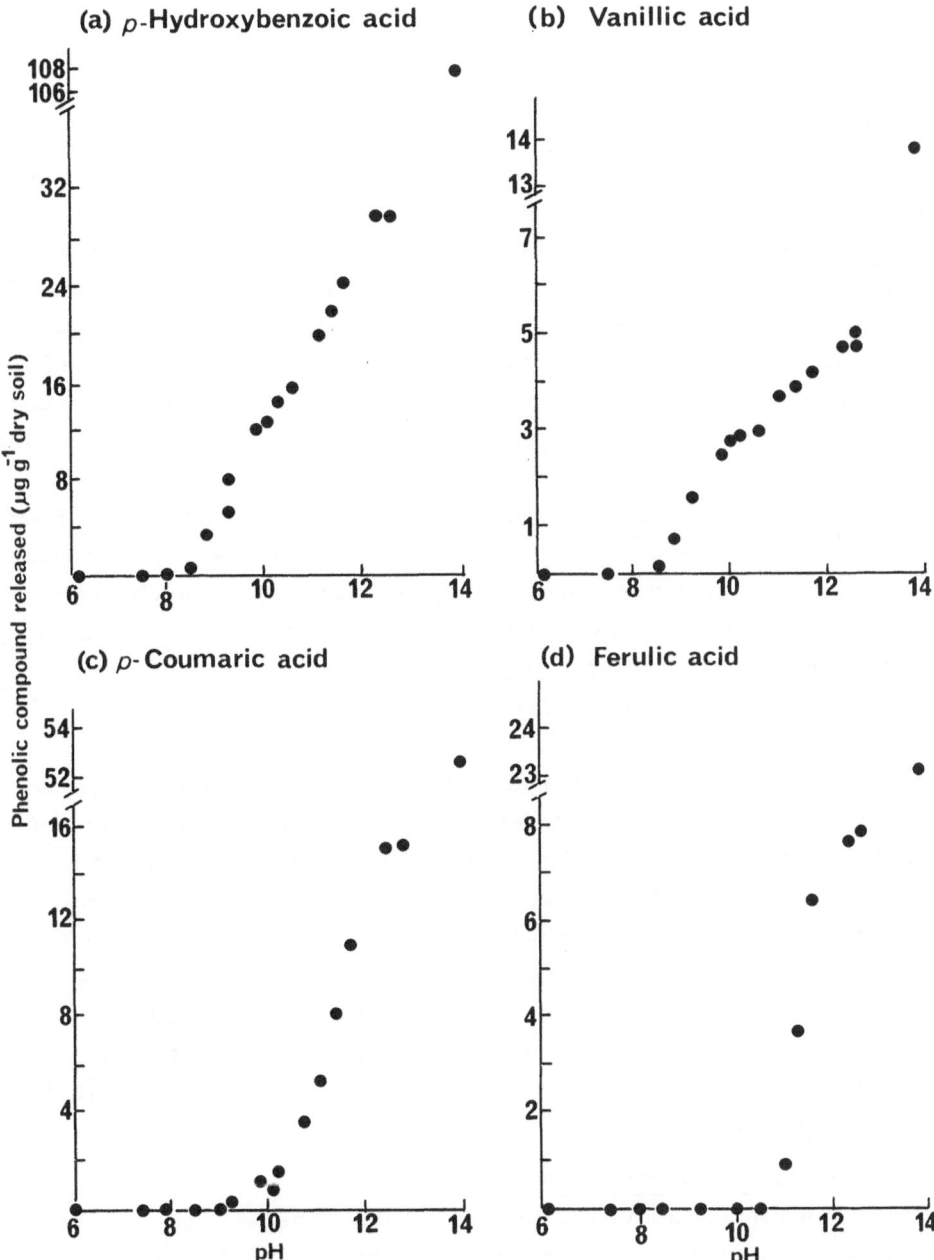

FIGURE 2.   The amounts of four phenolic acids released from a soil
under permanent grassland as influenced by extractant
pH.

but often more than 50%, of the phenolic acids and aldehydes were present in bound forms[134].

Ethylenediamine tetra-acetic acid (EDTA) at pH 7.5 has been proposed as a reagent for extracting the biologically active fraction of soil phenolics; it avoids the chemical changes that may occur in alkaline solutions[71,72]. Evidence was obtained that the amounts of *p*-hydroxybenzoic, *p*-coumaric and ferulic acids extracted by EDTA after their addition to soils, were correlated with the amounts biologically available. Availability was assessed by comparing the effects on seed germination of a range of concentrations in the absence and presence of soil[71]. Despite these apparently promising results, there appear to be no reports of the amounts of phenolic acids extracted by EDTA from unamended soils.

In another procedure, phenolic acids were extracted from soils with ethyl acetate[74,75]. The amounts were small compared with the total amounts assessed by extraction with 2M NaOH and, with *p*-coumaric and ferulic acids, large proportions of the amounts extracted were in bound form.

Autoxidation of phenolic compounds may occur during extraction, particularly at alkaline pH values, but can be prevented, or at least limited, by extracting under 'oxygen-free' nitrogen. It is advisable to treat reference compounds in a similar manner. After extraction with water, alkali or EDTA, extracts are acidified, usually to about pH 2.5, to convert the phenolics to the undissociated form. This enables them to be separated by partition into a suitable organic solvent (e.g. diethyl ether, ethyl acetate). In order to minimize the *cis-trans* isomerisation of the substituted cinnamic acids, these manipulations should be carried out in fluorescent light[57,70,96].

4.3. Sources of phenolic acids

As indicated above, plant residues constitute a major source of the phenolic compounds in soils. The influence of vegetation on the content of various phenolic compounds in the underlying soil has been shown, for example, in studies of

allelopathy in areas of 'chaparral' vegetation in California[12,95].
A striking characteristic of this type of vegetation is the
complete lack of herbaceous species under or between the shrubs,
an effect apparently due largely to various phenolic compounds
reaching the soil, in part by leaching from the foliage and,
in part, by decomposition of plant residues. The importance
of rice in the agriculture of Asian countries, including Japan
and Taiwan, has led to attention being given to the influence
of decomposing rice straw on the growth of subsequent crops.
In this context, studies have been made of the phenolic acids
in rice straw, and of the changes occurring during its decomposition[10,11,109]. The phenolic acids occurring in the
residues of other cereal crops have also been studied[39,90].

The influence of vegetation on the content of phenolic
compounds in soils has also been shown with various crop and
weed species. Thus, Wang *et al*.[129] reported that substantial
differences occurred in the contents of *p*-hydroxybenzoic,
vanillic and *p*-coumaric acids, apparently resulting from the
growth of sugar cane, pineapple, banana, sweet potato and
tobacco. More recently, Whitehead *et al*.[133] examined the
phenolic compounds extracted from soils of 14 plots at the
same site, on which different plant species had been grown
individually for 16 years. The plants comprised eight
dicotyledonous and four graminaceous species, and two species
of Pteridophytes; their roots were also examined for phenolic
compounds. The plant species were found to exert a marked
influence on the amounts of the phenolic compounds extracted
from the soils, the influence being apparent both with water
extracts, in which the amounts of phenolic compounds were small,
and with alkaline extracts in which the amounts were much
greater. The magnitude of the differences between the soils
of the 14 plots in their content of the phenolic compounds
varied both with the compound and with the extractant. For
any one compound and extractant, the highest contents were
often about 30 times the lowest. Part of the influence of
species appeared to result from a direct contribution of
phenolic compounds from plant material to soil. Thus, for

example, the grass roots contained much more ferulic acid
extractable by 2M NaOH than did the other species, and this
difference was reflected, though to a lesser extent, in the
soils. However, other factors appeared to be involved since,
although there was a large difference between the grasses and
the other species in the amount of *p*-coumaric acid extracted
from the roots, this was reflected much less closely in the
soils. Also the soil having the highest content of water-
soluble *p*-hydroxybenzoic acid was that under *Agropyron repens*,
but the roots of this species contained considerably less of
both water-soluble, and NaOH-soluble, *p*-hydroxybenzoic acid
than did those of several other species. The results suggested
that there was a trend towards uniformity of composition during
the transformation of plant residues into soil organic matter.
There was also a marked reduction in the content of most of
the phenolic compounds in the organic matter, as assessed by
extraction with 2M NaOH and expressed per unit of organic C.
However in some instances where the amounts extracted from
the roots were small, e.g. vanillic acid from *Pteridium
aquilinum*, the amounts in the soil organic matter were greater.

In addition to their release during the decomposition
of plant residues, phenolic compounds may also be released
from living plant roots. In a study of Bigalta limpograss
grown in sand with a circulating nutrient solution, twelve
aromatic compounds, including cinnamic, syringic and ferulic
acids, were identified as root exudates[120].

Microbial transformations, possibly including synthesis
as well as the decomposition of the plant residues, are
clearly an important factor influencing the content of phenolic
acids in soils. The microbial synthesis of phenolic compounds,
including several phenolic acids, has been demonstrated by the
work of Haider and Martin[41]. In this study, in which the
fungus *Epicoccum nigrum* was grown in a glucose-asparagine
medium, more than 20 phenolic compounds were identified within
a period of a few weeks. Two major intermediates appeared to
be 2,4-dihydroxy-6-methylbenzoic and 2-methyl-3,5-dihydroxy-
benzoic acids, and these compounds were apparently subject to

decarboxylation, to the introduction of additional hydroxyl groups and oxidation of methyl groups, to produce a range of other compounds including p-hydroxybenzoic acid[41]. In subsequent work, the saprophytic fungus, *Hendersonula toruloidea* was found to synthesise a wide range of phenolic compounds and to produce particularly high yields of phenolic polymers[89].

The ability to synthesise aromatic compounds from non-aromatic carbon sources is generally greater with fungi and actinomycetes than with bacteria[118].

4.4. Microbial decomposition of phenolic acids

Phenolic acids in soils are subject to various processes including decomposition and polymerization. Microbial decomposition of phenolic compounds can occur both aerobically and anaerobically, but in most soils aerobic pathways are likely to be the most important.

During the period 1950-1965, a considerable amount of work, reviewed by Dagley[16,17], was carried out on the microbial degradation of phenolic compounds, much of it in the context of the decomposition of lignin under aerobic conditions. This work showed that the simple aromatic compounds, including ferulic and vanillic acids, that are produced during the decomposition of lignin can be used as sources of carbon by many fungi and bacteria.

In general, the degradation of an aromatic compound by soil micro-organisms involves firstly the formation of a dihydroxyphenol, usually with the hydroxyl groups attached to adjacent carbon atoms of the nucleus. A structure of this type appears to be necessary before the aromatic ring can be split enzymically. The necessary reactions may entail the removal of substituent side chains, the conversion of methoxyl or other ether groups into hydroxyl, or the introduction of hydroxyl groups to the ring. A general scheme for the aerobic decomposition of phenolic acids is shown in figure 3, based on those developed by Martin and Haider[87,88] and Shindo and Kuwatsuka[109]. By specifically labelling individual C atoms of ferulic acid with ^{14}C, Martin and Haider[88] found, as in

124

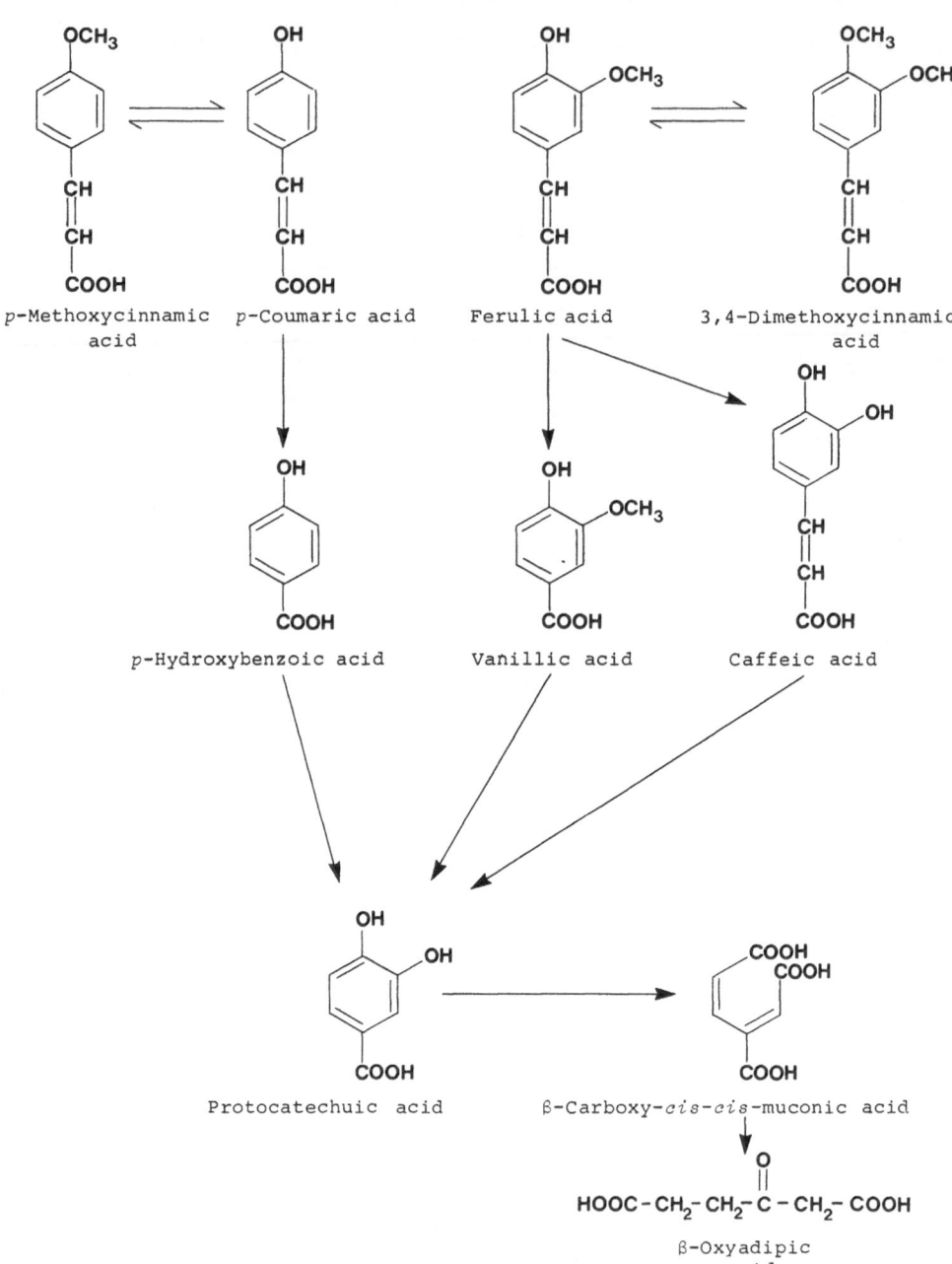

FIGURE 3. Probable decomposition pathways of phenolic acids.

earlier studies with other benzoic and cinnamic acid derivatives[42], that the whole molecule was readily utilised by soil
micro-organisms. However, compared with simple aliphatic
compounds, such as glucose and acetic acid, less of the ferulic
acid carbon was oxidised. For example, about 60% of the side
chain carbons of ferulic acid were evolved as CO_2, compared
with 71% and 87% for the methyl and carboxyl carbons of acetic
acid. It appeared that the methoxy group was utilised almost
completely for microbial energy or for microbial synthesis,
and that overall about 68% of the ferulic acid carbon was
evolved as CO_2. Huntjens *et al.*[69] reported that the rate of
decomposition, as assessed by evolution of carbon dioxide,
was lower for benzoic acid derivatives with OH groups in the
ortho position than for those with OH groups in the meta or
para position. The replacement of the COOH group by a C=C-
COOH group reduced decomposition. Turner and Rice[122] found
that *p*-coumaric, vanillic and *p*-hydroxybenzoic acids were
produced in soil during the decomposition of leaves containing
a high concentration of ferulic acid, and that ferulic acid
was not detectable after 300 days incubation. When ferulic
acid was added to a soil, there was an increase in the
evolution of CO_2 and in the total numbers of micro-organisms
in the soil. Eighteen micro-organisms were isolated which
were able to grow in soil amended with ferulic acid, of which
many belonged to the genus *Pseudomonas*, three were fungi, and
the others appeared to be actinomycetes. The fungi *Cephalos-
porium curtupes* and *Rhodotorula rubra* were particularly
effective in the degradation of ferulic acid. Considine and
Patching[15] reported that the aerobic decomposition of phenolic
acids by a species of *Penicillium* resulted in the production
of ethylene.

Three anaerobic pathways for the decomposition of simple
aromatic compounds by bacteria have been described by Evans[22].
These involve (i) photometabolism by organisms carrying out
photosynthesis, notably members of the *Athiorhodaceae*, (ii)
nitrate respiration by a limited number of organisms including

Pseudomonas and *Moraxella* species, and (iii) methanogenic
fermentation in which the benzene nucleus is first reduced
and then cleaved by hydrolysis. The third pathway apparently
requires a mixed culture of bacteria. Since the *Athiorhodaceae*
require both light and anaerobic conditions, it is unlikely
that photometabolism is an important pathway in soils, but
'nitrate respiration' reactions may well occur following the
diffusion of nitrate into anaerobic zones of soil. Micro-
organisms able to produce methane are widespread but generally
prefer, as substrate, cellulose to aromatic compounds[22]. It
is thought that all three anaerobic pathways involve reduction
and then cleaving of the aromatic ring, thus producing aliphatic
acids; subsequent reactions depend on the types of organisms
and energy metabolism involved. Healy *et al.*[64] reported that
ferulic acid could be degraded to methane by mixed bacterial
culture under anaerobic conditions, but the extent to which
this process occurs in soils is unknown.

4.5. Retention in the soil by sorption and polymerisation

The retention of phenolic acids in soils, so that they
are less susceptible to microbial decomposition and leaching,
occurs by both sorption and polymerisation. Sorption may
promote polymerisation.

Evidence of sorption was reported by Wang *et al.*[130], who
found that several phenolic compounds could not be recovered
quantitatively by extraction with ethanolic NaOH immediately
after their addition to soils. Ferulic and syringic acids
were sorbed more strongly than p-hydroxybenzoic, vanillic
and p-coumaric acids. Of several inorganic soil components,
the amorphous hydroxy iron and aluminium compounds appeared
to be particularly important, their capacity for sorption
being much greater than those of kaolinite, illite and
vermiculite[68]. Also the removal of the amorphous sesquioxide
components from the soils significantly decreased their sorption
of the acids. With synthetic hydroxy iron and aluminium compounds, there were only small differences in the extent of

sorption between *p*-hydroxybenzoic, vanillic, *p*-coumaric, ferulic and syringic acids. In apparent contrast to the results of Huang *et al.*[68], Shindo and Kuwatsuka[110] reported that the content of 'free' iron in eleven soils was not closely related to their sorption of phenolic acids, although allophanic soils sorbed more than did kaolinitic or montmorillonitic soils. They found that, in general, sorption increased in the order vanillic, ferulic, α-resorcylic (3,5-dihydroxybenzoic), *p*-hydroxybenzoic, *p*-coumaric and protocatechuic acid, the last named showing considerably greater sorption than the other compounds. The possibility that soil organic matter might be involved in the sorption of phenolic acids[130] was supported by the finding of Huang *et al.*[68] that the oxidation of four soils with $NaOAc-H_2O_2$ considerably reduced sorption. Microbiological stabilisation is also possible[4].

With soils having a low sorbtive capacity, phenolic acids are subject to rapid leaching[110] and those acids, such as protocatechuic and salicylic, which chelate polyvalent metal ions may accelerate the downward movement of iron, aluminium and manganese[111].

The polymerisation of phenolic acids is an oxidative process which may be catalysed by enzymes such as phenolase and peroxidase, as well as by inorganic soil components. The enzymic polymerisation of phenolic compounds, studied extensively by Flaig, Martin, Haider and their co-workers, in the context of the formation of humic substances, has been reviewed by Flaig *et al.*[27] and by Stevenson[118]. Various enzymic reactions that may be involved in the polymerisation of aromatic compounds have also been reviewed by Sjoblad and Bollag[114]. Polymerisation of the commonly occurring phenolic acids is enhanced by the presence of clay minerals, particularly montmorillonite[128].

Polymerisation may involve the formation of quinones and, in soils, many different quinones are likely to occur transitorally from the oxidation of phenolic compounds. Quinones are highly reactive and form various condensation products, especially in the presence of amino compounds. Enzymic oxidation of phenolic compounds in the presence of amino acids, peptides or protein yields mixed polymers that have many properties of natural humic acids[43,118]. Amino sugars may also react with quinones to form humic polymers[5].

The incorporation of several phenolic compounds into humified organic matter, and into the microbial biomass of soil, was studied by Kassim *et al.*[73]. They found that substantial proportions of the intact molecules of phenolic compounds, such as ferulic acid and catechol, were stabilised by incorporation into existing humus. Incorporation into the microbial biomass was greatest during the first two to four weeks of incubation, after which the amount declined with a corresponding increase in the proportion evolved as CO_2.

4.6. Amounts and distribution of phenolic acids in soils

As indicated above, the amounts of the phenolic acids extracted from soils are very dependent on the pH of the extractant. However it is of interest to examine the amounts extracted (a) by water and (b) by alkali at pH 13-14.

The amounts of *p*-hydroxybenzoic, vanillic, *p*-coumaric and ferulic acids, extracted by water from a permanent grassland soil within 20 hours of collection[132], are shown in table 1, the amounts being expressed per unit of dry soil, per unit of organic carbon and as the concentration in the soil solution.

Table 1. The amounts of four phenolic acids, in free form, extracted by water from a soil under permanent grassland.

	p-Hydroxy-benzoic acid	Vanillic acid	*p*-Coumaric acid	Ferulic acid
ng g^{-1} dry soil	60	5.5	1.4	< 0.6
ng g^{-1} organic C	980	90	23	< 10
Concentration in soil solution[a], nM	1400	110	30	< 10

[a]23% H_2O in moist soil.

However in assessing these results it should be appreciated
that there is often considerable sorption when phenolic acids
are added to a soil suspension[110], and that the amounts
extracted by water therefore represent an equilibrium concentration. Also it should be noted that, in undisturbed soils,
the compounds would not be uniformly distributed and that
localised concentrations close to fragments of decomposing
plant material might well be much greater than the measured
average concentration.

Extraction with 2M NaOH[132] yielded much larger amounts
of the four phenolic acids from the permanent grassland soil
(table 2). With this extractant the four phenolic acids

Table 2. The amounts of four phenolic acids extracted by 2M
NaOH from a soil under permanent grassland.

	p-Hydroxy-benzoic acid	Vanillic acid	p-Coumaric acid	Ferulic acid
µg g^{-1} dry soil	108	14	53	23
µg g^{-1} organic C	1770	230	870	380

amounted in total to about 0.2% of the soil organic matter.
In the soils from 14 plots on which individual plant species
had been grown[133], the same four phenolic acids plus two aldehyde derivatives amounted to between 0.03 and 0.33% of the
organic matter. Low concentrations of phenolic acids, amount-
ing to only 0.03 - 0.05% of the organic matter, were reported
by Guenzi and McCalla[40] as being extracted by 2M NaOH from
soils used for cereal cultivation.

Quantitative estimates of the phenolic acids extracted
by 0.75M NaOH from a number of soils in Japan were reported
by Shindo et al.[112,113] and are summarised in table 3. In
addition to the four phenolic acids mentioned above, salicylic, syringic and protocatechuic acids were determined.

Table 3. The amounts of total phenolic acids extracted by
0.75M NaOH from three categories of soil.

	Paddy soils (9)	Forest soils (15)	Greenhouse and field soils (45)
$\mu g\ g^{-1}$ dry soil (range)	10.4 - 25.8	34.1 - 632.1	9.5 - 62.0
$\mu g\ g^{-1}$ dry soil (mean)	21.3	210.3	26.0

p-Coumaric acid was usually present in the largest amount.
The mean values represented 0.06, 0.12 and 0.10% respectively
of the total soil organic matter in the paddy soils, forest
soils, and greenhouse and field·soils[112]. It should be noted
that the paddy soils, and also many of the greenhouse and
field soils, had received substantial applications of rice
straw. Shindo et al.[112] reported that the concentration in
soils of p-hydroxybenzoic, vanillic, syringic, p-coumaric
and ferulic acids, which do not form chelates with metal
ions, were positively correlated with soil organic matter
content, but that this relationship did not occur with
protocatechuic and salicylic acids. In a forest soil examined
at successive horizons down the profile, the concentrations
of the phenolic acids decreased with increasing depth[113].

5. INFLUENCE OF PHENOLIC ACIDS ON PLANT GROWTH
     It is well established that many phenolic acids are
toxic to plants at concentrations greater than about 1 mM.
It is also possible that, at concentrations lower than 1 mM,
some of the acids may stimulate plant growth[25], but there is
little evidence of such effects with intact plants. In
investigations reported on the effects of phenolic acids on
plant growth, the concentrations of the acids have always
been considerably greater than the concentrations found in
soil solutions by extraction with water. However in some
studies they have been comparable with the concentrations
extracted by solutions of $Ca(OH)_2$. Also, as noted above, in

undisturbed soils, relatively high concentrations of phenolic
acids may occur around decomposing plant fragments.

Insofar as phenolic acids synthesised by plants influence
the growth of other plants, the effects, whether toxic or
stimulatory, are allelopathic (i.e. they represent a biochemical
interaction between plants). The subject of allelopathy has
been reviewed extensively by Rice[104,105] and by Putnam and
Duke[101], and in this review only those effects involving
phenolic acids will be considered.

5.1. Influence on seed germination

Hennequin and Juste[65] examined the effects of five phenolic
acids on seed germination of several plant species including
those shown in table 4, and of these *p*-coumaric acid was the
most inhibitive.

Table 4. Inhibition of germination and seedling growth of
three plant species by five phenolic acids (from Hennequin
and Juste[65]).

Concentration (μg ml^{-1}) in nutrient solution causing 50%
inhibition in (a) germination and (b) growth.

	Ryegrass (a)	(b)	Lucerne (a)	(b)	Wheat (a)	(b)
p-Hydroxybenzoic acid	450	450	450	450	>1000	400
Vanillic acid	400	300	>1000	200	>1000	400
p-Coumaric acid	150	150	450	150	>1000	400
Ferulic acid	400	400	1000	450	>1000	400
Caffeic acid	300	450	>1000	550	>1000	600

Ferulic, caffeic and chlorogenic acids, each at a concentration of 100 μM, were found to considerably reduce seed
germination with radish, though not with *Kochia scoparia*[81].
In an examination of the effects of 12 phenolic compounds on
the germination of seeds of various plant species, Williams

and Hoagland[135] found that effec†s were negligible at a
concentration of 10 μM. At 1mM, coumarin, hydrocinnamic acid
(β-phenylpropionic acid), juglone and pyrocatechol inhibited,
or at least delayed, the germination of one or more species.
These authors concluded that the inhibition of germination
was unlikely to be an important allelopathic mechanism. However, synergistic effects may be important[102].

Endogenous phenolic acids, particularly ferulic, may well
be involved in the natural regulation of germination, as shown
by studies carried out by van Sumere *et al.*[125].

5.2. Influence on plant growth in solution culture

The influence of individual phenolic acids on the growth
of various plant species has been shown clearly in studies
based on solution culture. Thus, toxic effects of the phenolic
acids, at concentrations in solution in the range of 625 to
5000 mg l^{-1} (approximately 4 to 30 mM), on the growth of wheat
seedlings were reported by Guenzi and McCalla[39]. However, much
lower concentrations were reported as being inhibitory in the
study of Wang *et al.*[129]. A concentration as low as 1 mg l^{-1}
(6 μM) of *p*-hydroxybenzoic acid was reported to reduce the
growth of wheat, maize and soyabean, while a concentration of
300 mg l^{-1} (1.8 mM) resulted in the death of all plants over
a period of 24 days. Sugar cane, which was found to be somewhat less susceptible than the other plant species, was subjected
to several concentrations of five phenolic acids, and *p*-coumaric,
ferulic and syringic acids were found to be more toxic than
p-hydroxybenzoic and vanillic acids at concentrations over the
range 25-100 mg l^{-1} (approximately 140 to 580 μM) in the
nutrient solution. Hennequin and Juste[65] examined the effects
of five phenolic acids on the seedling growth of several plant
species, and their results with ryegrass and lucerne showed
that *p*-coumaric acid was the most toxic (table 4).

Chandramohan *et al.*[8] reported that cinnamic acid at concentrations greater than 100 μM reduced the growth of rice,
and Shindo *et al.*[113] found that *p*-hydroxybenzoic, vanillic and
p-coumaric acids all inhibited the growth of rice seedlings at

concentrations of more than 100 mg l^{-1} (approximately 640 µM) in solution culture.

Sparling and Vaughan[116] examined the effects of *p*-hydroxy-benzoic, *p*-coumaric, ferulic and caffeic acids at concentrations of zero, 10 µM and 1 mM on the growth of wheat seedlings grown in nutrient solution, under either sterile conditions or with a bacterial inoculum. The weight of the wheat plants was reduced by up to 70% at a concentration of 1 mM, the toxicity of the compounds decreasing in the order ferulic > *p*-coumaric > *p*-hydroxybenzoic > caffeic. Concentrations of 10 µM had small but variable effects on plant yield. Subsequent studies showed that the toxicity caused by *p*-hydroxybenzoic, *p*-coumaric and ferulic acids at 1 mM was considerably reduced by inoculation of the cultures with two soil micro-organisms[127].

Glass[34] studied the effect on barley of adding a mixture of *p*-hydroxybenzoic, vanillic, *p*-coumaric and ferulic acids to the nutrient solution. The concentrations of the acids were those found[131] to be extracted by $Ca(OH)_2$ from a soil under bracken. When barley seedlings were grown in a solution of $CaSO_4$, the phenolic acids severely inhibited root growth over a period of four weeks from germination although, when seedlings were grown in a balanced nutrient solution, the phenolic acids caused no significant effect. The growth of several other plant species, including perennial ryegrass, was also inhibited by the phenolic acids when the plants were grown in $CaSO_4$ solution, but the growth of couch grass (*Agropyron repens*) was increased. This is of interest since couch grass has been reported to inhibit the growth of other species[38]. Additional evidence that the toxicity of phenolic acids is greatest when nutrient supply is low has been reported by Stowe and Osborn[119]. Aqueous extracts of decomposing rice straw were found to have toxic effects on the growth of both lettuce and rice seedlings, and five phenolic acids (*p*-hydroxybenzoic, vanillic, *p*-coumaric, ferulic and *p*-hydroxyphenylacetic) were identified in the extracts[10].

5.3. Influence on the uptake of nutrients by plants

In order to provide a possible explanation for the toxic effects of phenolic acids on plant growth, several workers have examined their effects on the uptake of nutrients.

Glass[32] reported that the uptake of potassium by barley was significantly reduced by each of 12 naturally occurring phenolic acids at a concentration of 250 µM. For example, p-hydroxybenzoic acid reduced uptake by 52% during a 3-hour period. The inhibitory effect on potassium uptake was thought to be due to a reversible increase in membrane permeability towards inorganic ions. This was supported by measurements of membrane potentials of the root cells of barley following the addition of salicylic acid[36]. Harper and Balke[48] also examined effects on the uptake of potassium and found, with oat roots, that both salicylic and ferulic acids caused inhibition, the former having a greater effect. The effects tended to be greater at lower pH values.

The uptake of inorganic phosphate by barley roots was found by Glass[33] to be inhibited by 15 hydroxybenzoic acid derivatives. In general, the degree of inhibition was related to the lipid solubility of the compounds, although salicylate derivatives were generally more inhibitory than expected. The relationship with lipid solubility supported the view that the phenolic compounds caused an increase in membrane permeability. McClure et al.[91] found that ferulic acid inhibited the uptake of phosphate by soyabean at concentrations of 500 and 1000 µM but that inhibition was negligible at 100 µM. Although the inhibitory concentrations were greater than those reported to occur in soil solutions, the authors considered that prolonged exposure to naturally occurring concentrations might inhibit uptake.

Recent work has shown that certain plants respond to a deficiency of iron by acidification of the root environment and by the release of 'reductants' from the roots[98,99]. These mechanisms result in the conversion of insoluble and unavailable Fe^{3+} into the more soluble Fe^{2+} with a resultant increase in uptake by the plant. Caffeic acid, produced largely by the

hydroxylation of p-coumaric acid, appears to be particularly
important as a reductant in plants which are less susceptible
than others to iron deficiency (figure 4). In addition to
p-coumaric acid synthesised by the plant, that present externally may promote iron uptake. Olsen *et al.*[98] found that the
growth of iron-deficient chlorotic tomato plants was improved
by the addition of 100 µM p-coumaric acid to the nutrient
solution.

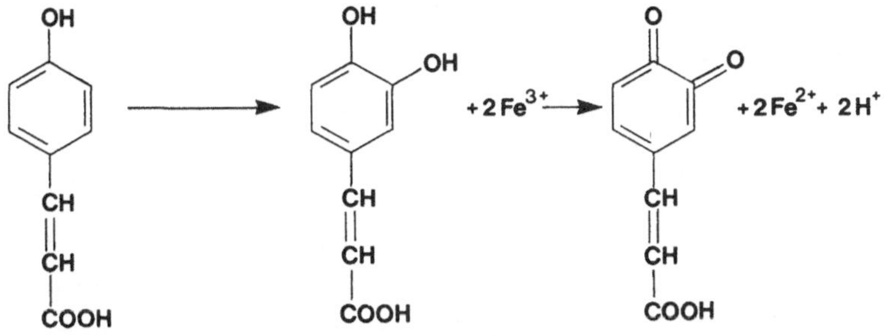

p-Coumaric acid   Caffeic acid   Caffeoyl-*o*-quinone

FIGURE 4. Possible mechanism in the root by which p-coumaric acid
may improve the availability of iron.

5.4. Interactions with enzymes, plant growth substances and chlorophyll

A considerable amount of work on the stimulatory effects
on plant growth of phenolic acids and related compounds, including quinones, was carried out by Flaig and co-workers, and
has been summarised by Flaig[25,26]. In general, he considered
that various lignin decomposition products, including ferulic,
vanillic and syringic acids, when present at concentrations
of about 100 µM in nutrient solution, could increase plant dry
weight by up to 10-15%. Increases occurred particularly when
other environmental conditions were far from the optimum for
plant growth. The effects of the 'physiologically active'

compounds, including phenolic acids and quinones, were thought
to be due to a weak uncoupling of phosphorylation, which in
turn influenced the production of various intermediary metabolites. The uncoupling of oxidative phosphorylation would
increase the content of inorganic phosphate, and consequently
increase metabolic processes such as glycolysis and the citric
acid cycle and, in turn, the production of carbohydrates and
proteins (see Chapter 2).

Other studies, often involving segments of plant tissue,
have indicated that certain phenolic acids influence various
fundamental plant processes, in addition to nutrient uptake,
in ways which may either inhibit or enhance plant growth.
These processes include respiration[19], synthesis of the enzyme
invertase[18], chlorophyll production and photosynthesis[20] and
the metabolism of the auxin, IAA (indole-3-acetic acid)[80].

Demos *et al.*[19] reported that 10 phenolic acids generally
inhibited the growth of mung bean hypocotyls, but the effects
on respiration varied considerably between the compounds. De
Kock and Vaughan[18] found that, at certain concentrations,
several phenolic acids, including cinnamic, ferulic, caffeic
and chlorogenic, enhanced the growth of excised segments of
pea root, and that the compounds enhancing cell elongation
also enhanced the synthesis of the enzyme invertase. With
regard to IAA metabolism, Horsley[67] has pointed out that certain
phenolic compounds may regulate endogenous levels of IAA by
their ability to stimulate or inhibit the activity of IAA
oxidase in plants. This ability has been demonstrated *in vivo*
for a variety of phenolic compounds in a few plant species.
In general, monohydroxy phenolic compounds stimulate IAA
oxidase, resulting in increased decarboxylation of IAA, while
dihydroxy phenolic compounds inhibit IAA oxidase. The interaction between phenolic compounds and IAA has also been studied
by Lee[80]. With segments of maize seedling stems, he found
that pre-treatment with certain compounds, including *p*-coumaric
and ferulic acids, decreased the binding of IAA applied
subsequently, and hence caused its accumulation. Caffeic and
protocatechuic acids had no such effect. Evidence was also

obtained that phenolic compounds influenced the enzymic oxidation of IAA.

A possible interaction between phenolic acids and chlorophyll was reported by Einhellig and Rasmussen[20] who found that vanillic, p-coumaric and ferulic acids, at concentrations of 1 mM and 500 mM, reduced the concentration of chlorophyll in leaves of soyabean, though not in sorghum.

5.5. Uptake of phenolic acids by plants

Studies of the uptake of phenolic acids and related compounds by plants were reviewed by Flaig[26], who concluded that uptake occurred readily with organic compounds having a molecular weight of less than about 1000. The results of studies on the uptake of ^{14}C-labelled p-hydroxybenzoic, vanillic and syringic acids by wheat seedlings have been reported by Flaig[26] and more extensively by Harms et al.[47]. An example of their findings, based on the distribution of the ^{14}C in roots and shoots and amounts of CO_2 evolved, is shown in table 5. The three compounds did not differ greatly in the amounts absorbed, but the amount of CO_2 evolved was greatest from syringic and least from p-hydroxybenzoic acid. At the end of the experiment, a proportion of each phenolic acid was still present in the nutrient solution, and the ^{14}C activity of the

Table 5. Distribution of ^{14}C from COOH-labelled p-hydroxybenzoic, vanillic and syringic acids after uptake by wheat seedlings over six days (from Flaig[26]).

	p-Hydroxybenzoic acid	Vanillic acid	Syringic acid
	% of amount added to nutrient solution		
Roots	7.8	8.6	7.7
Shoots	2.9	1.8	1.8
CO_2	0.9	6.4	7.3

roots was stable to treatment with dilute NaOH[26]. The roots
had a higher content, and the shoots a lower content, of
^{14}C when the label was in the methoxy group or in a ring
position, than when it was in the carboxyl group[47]. In comparison with the carboxyl group, smaller proportions of
methoxy and ring carbon atoms were evolved as CO_2. Some of
the phenolic compounds present in the plant material were
found as glycosides and glucose esters and, with vanillic and
syringic acids, small amounts of the corresponding hydroquinones were also identified[47].

The uptake of vanillic acid by wheat seedlings increased
with increasing pH of the nutrient solution, but transport
to the shoots was greatest when the pH was equal to the pK
value of vanillic acid *viz.* 4.4. Uptake was also greater with
low atmospheric humidity and high light intensity[26].

Studies of the uptake of hydroquinone, and of hydroquinone-
β-D-glucoside (arbutin), by excised barley roots indicated
that both diffusion and active transport were involved[35].

6. INFLUENCE OF PHENOLIC ACIDS ON SOIL MICRO-ORGANISMS
6.1. Influence on the microbial biomass

There is little information on the influence of phenolic
acids on microbial biomass (see Chapter 6). The results of
Knösel[77] indicated that a mixture of p-hydroxybenzoic, vanillic,
p-coumaric and ferulic acids, added to soils at the rate of
10 µg of each acid g^{-1} soil, had little effect on the microbial population. However in liquid culture media, soil
streptomycetes were inhibited by vanillic, p-coumaric and
ferulic acids added individually at a concentration of 100 mg
l^{-1} (approximately 600 µM). Turner and Rice[122] reported that
treatment of a soil with ferulic acid at a rate of 500 µg g^{-1}
increased microbial numbers over a period of several days,
and Sparling *et al.*[117] reported that p-hydroxybenzoic, vanillic,
ferulic and caffeic acids had no overall toxic effect on the
soil microbial biomass when added at the rate of 5 mg g^{-1}
soil. In the latter investigation, there were increases in

total biomass and in the rate of production of respiratory CO_2, indicating that the biomass was able to metabolise the acids.

While not directly related to soil micro-organisms, it is of interest to note that Chesson *et al.*[9] found that ferulic and *p*-coumaric acids suppressed the growth of three cellulo-lytic rumen bacteria when included in a growth medium at a concentration of > 5 mM: *p*-hydroxybenzoic and vanillic acids were less inhibitory.

6.2. Influence on microbial processes

Microbial processes that have been reported to be influenced by phenolic acids are those concerned with trans-formations of nitrogen, in particular nitrogen fixation and nitrification. The growth of *Rhizobium* was found to be in-hibited by *p*-hydroxybenzoic and ferulic acids as well as by several other phenolic acids at a concentration of 500 μM in a culture medium[100]; higher concentrations also caused inhibition but lower concentrations were not examined. Al-though not necessarily due to an effect on nitrogen fixation, the phytotoxicity of soil under *Erica australis* was greater with leguminous plants than with grasses[7].

Rice and Pancholy[106] studied the effects on nitrification of seven phenolic compounds (caffeic acid, chlorogenic acid, ferulic acid, isochlorogenic acid, isoquercitrin, myricetin and quercetin). They found with soil suspensions that *Nitrosomonas* was much more susceptible to inhibition than *Nitrobacter*, and reported that the nitrification of ammonium by *Nitrosomonas* was completely inhibited by ferulic acid at the extremely low concentration of 10 nM. However, since their procedure was based on a colorimetric test for the presence of nitrite, the possibility appears to exist that the result may have been due to the loss of nitrite rather than to inhibition of its formation. The other phenolic compounds tested were reported to cause inhibition of nitri-fication by *Nitrosomonas* at concentrations greater than 1 μM.

Lodhi and Killingbeck[82] reported that extracts of soils under *Ponderosa* pine inhibited *Nitrosomonas* in a soil suspension, and attributed the effect to various phenolic compounds, including caffeic and chlorogenic acids, present in the extracts.

7. CONCLUSION

Improved methods, particularly high performance liquid chromatography, are now available for the determination of phenolic acids in soils and plants but problems associated with the extraction of free and bound acids from the soil remain. The concentrations of the free acids extractable from most soils by water are very low i.e. of the order of 1 µM or less. These concentrations appear to be too low to have significant effects on plant growth, but further studies are required of the extent to which sorbed or bound phenolic acids may exert physiological effects. More information is also needed on the influence of the acids on microbial processes and, in particular, on nitrogen fixation and nitrification.

8. ACKNOWLEDGEMENTS

We wish to thank Mrs P. L. Burrows for the preparation of the typescript and Mrs K. M. Down for assistance with the References.

9. REFERENCES

1. ARMSTRONG G.M., ROHRBAUGH L.M., RICE E.L. and WENDER S.H. 1970. The effect of nitrogen deficiency on the concentration of caffeoylquinic acids and scopolin in tobacco. Phytochemistry, 9, 945-948.
2. BATE-SMITH E.C. 1962. The phenolic constituents of plants and their taxonomic significance. 1. Dicotyledons. Journal of the Linnean Society, 58, 95-173.
3. BATE-SMITH E.C. 1968. The phenolic constituents of plants and their taxonomic significance. II. Monocotyledons. Journal of the Linnean Society, 60, 325-356.

4. BATISTIC L. and MAYAUDON J. 1970. [Biological stabili-
    zation of C-14 labelled ferulic, vanillic and *p*-coumaric
    acids in soil]. Annales Institute Pasteur (Paris), 118,
    199-206.
5. BONDIETTI E., MARTIN J.P. and HAIDER K. 1972. Stabili-
    zation of amino sugar units in humic-type polymers.
    Soil Science Society of America Proceedings, 36, 597-602.
6. BROWN S.A. 1979. Biochemistry of the coumarins. In:
    Biochemistry of Plant Phenolics, pp. 249-286. Eds.
    Swain T., Harborne J.B. and Van Sumere C.F. Plenum Press,
    New York.
7. CARBALLEIRA A. and CUERVO A. 1980. Seasonal variation
    in allelopathic potential of soils from *Erica australis* L.
    heathland. Acta Oecologica, 1, 345-353.
8. CHANDRAMOHAN D., PURUSHOTHAMAN D. and KOTHANDARAMAN R.
    1973. Soil phenolics and plant growth inhibition.
    Plant and Soil, 39, 303-308.
9. CHESSON A., STEWART C.S. and WALLACE R.J. 1982. Influence
    of plant phenolic acids on growth and cellulolytic
    activity of rumen bacteria. Applied and Environmental
    Microbiology, 44, 597-603.
10. CHOU C.H. and LIN H.J. 1976. Auto-intoxication mechanism
    of *Oryza sativa*. 1. Phytotoxic effects of decomposing
    rice residues in soil. Journal of Chemical Ecology, 2,
    353-367.
11. CHOU C.H., LIN T.J. and KAO C.I. 1977. Phytotoxins
    produced during decomposition of rice stubbles in paddy
    soil and their effect on leachable nitrogen. Botanical
    Bulletin of Academia Sinica, 18, 45-60.
12. CHOU C.H. and MULLER C.H. 1972. Allelopathic mechanisms
    of *Arctostaphylos glandulosa* var. zacaenses. American
    Midland Naturalist, 88, 324-347.
13. CHOU C.H. and PATRICK Z.A. 1976. Identification and
    phytotoxic activity of compounds produced during decom-
    position of corn and rye residues in soil. Journal of
    Chemical Ecology, 2, 369-387.
14. CHOU C.H. and YOUNG C.C. 1975. Phytotoxic substances in
    twelve subtropical grasses. Journal of Chemical Ecology,
    1, 183-193.
15. CONSIDINE P.J. and PATCHING J.W. 1975. Ethylene production
    by micro-organisms grown on phenolic acids. Annals of
    Applied Biology, 81, 115-119.
16. DAGLEY S. 1967. The microbial metabolism of phenolics.
    In: Soil Biochemistry, pp. 287-317. Eds. McLaren A.D.
    and Peterson G.H. Dekker, New York.
17. DAGLEY S. 1971. Catabolism of aromatic compounds by
    micro-organisms. Advances in Microbial Physiology, 6,
    1-46.
18. DeKOCK P.C. and VAUGHAN D. 1975. Effects of some chelating
    and phenolic substances on the growth of excised pea root
    segments. Planta, 126, 187-195.

19. DEMOS E.K., WOOLWINE M., WILSON R.H. and McMILLAN C. 1975. The effects of ten phenolic compounds on hypocotyl growth and mitochondrial metabolism of mung bean. American Journal of Botany, 62, 97-102.
20. EINHELLIG F.A. and RASMUSSEN J.A. 1979. Effects of three phenolic acids on chlorophyll content and growth of soybean and grain sorghum seedlings. Journal of Chemical Ecology, 5, 815-824.
21. EL-BASYOUNI S.Z., NEISH A.C. and TOWERS G.H.N. 1964. The phenolic acids in wheat. III. Insoluble derivatives of phenolic cinnamic acids as natural intermediates in lignin biosynthesis. Phytochemistry, 3, 627-639.
22. EVANS W.C. 1977. Biochemistry of the bacterial catabolism of aromatic compounds in anaerobic environments. Nature, London, 270, 17-22.
23. EVANS L.J. 1980. Podzol development north of Lake Huron in relation to geology and vegetation. Canadian Journal of Soil Science, 60, 527-539.
24. FINCHER G.B. 1976. Ferulic acid in barley cell walls; a fluorescence study. Journal of the Institute of Brewing, 82, 347-349.
25. FLAIG W. 1965. Effect of lignin degradation products on plant growth. In: Isotopes and Radiation in Soil-Plant Nutrition Studies, pp. 3-19. Proceedings Symposium, IAEA, Vienna.
26. FLAIG W. 1968. Uptake of organic substances from soil organic matter by plant and their influence on metabolism. In: Organic Matter and Soil Fertility, pp. 723-776. Ed. Fernandez V.H. Wiley, New York.
27. FLAIG W., BEUTELSPACHER H. and RIETZ E. 1975. Chemical composition and physical properties of humic substances. In: Soil Components Volume 1, Organic Components, pp. 1-211. Ed. Gieseking J.E. Springer-Verlag, Berlin.
28. FRIEND J. 1979. Phenolic substances and plant disease. In: Biochemistry of Plant Phenolics, pp. 557-588. Eds. Swain T., Harborne J.B. and Van Sumere C.F. Plenum, New York.
29. FRIEND J. 1981. Plant phenolics, lignification and plant disease. Recent Advances in Phytochemistry, 7, 197-261.
30. FULCHER R.G., O'BRIEN T.P. and LEE J.W. 1972. Studies on the aleurone layer. I. Conventional and fluorescence microscopy of the cell wall with emphasis on phenol-carbohydrate complexes in wheat. Australian Journal of Biological Science, 25, 23-34.
31. GARRAWAY J.L. and RAMIREZ A.M.E. 1982. Phenolic acids in pig slurry subjected to various treatment processes. Journal of the Science of Food and Agriculture, 33, 697-705.
32. GLASS A.D.M. 1974. Influence of phenolic acids upon ion uptake. III. Inhibition of potassium absorption. Journal of Experimental Botany, 25, 1104-1113.
33. GLASS A.D.M. 1975. Inhibition of phosphate uptake in barley roots by hydroxy-benzoic acids. Phytochemistry, 14, 2127-2130.

143

34. GLASS A.D.M. 1976. The allelopathic potential of phenolic acids associated with the rhizosphere of *Pteridium aquilinum*. Canadian Journal of Botany, 54, 2440-2444.
35. GLASS A.D.M. and BOHM B.A. 1971. The uptake of simple phenols by barley roots. Planta, 100, 93-105.
36. GLASS A.D.M. and DUNLOP J. 1974. Influence of phenolic acids on ion uptake. IV. Depolarization of membrane potentials. Plant Physiology, 54, 855-858.
37. GROSS G.G. 1979. Phenolic acids. In: The Biochemistry of Plants, Volume 7, Secondary Plant Products, pp. 301-316. Ed. Conn E.E. Academic Press, New York.
38. GRÜMMER G. 1961. The role of toxic substances in the interrelationships between higher plants. In: Mechanisms in Biological Competition, pp. 219-228. Symposium of the Society of Experimental Biology No. 15, Ed. Milthorpe F.L. Cambridge University Press, London.
39. GUENZI W.D. and McCALLA T.M. 1966a. Phenolic acids in oats, wheat, sorghum and corn residues and their phyto-toxicity. Agronomy Journal, 58, 303-304.
40. GUENZI W.D. and McCALLA T.M. 1966b. Phytotoxic substances extracted from soil. Soil Science Society of America Proceedings, 30, 214-216.
41. HAIDER K. and MARTIN J.P. 1967. Synthesis and transfor-mations of phenolic compounds by *Epicoccum nigrum* in relation to humic acid formation. Soil Science Society of America Proceedings, 31, 766-772.
42. HAIDER K. and MARTIN J.P. 1975. Decomposition of specifi-cally carbon-14 labelled benzoic and cinnamic acid derivatives in soil. Soil Science Society of America Proceedings, 39, 657-662.
43. HAIDER K., FREDERICK L.R. and FLAIG W. 1965. Reactions between amino acid compounds and phenols during oxidation. Plant and Soil, 22, 49-64.
44. HARBORNE J.B. 1973. Phytochemical Methods. Chapman and Hall, London.
45. HARBORNE J.B. 1980. Plant phenolics. In: Secondary Plant Products, pp. 329-402. Eds. Bell E.A. and Charlwood B.V. Springer-Verlag, Berlin.
46. HARBORNE J.B. 1981. Introduction to Ecological Biochemistry. 2nd Edn. Academic Press, London.
47. HARMS H., SÖCHTIG H. and HAIDER K. 1971. Aufnahme und Umwandlung von in unterschiedlichen Stellungen C[14]-Mark-ierten Phenolcarbonsäuren in Weizenkeimpflanzen. Zeit-schrift für Pflanzenphysiologie und Bodenkunde, 64, 437-445.
48. HARPER J.R. and BALKE N.E. 1980. Inhibition of potassium absorption in excised oat roots by phenolic acids. In: Plant Membrane Transport: Current Conceptual Issues, pp. 399-400. Eds. Spanswick R.M., Lucas W.J. and Dainty J. Elsevier, Amsterdam.
49. HARRIS P.J. and HARTLEY R.D. 1976. Detection of bound ferulic acid in cell walls of the Graminae by ultraviolet fluorescence microscopy. Nature, London, 259, 508-510.

144

50. HARRIS P.J. and HARTLEY R.D. 1980. Phenolic constituents
    of the cell walls of monocotyledons. Biochemical Systema-
    tics and Ecology, 8, 153-160.
51. HARRIS P.J.,HARTLEY R.D. and LOWRY K.H. 1980. Phenolic
    constituents of mesophyll and non-mesophyll cell walls
    from leaf laminae of *Lolium perenne*. Journal of the
    Science of Food and Agriculture, 31, 959-962.
52. HARTLEY R.D. 1972. *p*-Coumaric and ferulic acid compon-
    ents of cell walls of ryegrass and their relationships
    with lignin and digestibility. Journal of the Science
    of Food and Agriculture, 23, 1347-1354.
53. HARTLEY R.D. 1973. Carbohydrate esters of ferulic acid
    as components of cell walls of *Lolium multiflorum*.
    Phytochemistry, 12, 661-665.
54. HARTLEY R.D. and BUCHAN H. 1979. High-performance liquid
    chromatography of phenolic acids and aldehydes derived
    from plants or from the decomposition of organic matter
    in soil. Journal of Chromatography 180, 139-143.
55. HARTLEY R.D. and DHANOA M.S. 1981. Rates of degradation
    of plant cell walls measured with a commercial cellulase
    preparation. Journal of the Science of Food and Agri-
    culture, 32, 849-856.
56. HARTLEY R.D. and HARRIS P.J. 1981. Phenolic constituents
    of the cell walls of dicotyledons. Biochemical Systematics
    and Ecology, 9, 189-203.
57. HARTLEY R.D. and JONES E.C. 1975. Effect of ultraviolet
    light on substituted cinnamic acids and the estimation
    of their *cis* and *trans* isomers by gas chromatography.
    Journal of Chromatography, 107, 213-218.
58. HARTLEY R.D. and JONES E.C. 1976. Diferulic acid as a
    component of cell walls of *Lolium multiflorum*. Phyto-
    chemistry, 15, 1157-1160.
59. HARTLEY R.D. and JONES E.C. 1977. Phenolic components
    and degradability of cell walls of·grass and legume species.
    Phytochemistry, 16, 1531-1534.
60. HARTLEY R.D. and JONES E.C. 1978. Phenolic components
    and degradability of the cell walls of the brown midrib
    mutant, *bm3*, of *Zea mays*. Journal of the Science of Food
    and Agriculture, 29, 777-789.
61. HARTLEY R.D., JONES E.C. and FENLON J.C. 1974. Prediction
    of the digestibility of forages by treatment of their cell
    walls with cellulolytic enzymes. Journal of the Science
    of Food and Agriculture, 25, 947-954.
62. HARTLEY R.D., JONES E.C. and WOOD T.M. 1976. Carbohy-
    drates and carbohydrate esters of ferulic acid released
    from cell walls of *Lolium multiflorum* by treatment with
    cellulolytic enzymes. Phytochemistry, 15, 305-307.
63. HARTLEY R.D. and KEENE A.S. 1984. Aromatic aldehyde
    constituents of graminaceous cell walls. Phytochemistry
    (in press).
64. HEALY J.B. Jr., YOUNG L.Y., REINHARD M. 1980. Methano-
    genic decomposition of ferulic acid, a model lignin
    derivative. Applied and Environmental Microbiology 39,
    436-444.

65. HENNEQUIN J.R. and JUSTE C. 1967. Présence d'acides phénols libres dans le sol. Étude de leur influence sur la germination et la croissance des végétaux. Annales agronomiques, 18, 545-569.
66. HIGUCHI T., ITO Y. and KAWAMURA I. 1967. *p*-Hydroxyphenyl-propane component of grass lignin and role of tyrosine-ammonia lyase in its formation. Phytochemistry, 6, 875-881.
67. HORSLEY S.B. 1977. Allelopathic interference among plants II. Physiological modes of action. Proceedings Fourth North American Forest Biology Workshop, 1976, pp. 93-136. Eds. Wilcox H.E. and Hamer A. Syracuse University Press, Syracuse, New York.
68. HUANG P.M., WANG T.S.C., WANG M.K., WU M.H. and HSU N.W. 1977. Retention of phenolic acids by non-crystalline hydroxy-aluminium and -iron compounds and clay minerals of soil. Soil Science, 123, 213-219.
69. HUNTJENS J.L.M., OOSTERVELD-van VLIET W.M. and SAYED S.K.Y. 1981. The decomposition of organic compounds in soil. Plant and Soil, 61, 227-242.
70. KAHNT G. 1967. *Trans-cis*-equilibrium of hydroxycinnamic acids during irradiation of aqueous solutions at different pH. Phytochemistry, 6, 755-758.
71. KAMINSKY R. 1980. The determination· and extraction of available soil organic compounds. Soil Science, 130, 118-123.
72. KAMINSKY R. and MULLER W.H. 1977. The extraction of soil phytotoxins using a neutral EDTA solution. Soil Science, 124, 205-210.
73. KASSIM G., STOTT D.E., MARTIN J.P. and HAIDER K. 1982. Stabilization and incorporation into biomass of phenolic and benzenoid carbons during biodegradation in soil. Soil Science Society of America Journal, 46, 305-309.
74. KATASE T. 1981a. Distribution of different forms of *p*-hydroxybenzoic, vanillic, *p*-coumaric and ferulic acids in forest soil. Soil Science and Plant Nutrition, 27, 365-371.
75. KATASE T. 1981b. The different forms in which *p*-coumaric acid exists in a peat soil. Soil Science, 131, 271-275.
76. KATO A., AZUMA J. and KOSHIJIMA T. 1983. A new feruloy-lated trisaccharide from bagasse. Chemistry Letters (Chemical Society of Japan), 137-140.
77. KNÖSEL D. 1959. Über die Wirkung aus Pflanzenresten freiwerdender phenolischer Substanzen auf Mikroorganismen des Bodens. II. Zeitschrift für Pflanzenernährung Düngung Bodenkunde, 85, 58-66.
78. KUWATSUKA S. and SHINDO H. 1973. Behavior of phenolic substances in the decaying process of plants. 1. Identification and quantitative determination of phenolic acids in rice straw and its decayed product by gas chromatography. Soil Science and Plant Nutrition, 19, 219-227.

79. LAU M.M. and VAN SOEST P.J. 1981. Titratable groups and
     soluble phenolic compounds as indicators of the digesti-
     bility of chemically treated roughages. Animal Feed
     Science and Technology, 6, 123-131.
80. LEE T.T. 1980. Effects of phenolic substances on meta-
     bolism of exogenous indole-3-acetic acid in maize stems.
     Physiologia Plantarium, 50, 107-112.
81. LODHI M.A.K. 1979. Germination and decreased growth of
     Kochia scorparia in relation to its autoallelopathy.
     Canadian Journal of Botany, 57, 1083-1088.
82. LODHI M.A.K. and KILLINGBECK K.T. 1980. Allelopathic
     inhibition of nitrification and nitrifying bacteria in a
     Ponderosa pine (*Pinus ponderosa* Dougl.) community.
     American Journal of Botany, 67, 1423-1429.
83. LYNCH J.M., HALL K.C., ANDERSON H.A. and HEPBURN A. 1980.
     Organic acids from the anaerobic decomposition of *Agropyron
     repens* rhizomes. Phytochemistry, 19, 1846-1847.
84. MARKWALDER H.U. and NEUKOM H. 1976. Diferulic acid as a
     possible crosslink in hemicelluloses from wheat germ.
     Phytochemistry, 15, 836-837.
85. MARTIN A.K. 1982a. The origin of urinary aromatic com-
     pounds excreted by ruminants 1. The metabolism of quinic,
     cyclohexanecarboxylic and non-phenolic aromatic acids to
     benzoic acid. British Journal of Nutrition, 47, 139-154.
86. MARTIN A.K. 1982b. The origin of urinary aromatic com-
     pounds excreted by ruminants 2. The metabolism of pheno-
     lic cinnamic acids to benzoic acid. British Journal of
     Nutrition, 47, 155-164.
87. MARTIN J.P. and HAIDER K. 1971. Microbial activity in
     relation to soil humus formation. Soil Science, 111,
     54-63.
88. MARTIN J.P. and HAIDER K. 1976. Decomposition of specifi-
     cally carbon-14-labelled ferulic acid: free and linked
     into model humic acid-type polymers. Soil Science Society
     of America Journal, 40, 377-380.
89. MARTIN J.P., HAIDER K. and WOLF D. 1972. Synthesis of
     phenols and phenolic polymers by *Hendersonula toruloidea*
     in relation to humic acid formation. Soil Science Society
     of America Proceedings, 36, 311-315.
90. McCALLA T.M. and NORSTADT F.A. 1974. Toxicity problems
     in mulch tillage. Agriculture and Environment, 1,
     153-174.
91. McCLURE P.R., GROSS H.D. and JACKSON W.A. 1978. Phosphate
     absorption by soybean varieties: the influence of ferulic
     acid. Canadian Journal of Botany, 56, 764-767.
92. McPHERSON J.K., CHOU C-H. and MULLER C.H. 1971. Allelo-
     pathic constituents of the Chaparral shrub *Adenostoma
     fasciculatum*. Phytochemistry, 10, 2925-2933.
93. MORITA H. 1975. Polyphenols in the lime water extractives
     of peat. Soil Science, 120, 112-116.
94. MORITA H. 1981. Changes in phenolic composition of a
     peat soil due to cultivation. Soil Science, 131, 30-33.

95. MULLER C.H. and CHOU C.H. 1972. Phytotoxins: an ecological phase of phytochemistry. In: Phytochemical Ecology, pp. 201-216. Ed. Harborne J.B. Academic Press, London.

96. NEISH A.C. 1961. Formation of *m*- and *p*-coumaric acids by enzymatic deamination of the corresponding isomers of tyrosine. Phytochemistry, 1, 1-24.

97. NEWBY V.K., SABLON R-M, SYNGE R.L.M., VANDE CASTEEL K. and VAN SUMERE C.F. 1980. Free and bound phenolic acids of lucerne (*Medicago sativa* cv. Europe). Phytochemistry, 19, 651-657.

98. OLSEN R.A., BENNETT J.H., BLUME D. and BROWN J.C. 1981. Chemical aspects of the Fe stress response mechanism in tomatoes. Journal of Plant Nutrition, 3, 905-921.

99. OLSEN R.A., BROWN J.C., BENNETT J.H. and BLUME D. 1982. Reduction of Fe^{3+} as it relates to Fe chlorosis. Journal of Plant Nutrition, 5, 433-445.

100. PURUSHOTHAMAN D. and BALARAMAN K. 1973. Effect of soil phenolics on the growth of *Rhizobium*. Current Science, 42, 507-508.

101. PUTNAM A.R. and DUKE W.B. 1978. Allelopathy in agroecosystems. Annual Review of Phytophathology, 16, 431-451.

102. RASMUSSEN J.A. and EINHELLIG F.A. 1977. Synergistic inhibitory effects of *p*-coumaric and ferulic acids on germination and growth of grain sorghum. Journal of Chemical Ecology, 3, 197-205.

103. RIBEREAU-GAYON P. 1972. Plant Phenolics. University Reviews in Botany pp. 254. Ed. Heywood V.H. Oliver and Boyd, Edinburgh.

104. RICE E.L. 1974. Allelopathy. Academic Press, New York.

105. RICE E.L. 1979. Allelopathy - an update. Botanical Review, 45, 15-109.

106. RICE E.L. and PANCHOLY S.K. 1974. Inhibition of nitrification by climax ecosystems. III. Inhibitors other than tannins. American Journal of Botany, 61, 1095-1103.

107. ROSTON D.A. and KISSINGER P.T. 1982. Liquid chromatographic determination of phenolic acids of vegetable origin. Journal of Liquid Chromatography, 5, (suppl. 1), 75-103.

108. SALOMONSSON A-C., THEANDER O. and ÅMAN P. 1978. Quantitative determination by glc of phenolic acids as ethyl derivatives in cereal straws. Journal of Agriculture and Food Chemistry, 26, 830-835.

109. SHINDO H. and KUWATSUKA S. 1975. Behavior of phenolic substances in the decaying process of plants. III. Degradation pathway of phenolic acids. Soil Science and Plant Nutrition, 21, 227-238.

110. SHINDO H. and KUWATSUKA S. 1976. Behavior of phenolic substances in the decaying process of plants. IV. Adsorption and movement of phenolic acids in soils. Soil Science and Plant Nutrition, 22, 23-33.

148

111. SHINDO H. and KUWATSUKA S. 1977. Behavior of phenolic
     substances in the decaying process of plants. V. Elution
     of heavy metals with phenolic acids from soil. Soil
     Science and Plant Nutrition, 23, 185-193.
112. SHINDO H., MARUMOTO T. and HIGASHI T. 1979. Behavior
     of phenolic substances in the decaying process of plants.
     X. Distribution of phenolic acids in soils of greenhouses
     and fields. Soil Science and Plant Nutrition, 25, 591-600.
113. SHINDO H., OHTA S. and KUWATSUKA S. 1978. Behavior of
     phenolic substances in the decaying process of plants.
     IX. Distribution of phenolic acids in soils of paddy
     soils and forests. Soil Science and Plant Nutrition,
     24, 233-243.
114. SJOBLAD R.D. and BOLLAG J.M. 1981. Oxidative coupling
     of aromatic compounds by enzymes from soil micro-organisms.
     In: Soil Biochemistry Vol. 5, pp. 113-152. Eds. Paul E.A.
     and Ladd J.N. Marcel Dekker, New York.
115. SMITH M.M. and HARTLEY R.D. 1983. Occurrence and nature
     of ferulic acid substitutuion of cell-wall polysaccharides
     in graminaceous plants. Carbohydrate Research, 118,
     65-80.
116. SPARLING G.P. and VAUGHAN D. 1981. Soil phenolic acids
     and microbes in relation to plant growth. Journal of
     the Science of Food and Agriculture, 32, 625-626.
117. SPARLING G.P., ORD, B.G. and VAUGHAN D. 1981. Changes
     in microbial biomass and activity in soils amended with
     phenolic acids. Soil Biology and Biochemistry, 13,
     455-460.
118. STEVENSON F.J. 1982. Humus Chemistry. Wiley, New York.
119. STOWE L.G. and OSBORN A. 1980. The influence of nitrogen
     and phosphorus levels on the phytotoxicity of phenolic
     compounds. Canadian Journal of Botany, 58, 1149-1153.
120. TANG C-S. and YOUNG C-C. 1982. Collection and identi-
     fication of allelopathic compounds from the undisturbed
     root system of Bigalta Limpograss (*Hemarthria altissima*).
     Plant Physiology, 69, 155-160.
121. TATE K.R. and ANDERSON H.A. 1978. Phenolic hydrolysis
     products from gel chromatographic fractions of soil humic
     acids. Journal of Soil Science, 29, 76-83.
122. TURNER J.A. and RICE E.L. 1975. Microbial decomposition
     of ferulic acid in soil. Journal of Chemical Ecology,
     1, 41-58.
123. VAN SOEST P.J., MASCARENHAS FERREIRA A. and HARTLEY R.D.
     1984. Chemical properties of fibre in relation to
     nutritive quality of NH_3-treated forages. Animal Feed
     Science and Technology (in press).
124. VAN SOEST P.J. and WINE R.H. 1967. Use of detergents
     in the analysis of fibrous feeds. IV. Determination of
     plant cell-wall constituents. Journal of the Associa-
     tion of Official Analytical Chemists, 50, 50-55.

125. VAN SUMERE C.F., COTTENIE J., DE GREEF J., KINT J. 1972.
Biochemical studies in relation to the possible germination regulatory role of naturally occurring coumarin
and phenolics. Recent Advances in Phytochemistry 4,
165-221.

126. VAN SUMERE C.F., VAN BRUSSEL W., VANTE CASTEELE, K. and
VAN ROMPAEY L. 1979. Recent advances in the separation
of plant phenolics. In: Biochemistry of Plant Phenolics,
pp. 1-28. Eds.Swain T., Harborne J.B. and Van Sumere,
C.F. Plenum Press, New York.

127. VAUGHAN D., SPARLING G.P. and ORD B.G. 1983. Ameliora-
tion of the phytotoxicity of phenolic acids by some soil
microbes. Soil Biology and Biochemistry, 15, 613-614.

128. WANG T.S.C., LI S.W. and FERNG Y.L. 1978. Catalytic
polymerization of phenolic compounds by clay minerals.
Soil Science, 126, 15-21.

129. WANG T.S.C., YANG T-K. and CHUANG T-T. 1967. Soil phenolic acids as plant growth inhibitors. Soil Science,
103, 239-246.

130. WANG T.S.C., YEH K.L., CHENG S.Y. and YANG T.Z. 1971.
Behaviour of soil phenolic acids. In: Biochemical
Interactions Among Plants, pp. 113-119. U.S. National
Committee for the International Biological Program,
National Academy of Sciences, Washington, D.C.

131. WHITEHEAD D.C. 1964. Identification of p-hydroxybenzoic,
vanillic, p-coumaric and ferulic acids in soils. Nature,
London, 202, 417-418.

132. WHITEHEAD D.C., DIBB H. and HARTLEY R.D. 1981. Extractant pH and the release of phenolic compounds from soils,
plant roots and leaf litter. Soil Biology and Biochemistry, 13, 343-348.

133. WHITEHEAD D.C., DIBB H. and HARTLEY R.D. 1982. Phenolic
compounds in soil as influenced by the growth of different
plant species. Journal of Applied Ecology, 19, 579-588.

134. WHITEHEAD D.C., DIBB H. and HARTLEY R.D. 1983. Bound
phenolic compounds in water extracts of soils, plant
roots and leaf litter. Soil Biology and Biochemistry,
15, 133-136.

135. WILLIAMS R.D. and HOAGLAND R.E. 1982. The effects of
naturally occurring phenolic compounds on seed germination.
Weed Science, 30, 206-212.

136. WILLIAMS R.T. 1964. Metabolism of phenolics in animals.
In: Biochemistry of Phenolic Compounds, pp. 205-248.
Ed. Harborne J.B. Academic Press, London.

137. WULF L.W. and NAGEL C.W. 1976. Analysis of phenolic
acids and flavonoids by HPLC. Journal of Chromatography,
116, 271-279.

CHAPTER 4

ORIGIN, NATURE AND BIOLOGICAL ACTIVITY OF ALIPHATIC SUBSTANCES AND
GROWTH HORMONES FOUND IN SOIL

J.M. LYNCH

Glasshouse Crops Research Institute, Littlehampton, West Sussex

CONTENTS

1. INTRODUCTION

Considering the diversity of soil organisms and their range of metabolic activities, there is potential for a vast range of products to be formed in soil. These may affect the growth of plants directly and indirectly by modifying the soil environment or the balance of other organisms, such as pathogens, present in the soil. In an earlier review [50], I catalogued some of the products and outlined some of the conditions which must be satisfied to be confident that substances extracted from the soil have ecological significance. However, many studies still do not consider this and therefore it seems prudent to reiterate those conditions here.

A major problem concerns extraction techniques. Humic substances are extremely complex polymers, but they are also relatively inert. They are a major fraction of the soil organic matter but they are retained with a half-life of at least 1000 years [64, 66]. However treatment with even mild extractants, such as dilute acid or alkali, can cause fragmentation of some of the end-groups of the polymer chains. These simple substances would not have been present in the soil solution, yet the extraction can produce an artefact in revealing their presence. Therefore, the first condition of ecological significance is that the substance must be biological active in the form in which it is actually present in soil. This can present problems in that whereas many substances can be recovered by aqueous extraction, other substances can only be recovered with more stringent procedures. For example, some substances are lipophilic and need a suitable organic solvent. Many substances are transported into the plant root in the aqueous phase, but others need to be lipophilic in order to enter cell membranes.

Having isolated a substance from the soil, it must be ascertained that it is biologically active in the concentration at which it was originally present. This can be difficult to determine because production is usually localized, at the site of available substrates and producer organisms. For example, substances produced on the surface of plant residue fragments in soil may attain very large concentrations, yet because they do not diffuse very far through soil or their transport is impeded by microbial degradation, they never attain very large concentrations in the soil itself (Fig. 1). This explains

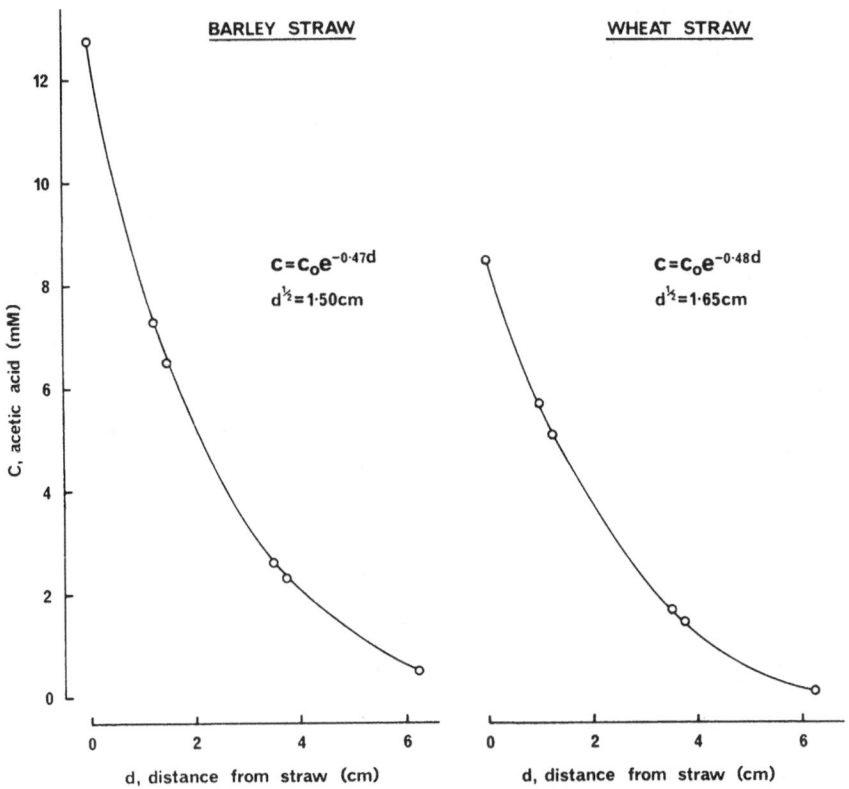

FIGURE 1. Movement of acetic acid from straw decomposing under anaerobic conditions [57].

why activity will only be detected when the plant roots come into direct contact with the substrate. Thus in order to determine the concentration of metabolite in the substrate it is necessary to perform a dilution analysis[57]. A sample of the substrate, such as a straw fragment, is taken from soil and the water content determined by removing a small part. A measured volume of distilled water is then added to the sample and shaken vigorously with glass

154

beads. The concentration (c mM) of the substance in the solution is then
determined from

$$c = A \frac{V + w}{w}$$

where A is the metabolite concentration determined (mM), V is volume of liquid
used for the extraction (ml) and w is the moisture content of the sample (ml).
If the water content of the substrate is expressed as W ml g^{-1} dry substrate,
the acid in the substrate (D μ moles g^{-1} dry substrate) can be calculated from

$$D = cW$$

In a similar way, metabolite concentration can be determined in soil itself.

In the determination of the biological activity of metabolites in soil, it
is essential that their effects be measured over a range of concentrations. It
is common for metabolites to stimulate plant growth in low concentrations but
to be inhibitory in higher concentrations (Fig. 2). Similar responses can also

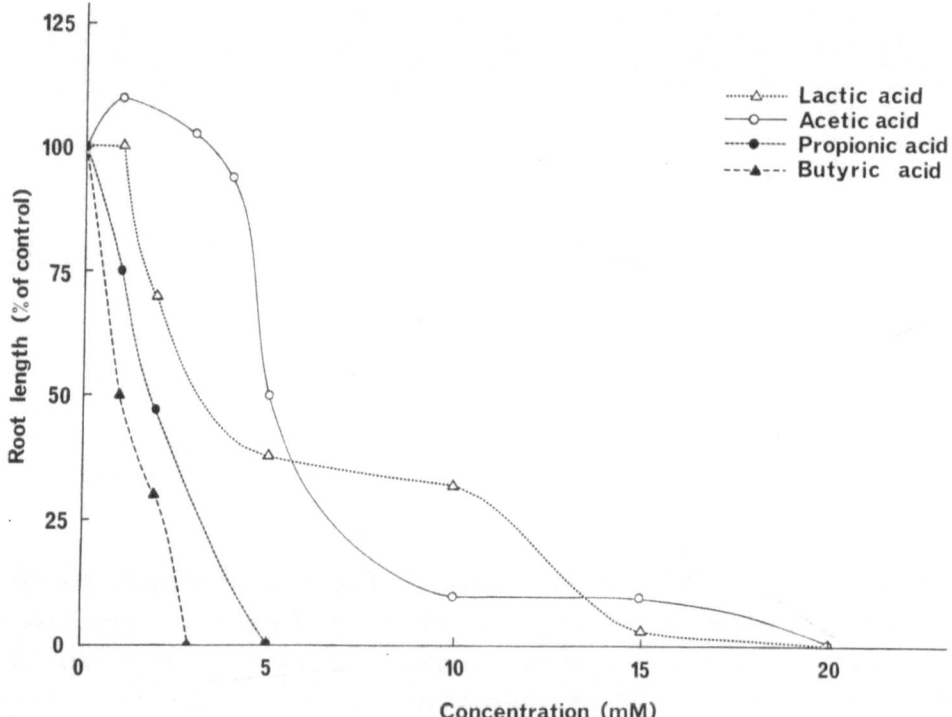

FIGURE 2. Effects of organic acids on the root extension of ryegrass[36].

be found with xenobiotics, such as herbicides. It should also be realised that the effect of the substrance will depend upon the number of contact points it makes with the root[36, 37].

The production of metabolites in soil is clearly linked to substrate availability. It has already been indicated that the humic/fulvic components of soil organic matter have long half-lives and therefore they cannot provide a significant contribution to the available substrate pool in soil. The major primary substrate input comes from living and dead plant roots and from plant residues. Table 1 identifies the substrate input to the surface layer of an arable soil. Similar budgets can be made for other soil horizons and for soils under different cultivation or natural vegetation. In addition to microbial usage, the soil fauna take a share of the available substrate. Both groups eventually die and provide substrates to secondary decomposers. Autotrophic micro-organisms are further sources of primary substrate input.

Table 1. Approximate annual input of substrates (primary productivity) to the surface 5 cm of an arable soil with an approximate microbial biomass of 400 kg C ha^{-1}.

Source	Amount kg C ha^{-1} year^{-1}
Root decomposition	400
Root exudation	240
Straw residues	2800
Autotrophic microbes	100
Total	3540

It is important to recognise that the accumulated pool of metabolites is measured in soil, which is usually less than the amount of metabolite produced. Thus there is a balance.

Accumulation = Production − Utilisation

Some metabolites only accumulate in soil by virtue of their utilisation being restricted. In this respect anaerobic soil conditions are particularly effective in blocking oxidative catabolic pathways. As fermentative metabolites can commonly be phytotoxic, wet soils are particularly significant for the study of the effects of microbial metabolites on plants.

The foregoing should be considered in assessing the significance of the presence of the following metabolites in soils.

2. ETHYLENE

At the Letcombe Laboratory, it was first demonstrated that ethylene could accumulate in phytotoxic concentrations in wet soils[81]. By autoclaving soil and leaving it to stand for a few weeks, the lack of subsequent ethylene production led to the conclusion that micro-organisms were responsible for the formation of the gas[80]. However, this evidence was equivocal because the treatment could have destroyed the necessary substrates for biotic or abiotic production of the gas. Indeed, immediately following autoclaving, there is a vast increase in the rate of ethylene production by the soil[73], probably resulting from chemical breakdown of soil organic matter.

It is generally recognized that ethylene biosynthesis is promoted by oxygen, both in micro-organisms[58] and in higher plants[1]. The accumulation of the gas in waterlogged soils is not paradoxical, but is the resultant of the relative rates of production and loss. Thus in waterlogged soil the gas may persist despite its slow generation, the anaerobic conditions retarding its metabolism and the presence of water preventing its diffusion to the atmosphere. The generation of such conditions depends on the size of soil aggregates (Fig. 3).

The formation of the gas in soil is heavily substrate-dependant. Methionine is generally considered to be the substrate for ethylene bio-synthesis in plants[2], with the intermediate being aminocyclopropane carboxylic acid. Methionine, in combination with glucose as an energy souce, has also been identified as the necessary substrate in a wide range of fungi, yeasts and bacteria[21, 49, 58, 59, 21, 85, 67, 69, 84]. This does not mean it is the universal substrate. Phenolic acids seem to be the substrate for some Penicillium spp.[18]. In P. digitatum, methionine is the precursor of ethylene in shaken cultures but in static cultures, glutamate and α-ketoglutarate are the precursors[8]. In P. cyclopium, biosynthesis of ethylene is dependent on availability of phosphate[65].

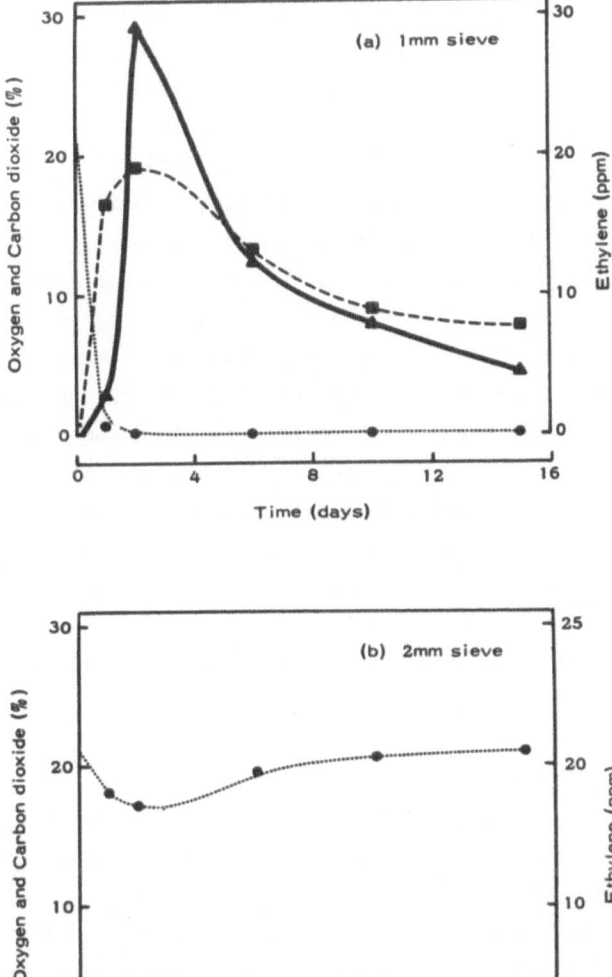

FIGURE 3. Oxygen, carbon dioxide and ethylene in a wetted chalk loam (Gore series) sieved to pass different screen sizes[60].

Soils can be categorized as substrate-rich or substrate-deficient in terms of their capacity to produce ethylene[60]. Crop residues provide natural substrates[60, 31] but there is no correlation between combined organic matter (humus) present in the soil and ethylene production[60]. To provide the substrate, localized variation in oxygen concentrations around crop residues probably favour the accumulation of methionine under anaerobic conditions. A similar reaction may take place at the anaerobic centres of crumbs with outward diffusion of methionine to aerobic micro-environments.

Smith and Cook[78] implied that fungi and aerobic bacteria could not be involved in ethylene production because the gas inhibited their growth; however, this hypothesis has since been disputed and there is much evidence that ethylene is not generally fungistatic[23]. It appears unnecessary at present to invoke anaerobes as the responsible organisms on the following grounds: (a) ethylene accumulates in soil in the presence of low concentrations of oxygen[80]; (b) aerobes can form small amounts of ethylene under totally anaerobic conditions[68]; (c) oxygen may prevent accumulation of ethylene simply by promoting its metabolism[19]; and (d) anaerobic conditions will initially retard the growth of aerobes, and low growth rates which are also caused by lack of available substrate and are usual for microorganisms in soil give rise to rapid rates of ethylene production[59]. When the aerobes die, additional amounts of ethylene may be formed, analogous to the ethylene produced by some senescing plant cells[1].

Fungi comprise the largest fraction of the microbial soil biomass[6]; accordingly, they are probably the main consumers of available substrates in soil and consequently may be important in the production of ethylene in soil. This does not preclude a contribution from bacteria; indeed bacteria may also be necessary to mobilize the substrates for ethylene formation which could be a rate-limiting step. However, the evidence that treatment of soils with anti-fungal agents fails to inhibit ethylene production[83] must be interpreted with caution as such agents are not generally antifungal, particularly to the ethylene-producing soil fungus, Mucor hiemalis[53].

Ethylene in samples of the soil atmosphere withdrawn from naturally structured, heavy, agricultural soils[79] in the spring and early summer can reach concentrations of 1-10 μl l^{-1}. Such concentrations cause a marked reduction in the extension of roots of barley, wheat, and corn in aerated solutions, but rapid extension is resumed when ethylene is no longer applied[81]. Root nodulation in legumes and nitrogen fixation in existing nodules are both inhibited by low concentrations of ethylene[32, 34].

Leaf chlorosis, suggesting premature senescence, has been correlated with increased internal concentrations of ethylene in aerial tissues of <u>Vicia faba</u> and tomato in waterlogged soil[25]. Leaf epinasty is a typical response to waterlogging in some species like tomato and sunflower, and this may in part be an effect of ethylene. Ethylene biosynthesis in submerged organs requires oxygen and in roots is curtailed when ambient oxygen concentrations become low[22, 43].

3. GIBBERELLINS, AUXINS, CYTOKININS AND ABSCISIC ACID

The formation, presence and activity of the non-volatile plant growth regulators in soil have received far less attention than that focussed on ethylene. Indeed the studies that have been undertaken have been somewhat less rigorous than those on ethylene. One of the major and fundamental differences is the ease of chemical analysis. Ethylene can be determined rapidly and unequivocally by gas chromatography. Determination of the non-volatile hormones usually involves a complex extraction procedure and more sophisticated chemical instrumentation. Even if only one of the gibberellins, gibberellic acid (GA_3), were to be determined, it would be desirable to label the material by incorporating $\left[{}^{14}C \right]$-mevalonate, a precursor, into the compound bio-synthetically, subsequent identification being by a radiochromatographic technique. If thin-layer techniques are used, several different solvent systems must be used for confident identification. For gas chromatography, analysis on two stationary phases is desirable. Alternatively combined gas chromatography-mass spectrometry as a single operation probably provides the most satisfactory information. The extensive literature on the presence of plant growth hormones in soils and micro-organisms is generally unsatisfactory in this respect because it is rare to find rigorous chemical analyses performed. Usually, the techniques used employ a mixture of paper chromatography and bioassay of components in selected Rf regions. It is therefore difficult to comment in detail on the validity of conclusions drawn. In the Handbook of Microbiology[55], I have catalogued some of these studies. Here I will give a brief background to each group. Specific citations on the presence of growth regulators in the named organisms can be found in the Handbook and in my earlier review[50].

3.1. Gibberellin

Gibberellic acid (GA_3)

The story of the identification of gibberellin as a plant growth regulator is now established as a classic example of the interactions between soil micro-organisms and plants. In 1926, Kurosawa in Taiwain noticed that a fungal infection of rice, which caused a foot rot, also caused the plants to become abnormally tall before they were killed. The spent culture medium in which the fungus Gibberella fujikuroi (an imperfect form of Fusarium moniliforme) had been grown also produced the effect on rice and other cereals in the laboratory, although under field conditions the effect only seemed to be produced on rice. The active principle was isolated in Japan and later in the UK and the USA. It was found to consist of three closely related compounds but the main component was found to be gibberellic acid (GA_3).

Gibberellins are isoprenoids, synthesized from mevalonic acid. Since the initial identification of the three gibberellins from G. fujikuroi, about 50 structurally similar substances have been found to occur naturally in plants. Although the gibberellins are proven endogenous stimulators of plant growth, it appears that neither developing nor mature roots exhibit a marked response to an exogenous supply of the growth regulator. Curtis[20] applied the filtrates from 1000 different fungi and 500 actinomycetes to the whorls of corn seedlings to look for gibberellin activity. It was found that only G. fujikuroi produced the activity. However, gibberellin activity has since been reported in the culture filtrates of a range of rhizosphere fungi[1], as well as in Azotobacter and in a blue-green alga[50]. The same concentration of authentic gibberellic acid as that apparently formed in cultures of Azotobacter chroococcum produces similar effects on tomato roots as the bacterium[13]; this has led to the speculation that gibberellin is the active principle in the bacterial culture. However, tomatoes are extremely sensitive to growth regulators and it has not

been possible to demonstrate a stimulation of growth when gibberellin was
applied to the roots of axenic barley seedlings[61], although Brown[12] did
demonstrate a stimulation in the growth of wheat seedlings by exongenous GA_3
and indole acetic acid (IAA).

3.2. Auxin

$$\text{indole ring} - CH_2 - COOH$$

indol-3yl-acetic acid (IAA)

Indole acetic acid, a biosynthetic product from tryptophan, is one of the
most common examples of the group of substances often structurally unrelated
but known generally as auxins. They were fist observed in fungi in 1934 to
1935 and it was not until 1945 that they were actually found in plants. Auxin
inhibits the elongation of roots with a concurrent thickening. Lateral growth,
which is stimulated, does not compensate for the dry matter lost. Low
concentrations or brief exposure to auxin can be stimulatory to root growth.
It has been shown that many effects of auxin are caused indirectly by a
stimulation in ethylene formation.

The production of auxin by fungi was reviewed by Gruen[35] and it is clear
that it is a common product of fungi; of the 81 different species of fungi
investigated to that date, 75 produce auxin. The inevitable problem is that
most fungi were investigated using the Avena curvature test, and although this
is probably the best bioassay available, it is still nonspecific and therefore
only "auxin-like" activity can be claimed. There appears to have been some
variation in the interpretation of the physiology of auxin production by fungi.
Although pathogens often increase the amount of IAA in the diseased plant
tissue, it is not clear whether the fungus or the plant is the source of the
increased concentration.

It could be that auxin production is the cause of the common inhibitions of
plant growth by fungi. However, it is also a product of soil bacteria,
including those which can stimulate plant growth.

Considering the widespread ability of soil microorganisms to form auxin, it is perhaps not surprising to find it in soil. Hammence[38] found 0.0025 mg to 0.008 mg per 100 g of soil. Autolyzing plant tissues probably contribute to the auxin pool in soil. However, Wainwright and Pugh[86] could find no indole auxins in soil but found phenolic acids with auxin activity. In this context it has been claimed that humic substances may act as auxins (see Chapter 2). Auxin in the soil can only be significant to plant growth if it is actually available to the plant and not metabolized by other soil microorganisms. It is metabolized by soil bacteria[50], but the difficult and critical experiments, which might show if rhizosphere production of auxin could proceed at a faster rate than its degradation to produce a "pool" of the regulator around roots, have not been done.

There is probably less opportunity for the microbial breakdown of IAA when it is formed by microorganisms in symbiotic association with the plant. Mycorrhiza[77] and root nodules produce[46, 75] auxin. However, it has been also demonstrated that IAA can be catabolized by soybean bacteroids under aerobic conditions even though IAA-oxidase seems unable to function anaerobic anaerobically[72].

3.3 Cytokinins

6-(γ,γ -dimethylallyl)-aminopurine (isopentyladenine)

6-(4-hydroxy-3-methylbut-2-enyl)-aminopurine (zeatin)

Cytokinins promote cell division in plant tissues. They are aminopurines and as such as linked to phosphate esters of sugars to form nucleosides and nucleotides.

The presence of cytokinins in microorganisms (11 bacteria and 4 fungi) has been reviewed[5]. Zeatin is the common cytokinin of plants and isopentyladenine that of microorganisms. However, this generalization is not completely valid because cytokinins other than zeatin occur in higher plants and zeatin occurs in the mycorrhizal fungus Rhizopogon roseolus. Since Van Andel's and Fuch's review, cytokinins have been reported in the bacteria Agrobacterium tumefaciens, Azotobacter chroococcum and Bacillus cereus mycoides, Lactobacillus acidophilus, and the fungi Taphrina spp., Taphrina cerasi and Rhizopogon ochraceorubens. There is little available evidence on the effect of these exogenously produced cytokinins on plants.

3.4. Abscisic Acid

abscisic acid (ABA)

Abscisic acid is primarily a plant growth inhibitor. Besides its involvement in abscission, it is involved in other physiological actions such as the dormancy of seeds and ion uptake by roots. It is formed biosynthetically from mevalonate, and whereas it could arise by fragmentation of carotenoids, the intermediates have not been established. It has not yet been reported as a microbial product, but this probably merely reflects that it has not been looked for. Mucoraceous fungi, which are common in soils, can produce trisporic acids when (+) and (-) sexual forms mate[7], these substances have superficial structural similarities to abscisic acid.

trisporic acid 'C'

Although trisporic acids appear to have abscisic acid-like activity on seed germination[82] and pea growth[26], these observations must be treated with some caution because the former are fully conjugated molecules and extremely unstable, and thus the activity may be due to breakdown products[61].

3.5. Microbial inoculants

Much of the interest in the manipulation of crop growth has been focussed on the introduction of nitrogen-fixing bacteria into the rhizosphere, usually by coating seeds or drenching roots on transplanting. It now seems very unlikely that nitrogen fixation is the principal process governing the response of the crop, certainly the claims appear to have been exaggerated. Barea and Brown[8] inoculated a variety of dicotyledons and monocotyledons growing in pots in natural soil with Azotobacter paspali. The large increases in early plant growth were not due to nitrogenase activity around the roots but were more characteristic of plant growth regulator responses. The culture filtrates were as active as the complete culture and they contained materials with IAA, gibberellin and cytokinin-like activity. Others[30, 70] made similar observations but it must be emphasized that the evidence is still circumstantial for the growth regulators which were tentatively identified being the factors responsible for the stimulation in plant growth.

A further type of microbial inoculant which has been considered is the phosphate-solubilizing bacteria. Of 50 phosphate-dissolving bacteria isolated from the rhizospheres of crop plants, twenty synthesized IAA, gibberellins and and cytokinins as assessed by a range of bioassays[9]. IAA was produced by 43, 20 formed gibberellins and 45 produced cytokinin-like substances. IAA was decomposed by 28 and some isolates produced plant growth inhibitors. As yet however, there is no scientifically proven application of the application of such inoculants, neither as solubilizers of rock phosphate nor as stimulators of plant growth, although there is a patent[45].

4. ALIPHATIC ORGANIC ACIDS

Short-chain aliphatic acids (acetic, propionic, and butyric) have been isolated from a range of anaerobic soils (for a review see Drew and Lynch[23]). Cellulose is the principle substrate for their formation and plant residues are the natural sources of this. In stirred soil slurry in a chemostat, the reaction is primarily linked to redox potential and consequently with the solubilization of Fe^{2+} and Mn^{4+} [130]; at very low redox potentials the acid itself serves as a substrate for methanogenic bacteria. Whereas the methanogenic bacteria of soil are well characterised the identity of the acetogens is less certain, although they must be ubiquitous.

Organic acids have been shown to reach phytotoxic concentrations in soil[17, 41, 51]. Injury to roots arises from the adverse effects of undissociated short-chain aliphatic or aromatic acids on cell membranes, causing leakage of cell contents[48]. At equimolar concentrations, butyric acid and the aromatic acids are more toxic than propionic acid, which is more toxic than acetic acid[52]. However, acetic acid is usually produced in the greatest amount and is responsible for most of the toxicity of the steam volatile acids. Problems arising from microbial production of organic acids during decomposition of plant residues with rice crops is well recognized, but damage to crops can also occur in cereal production in temperate climates[14].

166

5. HYDROGEN SULPHIDE

Hydrogen sulphide is produced by obligate anaerobic bacteria if there are abundant sulphate and organic substrates at redox potentials below zero. These bacteria are unable to use oxygen and require sulphate as a terminal electron acceptor. The sulphate-reducing genera listed in Bergey's manual are _Desulfotomaculum_, but recently _Desulforibrio_ and _Desulfomonas_ and _Desulfuromonas_, have been described. Substrate oxidation is coupled to reduction of sulphate, sulphite, thiosulphate, tetrathionate or sulphur to sulphide. The nutrition of the bacteria is limited to the oxidation of hydrogen, pyruvate, lactate, ethanol, formate or malate, which are converted to fatty acids (acetate) and CO_2; hence the bacteria form syntrophic associations with organisms which excrete these substrates. Furthermore, _Desulfomonas_ and _Desulfuromonas_ can also utilize acetate. The physiology of sulphate-reducing bacteria has been reviewed[47]. It has been claimed that _Pseudomonas putida_, a denitrifying organism involved in the suffocation disease of rice in Taiwan[91], in the absence of nitrate converts sulphate from ammonium sulphate to sulphide. However, in most soils any sulphide which is produced reacts with iron, is oxidized, or is converted to the nontoxic sulphydryl ion[28].

Sulphide production has been observed in the rhizosphere and spermosphere of plants[24, 27, 29], including rice where oxygen diffusing out of roots of the rhizosphere would be expected to produce a locally oxidizing atmosphere. In conditions where hydrogen sulphide persists, particularly in wet acidic soils containing sulphate, it can exceed concentrations (ca. 0.1 µg ml^{-1} water) that inhibit respiration and poison cytochrome oxidase in rice roots[3]. Damage to rice shoots attributed to hydrogen sulphide is known in different parts of the world as browning, bronzing, brusome, akiochi, hie-imochi, suffocation, and straighthead diseases[40]. Control of the disease is best achieved through increasing soil redox potential by application of nitrate to prevent sulphide formation, rather than by promoting binding of sulphide. Oxygen diffusing in intercellular spaces from the aerial tissues can move out of roots into the anaerobic soil to produce a locally oxidised atmosphere, thus contributing to plant tolerance to Fe^{2+} and H_2S[33, 74]. Rhizosphere microorganisms like _Beggiatoa_ on rice roots oxidize hydrogen sulphide extracellularly to sulphur[44]. In the presence of excess Fe^{2+}, the formation of insoluble ferrous sulphide usually detoxifies hydrogen sulphide[28], but injurious concentrations have been detected in soil containing excess iron[4].

6. ALCOHOLS

Lower aliphatic alcohols are formed by a wide range of microorganisms during the fermentation of carbohydrates and by the hydrolytic deamination of amino acids and aliphatic amines. Thus methanol, ethanol, and n-propanol can be formed in a range of soils when sugarcane leaf powder or fresh Crotalaria juncea are added to soils under waterlogged or aerobic conditions. n-butanol is also formed when C. juncea is added[88].

The occurrence of alcohols in waterlogged soil has been rarely examined, but in one study[88] in which plant residues were incubated with soil at $30^{o}C$, ethanol and other lower aliphatic acids reached concentrations in the soil solution that were damaging to seedlings of rice, sugarcane, and corn when tested in solution culture. Ethanol applied exogenously is not greatly toxic. Growth of seedlings of sugarcane, corn, and rice was depressed by only 10-20% by 4 mM ethanol[88]. Barley roots tolerate 5 mM without damage to cell membranes[46], and the survival of emerging wheat and rye seedlings was halved between 30 and 170 mM[10, 76]. From the available evidence, ethanol has yet to be identified as a factor in waterlogging injury to roots or shoots.

7. ANTIBIOTICS

In the 1950s, the ecological significance of antibiotic production by microorganisms in soil was investigated. Detailed reviews were given by Waksman [11] and Brian[87]. It was recognized that whereas antibiotics could be considered as inhibitors of plant growth, they could sometimes stimulate growth either directly or by the removal of pathogens which are inhibitory to growth. Antibiotics are often metabolic inhibitors of great specificity and potency and amongst those with inhibitory effects at 5 µg/ml or less are actidione, azaserine, alternaric acid, and polymyxin. The activity of antibiotics on seed germination and subsequent root growth was compared with those of coumarin and IAA in the range 0 to 25 ppm by Wright[90]. They generally had little effect on percentage germination and most were less toxic than IAA and more toxic than coumarin to root extension. The most toxic were alternaric acid, glutinosin, mycophenolic acid, and gliotoxin and the least toxic were griseofulvin, penicillin, and streptomycin.

Antibiotics do not appear to be produced in soils except when there is a source of energy, such as seeds or straw, for the microorganisms. Even if they are produced they can be inactivated by adsorption onto clay surfaces or

degraded by other microorganisms. An example of the latter is the production
of penicillin in maize seeds by <u>Penicillium</u> <u>chrysogenum</u> and its detoxification
by the β-lactamase of <u>Bacillus</u> <u>cereus</u>[39], this could well be important in the
survival of the seed in the soil.

It has been postulated that the antibiotic patulin, an unsaturated lactone
produced by <u>Penicillium</u> <u>urticae</u>, is one of the most likely phytotoxins arising
from the stubble mulch farming practice[63].

patulin

In the range 20 to 75 ppm in solution, sand, and soil, it can inhibit
germination and root and shoot growth by 50% and decrease weed emergence,
affecting dicotyledenous plants more than monocotyledons. However, its effect
in the field is uncertain because the concentrations which actually occur in
soil are not known. Furthermore, soils high in organic matter can reduce its
phytotoxicity[62]. Its action also depends on the type of straw present and on
temperature[63]. Patulin can be effective against Gram-positive and Gram-
negative bacteria and fungi but it can also decrease the resistance of a plant
to mildew and smut[63]. Although patulin can be produced by several fungi and
Nordstadt and McCalla recognized that its first isolation from <u>Pencillium</u>
<u>expansum</u> concerned its phytotoxicity, they have placed much emphasis on the
role of <u>P. urticae</u> in patulin production in soil as it apparently occurs in
high concentrations in stubble-mulch. However, subsequent studies in the USA
(A.C. Waiss, personal communication) and in my laboratory have failed to detect
patulin in soil containing large amounts of straw.

The question 'Are antibiotics produced in soil?' was posed again recently
by Williams[89]. It was indicated that convincing evidence is still lacking 40
years after the discovery of those substances, although there have been
relatively few recent attempts to detect them. The reasons postulated for the
failure to detect them were given as: (a) the instability of antibiotics in

soil conditions; (b) adsorption of antibiotics by soil colloids;
(c) insufficient sensitivity of detection methods; (d) insufficient nutrients
for widespread, frequent growth of producer microbes. Their detection would be
of interest to microbial ecologists, but also to plant pathologists,
particularly those concerned with the biological control of plant disease
because antibiosis is claimed to be one of the major modes of action. A recent
such example was the identification of pyrrolnitrin produced by _Pseudomonas_
fluorescens to control the pathogen _Rhizoctonia solani_[42]. However, although
the antibiotics must be produced at localized sites, no evidence was sought for
its presence in soil.

8. CONCLUSION

Of the range of microbial products surveyed, those which occur in soil in
ecologically significant concentrations can probably be regarded as allelo-
chemicals, according to the definition of allelopathy[71]. They may also be
responsible for 'soil sickness' and other loosely-described phenomena.
However, it must be recognized that microbial metabolites in soil have as much
potential to stimulate plant growth as to inhibit it. A target for soil
biotechnology[54] must be to optimise soil conditions for the stimulatory effects,
while minimising inhibitory effects.

9. SUMMARY

Soil micro-organisms can produce a wide range of substances which
influence plant growth. There is substantial evidence that ethylene,
hydrogen sulphide and organic acids are produced in ecologically significant
concentrations in soil but the evidence for the occurrence of non-gaseous
plant growth hormones and antibiotics in soil is weak. Whereas these latter
substances are probably produced at microsites where substrates and producer
organisms are present, it remains to be proven that this would be significant
to the growth of a crop.

170

10. REFERENCES

1. ABELES F.B. 1973. Ethylene in Plant Biology. Academic Press, New York.
2. DAMS D.O. and YANG S.F. 1979. Ethylene biosynthesis: Identification of 1-aminocylopropane-1-carboxylic acid as an intermediate in the conversion of methionine to ethylene. Proceedings of the National Academy of Sciences USA, 76, 170–174.
3. ALLAM A.I. and HOLLIS J.P. 1972. Sulfide inhibition of oxidases in rice roots. Phytopathology, 62, 634–639.
4. ALLAM A.I., PITTS G. and HOLLIS J.P. 1972. Sulfide determination in submerged soils with an ion-selective electrode. Soil Science, 114, 456–467.
5. ANDEL O.M. van and FUCHS A. 1972. Interference with plant growth regulation by microbial metabolites. In: Phytotoxins in Plant Disease. Eds. R.K.S. Wood, A. Ballio and A. Graniti, pp. 227–249. Academic Press, London.
6. ANDERSON J.P.E. and DOMSCH K.H. 1973. Quantification of bacterial and fungal contributions to soil respiration. Archives for Microbiology, 93, 113–127.
7. AUSTIN D.J., BU'LOCK J.D. and GOODAY G.W. 1969. Trisporic acids: sexual hormones from Mucor mucedo and Blakeslea trispora. Nature, London, 223, 1178–1179.
8. BAREA J.M. and BROWN M.E. 1974. Effects on plant growth produced by Azotobacter paspali related to synthesis of plant growth regulating substances. Journal of Applied Bacteriology, 37, 583–593.
9. BAREA J.M., NAVARRO E. and MONTOYA E. 1976. Production of plant growth regulators by rhizosphere phosphate-solubilizing bacteria. Journal of Applied Bacteriology, 40, 129–134.
10. BELETSKAYA E.K. 1977. Changes in metabolism of winter crops during their adaptation to flooding. Soviet Plant Physiology, 24, 750–756.
11. BRIAN P.W. 1957. The ecological significance of antibiotic production. In: Microbial Ecology. Eds. R.E.C. Williams and C.C. Spicer, pp. 168–188. Cambridge University Press, Cambridge.
12. BROWN M.E. 1972. Plant growth substances produced by micro-organisms of soil and rhizosphere. Journal of Applied Bacteriology, 35, 443–451.
13. BROWN M.E., JACKSON, R.M. and BURLINGHAM S.K. (1968). Effects produced on tomato plants, Lycopersicum esculentum, but seed or root treatment with gibberellic acid and indol-3yl-acetic acid. Journal of Experimental Botany, 19, 544–552.
14. CANNELL R.Q. and LYNCH J.M. 1983. Possible adverse effects of decomposing organic matter on plant growth. In: Organic Matter and Rice. Ed. F.N. Ponnamperuma, in press. International Rice Research Institute, Los Banos.
15. CHALUTZ E., LIEBERMAN M. and SISLER H.D. 1977. Methionine-induced ethylene production by Penicillium digitatum. Plant Physiology, 60, 402–406.
16. CHALUTZ E. and LIEBERMAN M. 1978. Inhibition of ethylene production in Penicillium digitatum. Plant Physiology, 61, 111–114.
17. CHO D.Y. and PONNAMPERUMA F.N. 1971. Influence of soil temperature on the chemical kinetics of flooded soils and the growth of rice. Soil Science, 112, 184–194.
18. CONSIDINE P.J. and PATCHING J.W. 1975. Ethylene production by micro-organisms grown on phenolic acids. Annals of Applied Biology, 81, 115–119.
19. CORNFORTH I.S. 1975. The persistence of ethylene in aerobic soils. Plant and Soil, 42, 85–96.
20. CURTIS R.W. 1957. Survey of fungi and actinomycetes for compounds possessing gibberellin-like activity. Science, 125, 646.

21. DASILVA E.J., HENRIKSSON E. and HENRIKSSON L.A. 1974. Ethylene production by fungi. Plant Science Letters, 2, 63-66.
22. DREW M.C., JACKSON M.B. and GIFFARD S. 1979. Ethylene-promoted adventitious rooting and development of cortical air spaces (arenchyma) in roots may be adpative responses to flooding in Zea mays L. Planta, 147, 83-88.
23. DREW M.C. and LYNCH J.M. 1980. Soil anaerobiosis, microorganisms and root function. Annual Reviews of Phytopathology, 18, 37-66.
24. DOMMERGUES Y. and JACQ V. 1972. Microbiological transformations of sulphur in the rhizosphere and spermosphere. Annales Agronomiques, 23, 201-215.
25. EL-BELTAGY A.S. and HALL M.A. 1974. Effect of water stress upon endogenous ethylene levels in Vicia faba. New Phytologist, 73, 47-60.
26. FEOFILOVA E.P., LOZHNIKOVA V.N., BEKHTEREVA M.N., SAMOKHVALOV G.I. and CHAILAKHAIN M.Kh. 1973. The influence of trisporic acids on the growth and pigment formation and respiration of pea sprouts. Doklady Academii Nauk SSSR, 208, 483-486.
27. FORD H.W. 1973. Levels of hydrogen sulphide toxic to citrus roots. Journal of the American Society of Horticultural Science, 98, 66-68.
28. GAMBRELL R.P. and PATRICK W.H. 1978. Chemical and microbiological properties of anaerobic soils and sediments. In: Plant Life in Anaerobic Environments. Eds. D.D. Hook and R.M.M. Crawford, pp. 375-423. Ann Arbor, Michigan.
29. GARCIA J.L., RAIMBAULT M., JACQ V., RINAUDO, G. and ROGER P. 1974. Activities microbiennes dans les sols de rizieres du Senegal: Relation avec les characteristiques physico-chemiques et influence de la rhizosphere. Révue Ecologie et Biologie du Sol, 11, 169-185.
30. GASKINS M.H. and HUBBELL D.H. 1979. Response of non-leguminous plants to root inoculation with free-living diazotrophic bacteria. In: The Soil-Root Interface. Ed. J.L. Harley and R.S. Russell, pp. 176-182. Academic Press, London.
31. GOODLASS G. and SMITH K.A. 1978. Effects of organic amendments on the evolution of ethylene and other hydrocarbons from soil. Soil Biology and Biochemistry, 10, 201-205.
32. GOODLASS G. and SMITH K.A. 1979. Effects of ethylene on root extension and nodulation of peat (Pisum sativum L.) and white clover (Trifolium repens L.). Plant and Soil, 51, 387-395.
33. GREEN M.S. and ETHERINGTON J.R. 1977. Oxidation of ferrous iron by rice (Oryza sativa L.) roots: A mechanism for waterlogging tolerance? Journal of Experimental Botany, 28, 678-690.
34. GROBELAAR N., CLARKE B., HOUGH M.C. 1971. The nodulation and nitrogen fixation of isolated roots of Phaseolus vulgaris L. III. The effect of carbon dioxide and ethylene. Plant and Soil Special Volume, pp. 215-221.
35. GRUEN H.E. 1959. Auxins and fungi. Annual Review of Plant Physiology, 10, 405-440.
36. GUSSIN E.J. and LYNCH J.M. 1981. Microbial fermentation of grass residues to organic acids as a factor in the establishment of new grass swards. New Phytologist, 89, 449-457.
37. GUSSIN E.J. and LYNCH J.M. 1982. Effect of local concentrations of acetic acid around barley roots on seedling growth. New Phytologist, 92, 345-348.
38. HAMMENCE J.H. 1946. The determination of auxins in soils. Analyst, London, 71, 111-116.
39. HILL P. 1972. The production of penicillins in soils and seeds by Penicillium chrysogenum and the role of penicillin β-lactamase in the ecology of soil bacillus. Journal of General Microbiology, 70, 243-252.

40. HOLLIS J.P., ALLAM A.I., PITTS G., JOSHI M.M. and IRRAHIM I.K.A. 1975. Sulfide diseases of a rice on iron-excess soils. Acta Phytopathologica Academiae Scientarium Hungaricae, 10, 329-341.

41. HOLLIS J.P., RODRIGUES-KABANA R. 1967. Fatty acids in Louisiana rice fields. Phytopathology, 57, 841-847.

42. HOWELL C.R. and STIPANOVIC R.D. 1979. Control of Rhizoctonia solani in cotton seedlings with Pseudomonas fluorescens and with an antibiotic produced by the bacterium. Phytopathology, 69, 480-482.

43. JACKSON M.B., GALES K. and CAMPBELL D.J. 1978. Effect of waterlogged soil conditions on the production of ethylene and on water relationships in tomato plants. Journal of Experimental Botany, 29, 183-193.

44. JOSHI M.M. and HOLLIS J.P. 1977. Interactions of Beggiatoa and rice plant: detoxification of hydrogen sulphide in the rice rhizosphere. Science, 195, 179-180.

45. JUDD RINGER CORPORATION. 1980. Improvements in or relating to the improvement of growing media for plants. British Patent 1562556.

46. KEFFORD N.P., BROCKWELL J. and ZWAR J.A. 1960. The symbiotic synthesis of auxin by legumes and nodule bacteria and its role in nodule development. Australian Journal of Biological Sciences, 13, 456-467.

47. LE GALL J. and POSTGATE J.R. 1973. The physiology of sulphate-reducing bacteria. Advances in Microbial Physiology, 10, 81-133.

48. LEE R.B. 1977. Effects of organic acids on the loss of ions from barley roots. Journal of Experimental Botany, 28, 578-587.

49. LYNCH J.M. 1972. Identification of substrates and isolation of micro-organisms responsible for ethylene production in the soil. Nature, London, 240, 45-46.

50. LYNCH J.M. 1976. Products of soil micro-organisms in relation to plant growth. CRC Critical Reviews in Microbiology, 5, 67-107.

51. LYNCH J.M. 1978. Production and phytotoxicity of acetic acid in anaerobic soils containing plant residues. Soil Biology and Biochemistry, 10, 131-135.

52. LYNCH J.M. 1980. Effects of organic acids on the germination of seeds and growth of seedlings. Plant, Cell and Environment, 3, 255-259.

53. LYNCH J.M. 1983. Effects of antibiotics on ethylene production by soil micro-organisms. Plant and Soil, 70, 415-420.

54. LYNCH J.M. 1983. Soil Biotechnology. Microbiological Factors in Crop Productivity. Blackwell Scientific Publications, Oxford.

55. LYNCH J.M. 1983. Plant growth regulators. CRC Handbook of Microbiology Vol. VII, in press. CRC Press, Boca Raton.

56. LYNCH J.M. and GUNN K.B. 1978. The use of the chemostat to study the decomposition of wheat straw in soil slurries. Journal of Soil Science, 29, 551-556.

57. LYNCH J.M., GUNN K.B. and PANTING L.M. 1980. On the concentration of acetic acid in straw and soil. Plant and Soil, 56, 93-98.

58. LYNCH J.M. and HARPER S.H.T. 1974. Formation of ethylene by a soil fungus. Journal of General Microbiology, 80, 187-195.

59. LYNCH J.M. and HARPER S.H.T. 1974. Fungal growth rate and the formation of ethylene in soil. Journal of General Microbiology, 85, 91-96.

60. LYNCH J.M. and HARPER S.H.T. 1980. Role of substrates and anoxia in the accumulation of soil ethylene. Soil Biology and Biochemistry, 12, 363-367.

61. LYNCH J.M. and WHITE N. 1977. Effects of some non-pathogenic micro-organisms on the growth of gnotobiotic barley plants. Plant and Soil, 47, 161-172.

62. NORSTADT F.A. and McCALLA T.M. 1963. Phytotoxic substance from a species of Penicillium. Science, 140, 410-411.

63. NORSTADT F.A. and McCALLA T.M. 1971. Effects of patulin on wheat grown to maturity. Soil Science, 111, 236-243.
64. PAUL E.A., CAMPBELL C.A., RENNIE D.A. and McCALLUM K.J. 1964. Investigations of the dynamics of soil utlizing carbon dating techniques. Transactions 8th International Congress of Soil Science, 3, 201-208.
65. PAŽOUT J., PAŽOUTOVÁ S. and VANČURA V. 1982. Effects of light, phosphate and oxygen on ethylene formation and conidiation in surface cultures of Penicillium cyclopium Westling. Current Microbiology, 7, 133-136.
66. PERRIN R.M.S., WILLIS E.H. and HODGE C.A.H. 1964. Dating of humus podzols by residual radiocarbon activity. Nature, London, 202, 165-166.
67. PRIMROSE S.B. 1976. Formation of ethylene by Escherichia coli. Journal of General Microbiology, 95, 159-165.
68. PRIMROSE S.B. 1979. Ethylene and agriculture: The role of the microbe. Journal of Applied Bacteriology, 46, 1-25.
69. PRIMROSE S.B. and DILWORTH M.J. 1976. Ethylene production by bacteria. Journal of General Microbiology, 93, 177-181.
70. REYNDERS L. and VLASSAK K. 1979. Conversion of trytophan to indole acetic acid by Azospirillum brasilense. Soil Biology and Biochemistry, 11, 547-548.
71. RICE E.L. 1974. Allelopathy. Academic Press, London.
72. RIGAUD J. and PUPPO A. 1975. Indole-3-acetic acid catabolism by soybean bacteroids. Journal of General Microbiology 88, 223-228.
·73. ROVIRA A.D. and VENDRELL M. 1972. Ethylene in sterilized soil: its significance in studies of interactions between microorganisms and plants. Soil Biology and Biochemistry, 4, 63-69.
74. SANDERSON P.L. and ARMSTRONG W. 1978. Soil waterlogging, root rot and conifer windthrow: Oxygen deficiency or phytotoxicity? Plant and Soil, 49, 185-190.
75. SARMA K.S.B., LAKSHMI-KUNARI M., APTE R. and SUBBA RAO N.S. 1973. Some physiological characteristics of Rhizobium meliloti and R. trifolii in relation to efficiency of symbiosis with lucerne and Egyptian clover. Plant and Soil, 38, 299-305.
76. SIEGEL S.M. and HALPERN L.A. 1964. The effect of branching at C-1 on the biological activity of alcohols. Proceedings of the National Academy of Sciences USA, 51, 765-768.
77. SLANKIS V. 1973. Hormonal relationships in mycorrhizal development. In: Ectomycorrhizae - Their Ecology and Physiology. Eds. G.C. Marks and T.T. Kozolowski, pp. 231-298. Academic Press, New York.
78. SMITH A.M. and COOK R.J. 1974. Implications of ethylene production for biological balance of soil. Nature, London, 252, 703-705.
79. SMITH K.A. and DOWDELL R.J. 1974. Field studies of the soil atmosphere. 1. Relationships between ethylene, oxygen, soil moisture content and temperature. Journal of Soil Science, 25, 217-230.
80. SMITH K.A. and RESTALL S.W.F. 1971. The occurrence of ethylene in anaerobic soil. Journal of Soil Science, 22, 430-443.
81. SMITH K.A. and RUSSELL R.S. 1969. Occurrence of ethylene and its significance in anaerobic soil. Nature, London, 222, 469-471.
82. SPALLA C. and BIFFI G. 1971. Abscisic acid-like activity of trisporic acids. Experienta, 27, 1387-1388.
83. SUTHERLAND J.B. and COOK R.J. 1980. Effects of chemical and heat treatments on ethylene production in soil. Soil Biology and Biochemistry, 12, 357-362.

84. SWANSON B.T., WILKINS H.F. and KENNEDY B. 1979. Factors affecting ethylene production by some plant pathogenic bacteria. Plant and Soil, 51, 19-26.
85. THOMAS K.C. and SPENCER M. 1977. L-methionine as an ethylene precursor in Saccharomyces cerevisiae. Canadian Journal of Microbiology, 23, 1669-1674.
86. WAINWRIGHT M. and PUGH G.J.F. 1975. Phenol auxins and Ehrlich reactors in soils. Soil Biology and Biochemistry, 7, 287-289.
87. WAKSMAN S.A. 1956. The role of antibiotics in natural processes. Giornale di Microbiologica, 2, 1-14.
88. WANG T.S.C. and CHUANG T.-T. 1967. Soil alcohols their dynamics and effect on plant growth. Soil Science, 104, 40-45.
89. WILLIAMS S.T. 1982. Are antibiotics produced in soil? Pedobiologia, 23, 427-435.
90. WRIGHT J.M. 1951. Phytotoxic effects of some antibiotics. Annals of Botany (London), 15, 493-499.
91. WU M.M.H, WU C.S., CHIANG M.H. and CHOU S.F. 1972. Microbial investigations on the suffocation disease of rice in Taiwan. Plant and Soil, 37, 329-344.

CHAPTER 5

SOIL ENZYMES

J.N. LADD

Division of Soils, CSIRO, Glen Osmond, South Australia

CONTENTS

1. INTRODUCTION

Soils are enzymically active. Enzymes in soils are of plant, animal and microbial origin, and collectively their activities express the metabolic status of soils at a given time. Many assays have been devised to measure the activities of a wide range of enzymes in soils, usually under conditions which minimize or eliminate the proliferation of microorganisms during the assay period. Such activities are thought to be a more direct expression of soil biological activity generally , or of the activities of specific processes of nutrient cycling and organic matter turnover, than say measurements of microbial numbers.

Interest in soil enzymes derives partly from the notion that their potential activities, as assayed under artificial but relatively controlled conditions, relate to their activities as pertaining in the natural soil environment with varying pH, moisture and substrate supply. Such a relationship would permit valid comparisons of soil biological activities according to soil properties and treatments, e.g. fertilizer and pesticide use, return of plant residues etc. Indeed potential activities of individual or groups of enzymes have been correlated with plant growth in different soils or in soils with different crop histories, and are thought by some workers to serve as useful indices of soil fertility. Since 1978, when the subject of soil enzymes was comprehensively reviewed[18], the majority of studies has continued to deal with the activities of whole soil suspensions, and to be environmental in character. Most studies have been concerned with the effects of pesticides, heavy metal pollutants, etc., on enzymically-catalysed reactions and emphasize those reactions directly involved with the cycling of plant nutrients.

Soil enzymes function in complex and varying physico-chemical environments. Assays generally measure the activities of accumulated enzymes[57,58] which include (i) enzymes which function exocellularly, either free in the soil solution, or bound to inorganic and organic soil constituents, or present in particulate cell debris, and (ii) enzymes which are present either in dead cells or in live cells. Burns[20] envisaged ten distinct categories of enzymes according to their location in soils. He distinguished enzymes of proliferating microbial, animal and plant cells from enzymes of non-proliferating cells such as microbial spores, protozoan cysts etc., and further categorized enzymes of live cells into cytoplasmic enzymes, enzymes in the periplasmic space of Gram-negative bacteria, and enzymes attached to the outer surfaces of cells. Other categories included truly exocellular enzymes and enzymes leaked from lysed cells, enzymes attached to dead cells and cell debris, and enzymes adsorbed to clay minerals or associated with humic colloids. Thus the measured activity of a soil suspension will be due to enzymes from a variety of sources, existing in perhaps several states in the soil, and operating under a range of microenvironmental conditions, all of which will affect the kinetics of the reaction catalysed.

Skujins[127] coined the term abiontic enzymes to describe all enzymes in soil, exclusive of those within live cells. Some enzymes, eg. proteinases, nucleases, when active against externally added substrates of high molecular weight, may be considered reasonably to be functioning abiontically. Others, eg. nitrogenase, probably dehydrogenases, because of their mechanisms of action, are likely to exist in active forms in soils only when they are present in live cells. Such simple distinctions are deduced from knowledge of the nature of the reactions catalysed and not from the establishment of suitably selective assay conditions. In practice, enzyme assays of whole soil suspensions cannot be made sufficiently selective to distinguish the activities of live, non-proliferating cells in soils from those of abiontic enzymes. Even techniques designed to suppress microbial growth during the assay period, and hence increased activity of proliferating cells, have been questioned. McGarity and Myers[89] found that toluene, added to soils to suppress microbial growth, actually decreased urea-hydrolysing activities of soils. They considered that in the presence of toluene, activities assayed were due to exocellular ureases adsorbed to soil colloids, whereas in the absence of toluene measured activities included those of ureolytic microorganisms. Also measurements of changing kinetic properties of soil ureases during periods of active microbial growth have formed the basis of a technique to apportion urea-hydrolysing activities to the actions of exocellular and cell-bound ureases[104]. Both approaches involve assumptions; other interpretations are possible (see Section 2.1).

The persistence of some enzymic activities in soils, and the possibility that active abiontic enzymes may be stabilized by combination with soil colloids, has led a minority of workers to attempt the characterization of enzymes in soil extracts and in other fractions. Elucidation of the mechanisms by which abiontic enzymes are stabilized in soil may be important in the wider context of understanding the processes which confer biological resistance on soil organic N.

During the past twenty years, many reviews[17,37,58,64,72,93,111,124,127 128,152,153] have detailed the activities and properties of soil enzymes. A book devoted entirely to the subject was published in 1978 and included several chapters which dealt with the broad range of enzymes in soil[66] and with specific groups of enzymes[13,59,136]. This present review is necessarily selective. It emphasizes those studies published since 1978 and those concerned with the persistence and stability of enzymes in soil and the properties of extracted enzymes.

2. RANGE AND ASSAY OF SOIL ENZYMES

2.1. Range

Soil enzymes most frequently studied are oxidoreductases (especially dehydrogenases, catalase and peroxidase) and hydrolases (especially invertase, proteinases, phosphatases and urease). A few studies measure activities attributable to transferases and lyases; no investigations of isomerases and ligases have been reported (Table 1).

Because most soil enzyme studies deal with crude soil suspensions, or at best partially-purified soil extracts, care must be exercised in assigning an activity to the action of a particular enzyme. This note of caution is perhaps justified by considering the conversion in soils of urea to NH_4^+ and CO_2. Urea is an important nitrogenous fertilizer and is a major constituent of the urine of grazing animals, and studies of urea hydrolysis have dominated the soil enzyme literature. It is widely assumed that urea hydrolysis is catalysed by urease (urea amidohydrolase, EC 3.5.1.5). Studies have compared the kinetics and inhibition of urea hydrolysis by soils and by purified ureases, mainly from jack bean, *Canavalia ensiformis*. Jack bean urease catalyses urea hydrolysis by a pathway in which carbamate is an intermediate.

Table 1. Range of enzymes assayed in soils.

Group and Subgroup No.	Enzyme
Oxidoreductases	
1.1	dehydrogenases, glucose oxidase
1.2	aldehyde oxidase
1.7	urate oxidase
1.10	catechol oxidase, p-diphenol oxidase, ascorbate oxidase
1.11	catalase, peroxidase
1.12	hydrogenase
1.14	phenol o-hydroxylase
Transferases	
2.4	dextransucrase, levan sucrase
2.6	aminotransferase
2.8	rhodanese
Hydrolases	
3.1	carboxylesterase, arylesterase, lipase, phosphatase (mono- and diester), nuclease, nucleotidase, phytase, arylsulphatase
3.2	amylase, cellulase, laminarinase, inulase, xylanase, dextranase, levanase, polygalacturonase, α-glucosidase, β-glucosidase, α-galactosidase, β-galactosidase, invertase
3.4	proteinase, peptidase
3.5	asparaginase, glutaminase, amidase, urease
3.6	inorganic pyrophosphatase, polymetaphosphatase, adenosine triphosphatase
Lyases	
4.1	aspartase decarboxylase, glutamate decarboxylase, aromatic amino acid decarboxylase

$$O = C\begin{array}{c} NH_2 \\ NH_2 \end{array} \xrightarrow[\ \ H_2O \quad\quad NH_3\ \ ]{} O = C\begin{array}{c} OH \\ NH_2 \end{array} \xrightarrow[\ \ H_2O\ \ ]{} H_2CO_3 + NH_3$$

An alternative mechanism of urea hydrolysis, also involving a two-step reaction sequence, has been demonstrated in yeast[114], algae[53,74,123] and some fungi[125]. Urea is first carboxylated to allophanate by urea carboxylase (urea: CO_2 ligase, EC 6.3.4.6), a reaction requiring ATP. Allophanate is then hydrolysed by allophanate hydrolase (allophanate amidohydrolase, EC 3.5.1.13) to CO_2 and NH_3.[155,156]

$$O = C \underset{NH_2}{\overset{NH_2}{\diagup}} + HCO_3^- + H^+ \xrightarrow[\text{ATP}]{\text{ADP} + P_{in}} O = C \underset{NH_2}{\overset{NH\diagdown COOH}{\diagup}}$$

$$\xrightarrow{H_2O} 2CO_2 + 2NH_3$$

The requirement for ATP in the ligase-catalysed reaction suggests that if in soils urea were to be hydrolysed partly by the allophanate pathway, then this reaction would take place within live cells only (ATP is unlikely to persist in dead cells or to remain free in soil[55]). Thus it is conceivable that the decreases in urea-hydrolysing activities of soils which frequently have been observed following toluene addition may have resulted from cell death, loss of ATP and hence elimination of the allophanate pathway, rather than from the destruction of urease activities of live cells.

As yet there is no direct evidence that in soils urea is hydrolysed even in part via the allophanate pathway. Nevertheless there is a risk in assuming that urea-hydrolysing activities of soils are due entirely to the action of soil ureases. Soils differ appreciably in the stability of their urea-hydrolysing enzymes following incubation or storage; and in their activity responses to air-drying, or to toluene addition, or to irradiation to induce soil sterilization. Such treatments may actually increase rather than decrease activities in some soils. Clearly the situation in unfractionated soils is complex even for urea hydrolysis, and may involve different enzymes and enzymic mechanisms as well as enzymes acting in different microenvironments.

Many enzymic activities in soils can be assigned at the group or subgroup level only. For example dehydrogenases, which in soils may act exclusively within intact cells, are assayed by measuring rates of formation of a reduced dye which can be extracted and estimated spectrophotometrically. In some assays organic substrates eg. succinate, sugars, are added as part of the incubation mixture and their oxidation may contribute to the formation of reduced dye. In other cases, total reliance is placed on the oxidation of endogenous substrates, of unknown chemical nature and concentrations.

As a second example, proteinase activities in soils have been measured using proteins, protein and peptide derivatives, and substituted amides and esters as assay substrates. Ladd and Butler[70] tested a range of N-benzyloxycarbonyl (Z) dipeptide derivatives, including those with amino acid residues with either aromatic, acidic, or non-polar aliphatic (long or short) side chains. Soils differed markedly in their activities towards a given substrate but were consistent in their preferential hydrolysis of those anionic dipeptide derivatives containing amino acid residues with hydrophobic side chains. Z-phenylalanyl leucine (ZPL) was hydrolysed most rapidly by all soils. The soils displayed a specificity of action similar to that of chymotrypsin and carboxypeptidase A. The soils also hydrolysed the cationic substrate N-benzoyl L-arginine amide (BAA). However, their rates of hydrolysis of this substrate, relative to their hydrolysis of the Z-dipeptide derivatives, varied with each soil. Relative activities towards BAA and ZPL ranged from 0.2:1 to 15:1 for the topsoils tested[70,72]. It seems clear that soils contain different amounts of proteinases with different substrate preferences.

The cationic substrate BAA is hydrolysed by proteinases purified from plant, animal and microbial sources. Of those listed by Barman[6] for which rate constants (k_o) are available papain was the most active; 11.0 molecules of BAA were hydrolysed per second at 38°C, pH 6 per molecule of enzyme. This is equivalent to 11.1×10^3 µ moles of NH_4^+ released per hour per mg of enzyme N. Trypsin had a k_o of 0.18 molecules of BAA hydrolysed per second at 25°C, pH 7.8 per molecule of enzyme, corresponding to 175 µ moles of NH_4^+ released per hour per mg of trypsin N. Ladd and Butler[70] showed that BAA hydrolysis by soils was highly correlated with soil organic C and N contents. Soil activities for a range of Australian and New Hebrides soils approximated 1.5 µ moles of NH_4^+ released per hour at 40°C, pH8.1 per mg of soil N. If the enzymes hydrolysing BAA in soils were of the same specific activity as that of papain solutions, then they accounted for approximately 0.01% of the soil N; and for about 0.4% of the soil N if they were of the same specific activities as that of trypsin solutions, allowing for a doubling of specific activities between 25° and 40°C. Given that only 30-45% of soil N is represented by amino acid N after hot acid hydrolysis[139] and that this may be derived in part from peptides rather than proteins in soil, the enzyme studies suggest that BAA is hydrolysed by soil proteinases (peptidases) with considerably higher mean catalytic activities than that of pure trypsin.

By the mid to late seventies[58,66,127] the activities of approximately fifty enzymes purportedly had been demonstrated in soils. In recent years the list has been extended to include amidase[43], phosphodiesterase [15,39,54], adenosine triphosphatase[47], phenol o-hydroxylase[154], hydrogenase[32] and nitrogenase[61]. In addition new assays have been reported for enzymes described previously eg. inorganic pyrophosphatase[34], urease[82], 1,

3-β-glucanase[75], esterases[122] (using fluorescein diacetate as substrate), and proteinases[24] (using an azo dye-protein as substrate). Short-term rates of denitrification in soils have been measured using slurries or moist crumbs. The assays included acetylene as an inhibitor of N_2O reduction[130].

2.2 Assay

Individual enzymes or classes of enzymes in soils have been assayed under a variety of conditions, thus limiting the scope for comparing activities reported from different laboratories. Galstyan[47] recently has described assays for five enzymes considered important for assessment of soil biological activity and has called for a standardization of methods. Roberge[113] and Burns[19] have reviewed aspects of methodology regarding pH control, choice of buffer if any, duration of assay, substrate concentration, and soil sterilization.

Valid measurements of the activities of soil enzymes accumulated at the commencement of assay require that both enzyme production by growing populations of microorganisms and assimilation of reaction products are excluded during the assay period. Further, methods adopted to achieve these ends should not influence the activities of accumulated enzymes. Soils have been sterilized (irradiated) or treated with bacteriostatic agents (toluene). Toluene is still widely used despite the observations that (i) some fungi and bacteria can utilize toluene as a carbon source[56,128] (ii) toluene does not sterilize soils nor, in some treatments, eliminate completely dehydrogenase or respiratory activities, and thus there may remain the potential for energy generation and enzyme synthesis (iii) toluene may affect the activities of accumulated enzymes by modes of action which cannot be unequivocally interpreted. For example, the increases in urea-hydrolysing activities which sometimes have been observed following toluene addition to soils may have been due to ureases released from plasmolysed microbial cells[128]. More usually toluene addition has decreased urea-hydrolysing activities of soils, possibly by destruction of cell-bound ureases or urea carboxylases, or by inhibition of exocellular ureases (Thente[149] found that toluene decreased the activity of jack bean urease). The effects of toluene have varied with soils, conditions of assay, and enzyme assayed. Dehydrogenases are sensitive to toluene addition, and complete or very extensive losses of activity in soils have been interpreted as evidence for their occurrence in live cells only.

An alternative treatment to the use of bacteriostatic agents is to irradiate soils at sterilizing or lower dosage levels. This approach however, suffers from the same objections raised over the use of toluene in that soil enzyme activities can be influenced by the treatment, the responses differing with soils and radiation dose. The overall effects are difficult to interpret. Clearly it is preferable to avoid such treatments if possible, and to use assays sufficiently sensitive to reduce assay times to less than a few hours, thus minimizing opportunities for enzyme synthesis, microbial growth and product assimilation. Reaction rates usually remain constant for several hours. Nevertheless Ahmed et al.[1] found that the amounts of ATP extractable from soil increased markedly within less than an hour after remoistening air-dried soils. The increases may have been due partly to more effective extraction of residual ATP and partly to synthesis of ATP, the latter being indicative of renewed activities of energy-generating systems and perhaps the potential for de novo enzyme synthesis.

As in all enzyme assays, concentrations of substrate in the soil reaction mixtures should remain sufficiently high to maintain zero order kinetics. This may require initial concentrations to be as high as 100 times those of the Km values of the reactions. Initial velocities of reaction should form the basis of comparisons and these are most easily obtained if it can be demonstrated that reaction rates remain constant over the assay period.

The choice of substrate is often arbitrary and may involve a compound that does not occur naturally in soil. To meet kinetic requirements, the substrate is supplied in concentrations higher than would occur normally in soils. Further, other assay conditions eg. temperature, moisture content, will also depart from the natural situation. It seems of doubtful merit therefore to debate whether or not soil suspensions should be buffered, perhaps at a pH different from the measured pHs of soil-water suspensions, on the grounds that to do so may alter their activities in ways that they may no longer reflect the activities under natural conditions. The impact of such an argument is further diminished when it is considered that the pH of the bulk soil may differ considerably from that at the sites of enzymic action, whether within cells or tissues or within the microenvironments of enzymes sorbed to soil surfaces. Clearly the pH of the assay mixture and the nature and concentration of the buffer do affect reaction rates. These, like other factors, should be chosen to obtain controlled conditions of assay and to allow expression of maximal potential reaction rates. In practice, soil enzymes are assayed under conditions where substrate concentrations do not limit activities, where the choice of buffer is based on that giving maximal activity of 2 or 3 tested, where pH is optimal but where assay temperatures are less than optimal. The latter choice is often for convenience, but to some extent is based on the precision of assay, especially in some cases where higher temperatures significantly increase non-enzymic activities. It is of course hoped that potential reaction rates may directly reflect soil biological activities (see section 3). If however the purpose of an assay is to obtain measurements of soil activities closely related to those pertaining naturally, then reasonably, conditions should allow for the impact of biotic factors (growth and turnover of microbial biomass, synthesis and turnover of enzymes) as well as abiotic factors.

In devising assays for soil enzymes it is necessary to ensure that the reaction catalysed is biochemically mediated, and that the rates of reaction, whether based on substrate disappearance or product formation, are not influenced by their participation in other reactions, biological or physico-chemical. Enzyme activities may also be influenced by the time and conditions of soil storage between sampling from the field and assay. The basic requirements for establishing soil enzyme assays, the advantages and disadvantages of selected assay conditions, and the responses of different enzymes to soil pretreatments have been reviewed in detail elsewhere[13,19,113,127,128,149] etc.

The basis for assay and properties of a broad range of enzymes was reviewed[66] in 1978 and are not repeated here. The following specific comments are directed at more recent reports.

Frankenberger and Tabatabai[43] have described an assay for soil *amidases*, which catalyse the hydrolysis of aliphatic amides to carboxylic acids and ammonia. Their assay requires toluene-treated soils to be incubated at 37°C with 0.05M substrate (formamide, acetamide, propionamide) for either 2 or 24h. The suspensions were buffered at pH 8.5; the choice of buffer influenced the amounts of NH_4^+ released, which formed the basis for estimating activities. The pH optimum for hydrolysis of all substrates was about pH 8.5; the optimum temperature was about 65°C. Reaction rates in the presence of toluene were linear both with time and with the amounts of soil when substrate concentrations were 0.05M. At this concentration, and above, the reactions exhibited zero-order kinetics for the recommended assay periods.

In the absence of toluene, soils exhibited exponentially increasing amidase activities in longer term incubations, including the 24h period required for assay of activities towards acetamide and propionamide. Toluene addition not only maintained a constancy of reaction rates but prevented increases in the numbers of soil microorganisms capable of utilizing amides as sole C and N sources[45]. The action of toluene was considered to be due to the prevention of amidase induction by the soil microflora. Nevertheless, even for short term (2h) assays of formamide hydrolysis the addition of toluene decreased amidase activities of all soils tested by more than 50%. Rates of formamide hydrolysis in the absence of toluene and throughout the 2h assay were not available to permit an assessment of the constancy of reaction rates. The question whether the effect of toluene on formamide hydrolysis is due to the elimination of some *de novo* amidase synthesis in the 2h period, or solely to inhibition or destruction of existing soil amidases, is not resolved. Whatever the explanation, the effect of toluene illustrates the difficulties in using this agent to establish satisfactory conditions of assay for soil amidases and some other soil enzymes.

Activation energies and temperature coefficients for soil amidase action against the three substrates were determined for a range of soils. Q_{10} values for temperatures between 10 and 60°C averaged about 1.7 for formamide and acetamide, and 1.4 for propionamide. Amidase activities decreased with depth of soil profile, and were correlated with topsoil organic C and N contents, but not soil pH.

Phenol o-hydroxylase is an NADPH-dependent flavin mono-oxygenase which catalyses the oxidation of phenols to *o*-diphenols. Wainwright[154] described a soil enzyme assay based on the colorimetric measurement of substrate (phenol) disappearance. Soils were incubated for 1h with phenol (0.01M) in a citrate -phosphate buffer pH 4.0 (optimum) at 37°C (temperature optimum 40-45°C). Under these conditions soil activities were linear with time and the rates of reaction were proportional to the amounts of soil. Soil activities were stimulated by Cu^{2+} (purified phenol *o*-hydroxylase has a Cu^{2+} requirement).

An assay[54] for *phosphodiesterase* activity in soils based on the colorimetric measurements of *p*-nitrophenol released from *bis*-(*p*-nitrophenyl) phosphate has been developed further by Browman and Tabatabai[15]. Toluene-treated soils were incubated with substrate (0.001M) at 37°C for 1h. The suspensions were buffered at pH 8.0 with THAM. The reagents used to terminate the reaction and to extract the released *p*-nitrophenol did not themselves hydrolyse unreacted substrate. Soils were most active as suspensions buffered in THAM than in other buffer solutions. The pH optimum was 8.0, higher than that (pH 6.0) of phosphodiesterases extracted from soil[52], and more distinctive than the pH optima obtained for another suite of soil suspensions by Ishii and Hayano[54]. The temperature optimum for the reaction in the soils studied by Browman and Tabatabai[15] was consistently at about 70°C for a range of activities. The enzymes in soil appeared to be less affected by heat denaturation than enzymes in soil extracts[52]. Toluene was added to soils to prevent enzyme synthesis during the assay despite the brevity of the assay period, and actually increased phosphodiesterase activities by about 26-29% in all soils[15].

Under the described conditions, the rates of release of p-nitrophenol obeyed zero order kinetics until about 50% of substrate was hydrolysed. Q_{10} values averaged 1.7 for six soils between 10° and 60°C. Km and Vmax values ranged with the soil and method of estimation (Table 2). Phosphodiesterase activity was inhibited competitively by orthophosphate. Soil activity was also inhibited by chelating agents (EDTA, citrate), and thus was consistent with studies with purified phosphodiesterases. Phosphodiesterase activities were significantly correlated with soil organic C contents. Earlier studies by Eivazi and Tabatabai[39] compared the kinetics and properties of phosphomonoesterases, phosphodiesterases and "phosphotriesterases" using respectively p-nitrophenyl phosphate and the *bis*- and *tris*- p-nitro derivatives as assay substrates. The soil phosphatases varied in the nature of their response to air-drying, steam sterilization and toluene treatments of soils.

Dick and Tabatabai[34] have described an assay for *inorganic pyrophosphatase* activity in soils based on a specific colorimetric determination of extracted orthophosphate after its release from inorganic pyrophosphate (0.05M) incubated with soils for 5h at 37°C. Soil suspensions were buffered at pH 8.0. In the absence of toluene, the amounts of orthophosphate released g^{-1} soil were proportional to incubation time (up to 7h), and for a given time to the amounts of soil added. Activities were unaffected by toluene addition to soils and toluene was omitted from the recommended assay procedure.

Activation energies and kinetic constants (Km, Vmax) were calculated for inorganic pyrophosphatases in several soils. Some residual activity remained after steam sterilization and formaldehyde treatment of soils. Enzyme activities were significantly correlated with organic C of soil profiles, and with organic C and clay contents of acidic top soils[146].

Conrad and Seiler[32] have shown that moist soils utilized atmospheric H_2 at rates which followed first order kinetics. Irrespective of the fate of the fixed H_2, the initial reaction was assumed to be catalysed by soil *hydrogenases*. Activities were abolished by autoclaving soils, but were only partly removed by fumigation with $CHCl_3$, NH_3 or acetone, by UV irradiation, by addition of toluene, antibiotics or NaCN and NaN_3. Activities declined with prolonged storage of dried soils. The utilization of H_2 by soils was thought to be due to the activities of both live cells and abiontic enzymes.

Lethbridge *et al.*[75] have described an assay for *laminarinase* (1,3-β-glucanase) by measuring laminarin hydrolysis to glucose by shaken soil suspensions buffered at the pH optimum (pH 5.4) and incubated at 25°C for 17h. Sodium azide was added to prevent microbial growth and glucose metabolism. Rates of glucose formation were linear with time. Michaelis-Mentin kinetics were applied to obtain Km and Vmax values for enzymes of a soil suspension (Table 2). The Q_{10} value in the range 4-50°C was 1.93, the activation energy 49kJ mole^{-1}.

Activities of soil *proteinases* have been determined by measuring spectrophotometrically the release of a soluble red azo dye from Azocoll, an insoluble protein containing adsorbed dye and marketed as a substrate for a range of proteinases[24]. The assay required that unshaken suspensions of soil in phosphate buffer, pH 7, were incubated with substrate for 5h at 37°C. The kinetics of the hydrolysis were not established. Autoclaved soils were inactive. Soil proteinases were increasingly inhibited by $HgCl_2$ in final concentrations ranging from 50 to 1000 µg ml^{-1}.

A wide range of bacteria and fungi, and some protozoan, algal and mammalian cells have the ability to hydrolyse fluorescein diacetate. Recently Schnurer and Rosswall[122] have used this substrate to develop an assay for *esterase* activity in soil and straw based on the spectrophotometric determination of extracted fluorescein. The amounts formed increased linearly with time (up to 3h). The rates of hydrolysis were directly proportional to the amounts of soil used, to the amounts of fungal mycelium (*Fusarium culmorum*) or bacterial biomass (*Pseudomonas denitrificans*) in buffered (pH 7.6) pure culture suspensions, and, in the case of fungi, to the amounts added to autoclaved soil, which alone was inactive. Esterase activities correlated with respiratory (O_2 uptake) activities for samples from different depths in the soil profile.

3. ACTIVITY OF SOIL ENZYMES
3.1 Effect of soil types and management, and environmental conditions
    3.1.1. Kinetic constants. The activities of many enzymes in soils have been shown to follow Michaelis-Menten kinetics. Values for both K_m and

Vmax vary with soil, with soil physical fraction, and with the changing distribution of enzymes in soil, eg. when accompanying turnover of microbial biomass following organic amendments. Values are also influenced by assay conditions, eg. choice of substrate and buffer, the use of shaken or unshaken soil suspensions. Table 2 shows the ranges of values for kinetic constants calculated for different enzymes in soils and soil fractions.

The utilization of atmospheric H_2 catalysed by hydrogenases of moist, unbuffered soils exhibited Km values of 24 to 32 μl H_2 l^{-1} for two of the soils tested[32]. Km values were independent of soil water contents but were affected by exposing soils to progressively decreasing concentrations of H_2; for a sand, Km values rose from 32 to 60 μl H_2 l^{-1}. The Km values for soil hydrogenases corresponded to H_2 concentrations in water 10 to 1000 times less than Km values for soluble hydrogenases. However, the kinetic constants of soil hydrogenases were established from rates obtained with relatively low H_2 concentrations (μl per litre). Further, the electron acceptors, although unknown in the soil systems, were considered highly likely to be in lower concentrations than those employed for *in vitro* measurements of reaction rates of soluble hydrogenases (decreasing the concentrations of an electron acceptor decreased the Km values of a soluble hydrogenase).

Values for Vmax were dependent upon soil moisture levels and were achieved at substrate concentrations of about 200 μl H_2 l^{-1}. For a sand, the Vmax was approximately doubled when the water content was decreased from 16.3 to 11.3%. The first-order rate constants for H_2 utilization in soils were directly proportional to the amounts of soil used. Decreased activities at increasing soil water contents were considered due more importantly to decreased concentrations of suitable electron acceptors rather than to effects of anaerobiosis or to effects on H_2 diffusion.

Shaking soil suspensions during assay decreased Km values and increased Vmax values for soil ureases, phosphatases and arylsulphatases[141,144]. Frankenberger and Tabatabai[44] showed that Km and Vmax values for soil amidases, active against formamide, acetamide and propionamide, ranged with soil type as well as assay substrate. Km values also varied with pH of assay, being lowest at the pH optimum (pH 8.5).

For soil ureases, Km values decreased as the proportions of soil activity due to the microbial biomass increased following substrate addition, and later increased as the contributions from intact cells diminished. Km values for cellular and adsorbed ureases of a soil were estimated to be 57 mM and 252 mM respectively[104,105].

Table 2. Kinetic constants of soil enzymes.

Enzyme	Soil or soil fraction, buffer	Km*	Vmax**	Reference
'Hydrogenase'	Moist soil, unbuffered	24-60	2-4	32
Phosphomonoesterase				
(a) acid phosphatase	Soil suspension, maleate buffer, pH 6.9	0.32	213	106
	" " , universal buffer, pH 6.5	1.1 -3.4	200-625	39
	" " , " "	0.9 -1.8	233-735	144
(b) alkaline phosphatase	" " , unbuffered	0.35-5.4	-	26
	" " , universal buffer, pH 10	0.44-4.9	124-588	39
Phosphodiesterase	" " , tris (hyroxymethyl) amino methane	0.25-1.3	46-127	39
	" " (THAM) buffer, pH 8.0	1.3 -2.0	52-530	15
Arylsulphatase	" " , acetate buffer, pH 5.8	5.5	131	106
	" " , " "	2.0 -3.3	133-385	144
Laminarinase	" " , maleate buffer, pH 5.4	0.21	0.40	75
Amidase	" " , THAM buffer, pH 8.5, formamide	6.7 -17.9	138-438	43
	" " , " " " , acetamide	4.0 -5.1	13-43	
	" " , " " " , propionamide	10.1 -20.2	35-105	
Urease	moist soil, unbuffered	213	19.9	104,105
	soil microorganisms	57	2.6	
	adsorbed, exocellular soil colloids	252	21.8	
	soil suspension, THAM buffer, pH 9.0	1.1 -3.4	7.3-87	141
	clay fraction, " "	2.0 -4.9	4.3-8.5	
	silt fraction, " "	1.1 -4.1	7.0-4.3	
	sand fraction, " "	3.8 -1.3	9.5-1.3	
	soil suspension, " " , pH 7.0	52.3	-	107
	" " , phosphate buffer, pH 7.0	62.5		
	" " , THAM buffer, pH 7.2	10.4 -22.2	0.8-2.8	11
	" " , " "	2.0 -32.0	3.5-10.0	10
Inorganic pyrophosphatase	" " , phosphate buffer, pH 7.0	31.8 -62.3	-	28
	" " , pyrophosphate buffer, pH 8.0	20-51	130-830	34

* Values for Km expressed as mM solutions, except for (a) hydrogenase where expressed as μl hydrogen l^{-1} and (b) laminarinase where expressed as mg laminarin ml^{-1}.

** Values for Vmax expressed as follows (a) hydrogenase, μl H_2 consumed $100g^{-1}$ moist soil min^{-1} (b) acid and alkaline phosphatases, phosphodiesterase, arylsulphatase, μg p-nitrophenol released g^{-1} soil h^{-1} (c) laminarinase, $\mu moles$ glucose formed g^{-1} soil h^{-1} (d) amidase, μg NH_4^+-N released g^{-1} soil $24h^{-1}$ (e) urease, μg urea N hydrolysed g^{-1} soil (or soil fraction) h^{-1} (f) inorganic pyrophosphatase, μg inorganic P released g^{-1} soil $5h^{-1}$.

Km and Vmax values for ureases of different particle size fractions of soils differed respectively from each other and from those of unfractionated soil[141]. Generally the Km values of fractions were greater than those of the respective unfractionated soils, but there was no trend with particle size. Vmax values of all fractions were considerably less than those of the unfractionated soils, indicative of urease destruction perhaps during the sonic vibration treatment used to disperse the soils before isolating the fractions. Differences in assay conditions and in the bases for determining urea hydrolysis may be as important in accounting for the large differences in Km values reported by different investigators as different soil properties.

3.1.2. <u>Soil properties</u>. Soil enzyme activities decrease with depth of sampling, and vary with season and with the type of vegetative cover. Activities of topsoils are directly correlated with organic matter (C and N) contents, and for soils which have received no recent organic substrate amendments activities are poorly correlated with microbial numbers. Although not without exceptions, the pattern of results has been remarkably consistent, given that the enzymes assayed have ranged widely, as have the type and geographical spread of the soils used, and also the assay conditions employed by different investigators. Soil phosphatase activities have correlated positively with soil organic P contents, but negatively with soil available P[38]. Soil proteinase activities have correlated with organic N contents, but not with net rates of inorganic N formation[117]. Soil sulphatase activities have correlated with organic S contents[33,48] but not necessarily with rates of sulphur mineralized[63,145].

Enzyme activities also tend to be positively correlated with clay contents of soils (and cation-exchange capacities and surface areas). Correlations with soil pH have not been achieved consistently, even for the same enzyme assayed.

3.1.3. <u>Management</u>. The majority of more recent studies have been devoted to measuring the effects of various management practices (choice of vegetation, mulching, tillage, prescribed burning, application of fertilizers and pesticides, etc. on the activities of selected suites of enzymes. In some studies the effects of the nature of the plant cover have been offered as indirect evidence for the persistence of plant enzymes in soils (see section 3.2.1).

Tillage decreased enzymic activities of topsoils[36,60]. Phosphatase and dehydrogenase activities of a range of surface soils which had received long term conventional tillage and no tillage treatments, were positively related to organic C and N and water contents, and to the populations of several classes of microorganisms. The trend for greater enzymic activities of the no-tilled surface soils was reversed for subsoil samples[36]. Changes in the relative enzymic activities in the profiles of tilled and no-tilled plots may be a consequence of the large relative changes in the populations of aerobic and facultative anaerobic microorganisms. The biochemical environments of no-till soils were indicated to be less oxidative than those of soil receiving conventional tillage practices.

The effects of flooding of soils on enzymic activities varied with the enzyme assayed. For example, flooding increased dehydrogenase activity but decreased both invertase activity[30], and also the aerobic oxidative activities against a range of substrates, aromatic ring metabolism being the most affected[148]. Submergence of soils increased phosphatase activity and bacterial populations, each of which were negatively correlated with decreasing hydraulic conductivities[46]. Smith and Tiedje[131] have followed changes in denitrification rates in oxygen depleted soils. They recognized two phases of activity in response to reduce aeration. Initial activities were attributed to the existing soil microflora and were uninfluenced by chloramphenicol or organic amendment (1-3h). During the second phase (4-8h), activities increased due to the derepression of denitrifying enzyme synthesis but without growth of denitrifying organisms.

3.1.4. Fertilizers. The effects of inorganic fertilizers on soil enzyme activities vary with soil and with the enzyme assayed. In long term experiments the effects may be indirect, due perhaps to changes in plant yields, and thus in soil moisture regimes, soil solution nutrient concentrations and in the return to soil of organic residues, etc. Spiers and McGill[138] showed that P fertilization over 5 years decreased phosphatase activities of a soil initially of high organic matter content and of high phosphatase activity, but increased the phosphatase activity of a soil initially of low organic matter content and activity. In the latter soil increased activity was attributed to a greater return of plant residues and thus increased potential for microbial syntheses. Orthophosphate (0.55 M) in assay mixtures inhibited soil phosphatase activities. Orthophosphate incubated with soils amended with glucose completely repressed the synthesis of phosphatases by the newly-formed microbial populations and slightly inhibited phosphatases already present. Clearly P concentrations of the soil

192

solution, the availability of energy sources, and the accumulation of soil organic matter will be factors in determining the levels of soil phosphatase activities and the response to added P fertilizers. Applications of inorganic fertilizers without P generally has increased soil phosphatase activity, although the inclusion of trace elements in an NPK fertilizer applied to plots over eight years reduced phosphatase activity[79].

3.1.5. Metal ions. The relative effectiveness of trace elements, including heavy metal ions, as inhibitors of enzyme activities, has varied with the soil, the concentration and form of the added element, and the enzyme assayed. Enzymes tested have included urease[12,142] arylsulphatase[2,157], phosphatase[157], dehydrogenase[35,157], amylase[31], α-glucosidase[31], and others. Environmental studies have demonstrated negative correlations between enzymic activities of soils and their contents of heavy metals, especially Cu^{2+}, Pb^{2+} and Zn^{2+}[78,80,150,151]etc.

3.1.6. Pesticides. The potential for environmental pollutants and agrochemicals to influence the rates of soil biochemical processes and hence plant growth has provided an impetus for studies of their effects on soil enzyme activities. Studies of soil pesticides dominate, and have been the subject of recent reviews[29,153]. Wainwright[153] concluded that with some exceptions eg. fumigants, pesticides applied at normal field rates do not seriously affect biochemical activities of soils. Recent evidence obtained with a variety of compounds tested against a range of enzymes continues to support this general conclusion. Soil enzyme activities may be both inhibited and enhanced by pesticides. In many cases the effects are temporary and within several months activities may return to levels similar to those of untreated soils. The effects may be indirect. Long term effects of herbicides may be due more to the removal of vegetative cover and the changes in plant residue return to soils than to a direct effects on enzyme catalysis.

3.1.7. Urease inhibitors. Controlling the rates of transformation of urea in soils is of great practical importance because of the potential for large gaseous losses of fertilizer nitrogen due to ammonia volatilization and for toxic effects on germinating crops. The rationale for the successful use of urease inhibitors and the results obtained from testing a very wide range of compounds have been comprehensively reviewed by Mulvaney and Bremner[13,93]. Some of the more effective inhibitors of urea hydrolysis in soils also are inhibitors of purified jack bean urease, eg. heavy metal ions[12,142] substituted urea derivatives[27,28], and 1:4 benzoquinones and related phenols[12,16,51,91]. Of the latter methyl- and halogen- substituted p-benzoquinones were the most effective. Heavy metal ions and quinones probably inhibited by reacting with enzyme SH groups shown to be essential for urease activity, whereas inhibition by urea derivatives was of the mixed type. Inhibition by p-benzoquinone and hydroquinone was inversely related to the organic C, organic N, clay and silt contents of surface soils[92].

Other effective inhibitors of urea-hydrolysing activities in soils have included hydroxamates[198,109], heterocyclic mercaptans and disulphides[51], xanthates[5], and phosphoric acid phenyl ester diamide[81]. EDTA inhibits jack bean urease by removing essential Ni^{2+} ions but this compound, and other chelating agents, had little if any effect on urea hydrolysis in soil[119].

3.2. Persistence and stability of soil enzymes

Many factors influence the turnover of enzymes in soils. They include (a) direct contributions of plant enzymes to soil, (b) growth conditions and the availability of energy sources for de novo syntheses by a variety of soil animals and microorganisms, and (c) the availability of substrate for specific induction of enzymes by particular organisms. They include too the properties of the enzymes themselves, whether they are free or complexed, and the conditions pertaining in the microenvironments in which enzymes are located in soils, thus affording varying measures of protection against denaturation or biological degradation. Undoubtedly the persistence of enzyme activities in soils is due in part to enzymes which turnover relatively rapidly, their renewal being dependent essentially upon a continuing or spasmodic return to soil of energy-rich plant residues. By contrast, some of the enzymic activities of soils are due to stable or stabilized enzymes. For example, urease activity has been detected in stored, geologically preserved soils, carbon dated to be about 9000 years old[129].

3.2.1. Origin. Evidence for the persistence of plant enzymes in soils is mainly indirect. Many studies have related the activities of a range of soil enzymes to the nature of the plant cover. For example, Pancholy and Rice[102,103] showed that the activities in soils of some carbohydrases decreased, whereas dehydrogenase and urease activities increased, during stages of revegetation with prairie grass, and with oak, and with oak-pine forests. The types of plant residues rather than the total residue return were considered to be important in determining the activity gradients of the enzymes assayed. Other studies have contrasted the changes in enzyme activities of planted and unplanted soils. Speir et al.[135] showed that the urea-hydrolysing activities of moist, unplanted soils progressively declined during a 5 month period, whereas soils planted to ryegrass either increased their activities or exhibited lower rates of decrease. Similar trends were established for soil sulphatases and proteinases. Other approaches have involved comparisons of enzyme activities of soils and soil fractions (extracts, light-fraction material), with those of the associated vegetation[65,115,116,133,134]. Such experiments implicate plant enzymes as contributors to soil enzyme activities but do not eliminate an alternative secondary role of the plant residues in promoting conditions in soil for microbial growth and hence for enzyme synthesis and degradation.

The most direct demonstration of the persistence of plant enzymes in soil has been achieved by examining the soil ultrastructure using a combination of histochemical and electron microscopic techniques[41]. Foster, as quoted by Ladd[66], examined the light fraction of a sandy topsoil, sampled 11 months after the last accession of living plant tissue, and found phosphatase activity to be associated with intact plant cell walls and plant cell wall fragments, as well as with amorphous organic material of unknown origin. The method is not quantitative, and is restricted in its use in soils to a few enzymes only. There is a general lack of suitable techniques which fix, and render electron dense, reaction products at the sites of their formation, without also denaturing the enzymes or inhibiting their activities. The technique has demonstrated that plant phosphatases can persist for nearly a year at least in soil within decomposing plant tissue; it did not allow any assessment of the contribution of the plant phosphatases to the total phosphatase activity of the soil, or of the stability of the plant residue enzymes relative to phosphatases from other sources or in other locations.

Although soil enzyme activities do not consistently correlate with the numbers of microorganisms in soil, or with soil respiratory activities, the contributions from enzymes of microbial origin are considered to be of great importance, and are enhanced under conditions favouring microbial growth and turnover.

3.2.2. Locus. Soil exocellular enzymes active against, and therefore accessible to, externally added substrates of high molecular weight, may themselves be vulnerable to biological degradation[71,72]. Such enzymes are expected to turn over relatively rapidly, and thus their continued activity would be dependent on the availability of energy sources for *de novo* enzyme synthesis. A portion of the active enzymes, probably adsorbed to soil surfaces, may turn over more slowly having survived the action of destructive proteinases which themselves are subject to degradative reactions (including autodigestion) and inactivating adsorption reactions. A portion of the exoenzymes, through perhaps adsorption to soil colloids, or covalent bonding to resistant organic materials, or physical entrapment, may become protected from degradation by soil proteinases. It is likely that concomitantly the protected exocellular enzymes would be rendered inaccessible to their substrates of high molecular weight. Their activities towards low molecular weight substrates would be far less affected. In these cases, changes in activities due to enzyme protection may result from altered affinities of bound enzymes for their substrates, or from decreased rates of diffusion of substrates to the enzymes' active sites, but not from steric exclusion of substrates. To take a specific example, soil proteinases active towards both protein and peptide substrates are believed to be susceptible to degradation by other proteinases, and, if free in the soil solution, to autodigestion reactions. Any protection conferred on the proteinases may limit their substrate range to peptides only. Some mechanisms of protection from biological degradation may also enhance exocellular enzyme stability to physical denaturing agents.

The persistence of soil enzymes active against low molecular weight substrates only is less predictable in that reactions may be catalysed solely by intracellular enzymes, or partly by intracellular and partly by exocellular (abiontic) enzymes. Activities by exocellular enzymes may be less dependent on energy source availability because of contributions from accumulated, protected enzymes.

3.2.3. <u>Organic amendment</u>. Ladd and Paul[73] have followed changes in microbial enzyme activities accompanying glucose metabolism in soil (Fig.1). Enzymes assayed included (a) proteinases, presumed to have acted exclusively exocellularly against casein as substrate but not necessarily so against low molecular weight peptide and amide derivatives, (b) dehydrogenases, presumed to have acted exclusively intracellularly, and (c) phosphatase. The greatest relative changes occurred with casein hydrolysing activities; net activities remained low during a period of rapid microbial growth, but then increased rapidly during a period of net decline in bacterial numbers but of intense metabolism of microbial products. Caseinase activity fell markedly during further incubation as microbial substrates were depleted. Dehydrogenase activity increased with cell numbers, reaching a maximum during a period of

196

active metabolism of microbial products, then decreased more slowly than did caseinases and approximately in parallel with the decline in bacterial numbers. The behaviour of the caseinase and dehydrogenase enzymes were predictable for exocellular and intracellular enzymes respectively.

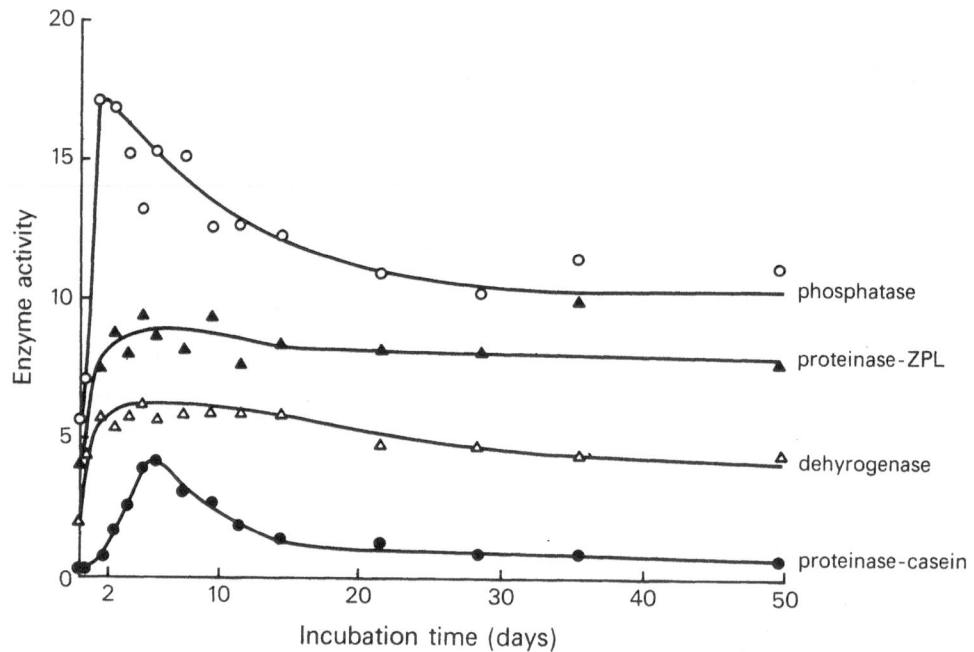

FIGURE 1. Enzyme activities of a soil incubated with glucose and NO_3^-. For phosphatase, proteinase (casein), and proteinase (ZPL) respectively, activities are expressed as μmoles of p-nitrophenol, tyrosine, and leucine released g^{-1} soil h^{-1}. For dehydrogenase, activities are expressed in absorbance units x15 (485 nm) g^{-1} soil h^{-1}. Conditions of assay have been described[73].

The behaviour of enzymes active against low molecular weight substrates was not so predictable. Net increases in the activities of phosphatases and of enzymes hydrolysing the peptide, benzyloxycarbonyl phenylalanyl leucine (ZPL) coincided with rapid increases in bacterial numbers. However, whereas phosphatase activities then declined rapidly coinciding with the initial rapid decline in cell numbers, and continued to decline at a rapid rate (similar to casein activities), ZPL- hydrolysing activities increased slightly to remain at a relatively high constant level for the remainder of the incubation period. Clearly the initial gains in activities of both types of enzyme were closely related to microbial growth. The persistence of ZPL-hydrolysing activities long after the decline in both bacterial numbers and metabolic activity suggested that either exocellular proteases may have become stabilized in the soil, retaining their activity against low molecular weight peptides, or that the proteases were acting intracellularly and were present in a stabilized section of the microbial biomass not revealed by the counting techniques. After 50 days incubation, neither cell numbers nor activities of any of the enzymes assayed had returned to the respective values before amendment of the soil.

Studies by Nannipieri et al.[99] confirmed that newly-formed caseinase activities were short lived in soils, and that the initial gains in phosphatase activities were the more closely linked to bacterial biomass accumulation, their activity peak preceding that of caseinase. By contrast, newly-formed phosphatases were more stable in their soil than in the soil used by Ladd and Paul[73]. Activities of newly-formed ureases also correlated with increases in bacterial biomass, the subsequent rate of urease decline being intermediate between those of proteinase and phosphatase[99].

Zantua and Bremner[159,160] also have demonstrated the rapid production and decline of urease activities in a range of soils accompanying the metabolism of added glucose or other organic materials (addition of urea itself did not induce urease activity). The initial urease activities of moist soils and the persistence of the increased urease activities varied with the soil. For most, urease activities had declined within 60 days of amendment to values equal to the activities of incubated unamended soils and equal to the respective initial activities. For such soils, urease activities declined no further on continued incubation. (Urease activities of the soils used by Zantua and Bremner[159] remained constant when moist unamended soils were incubated for several months, in contrast with activity decreases obtained by Speir et al.[135] and others). Further, Zantua and Bremner[160] showed that air drying of soils, followed by remoistening and incubation for several days, caused a partial loss of urease activity. Addition of glucose or of urease itself to these soils led to temporary increases in soil urease activities and then a return to the initial levels pertaining immediately before the air drying treatment. The results suggested that soils varied in their specific capacities to protect ureases from biological degradation and other destructive or destabilizing processes; and in the field moist soils of Zantua and Bremner the characteristic stable levels for maximal potential urease activity had been reached before any organic amendments or soil pretreatments were introduced.

198

By assuming a simple division of soil ureases into two groups, those associated direcly with microorganisms and those adsorbed to soil colloids, Paulson and Kurtz[104,105] apportioned activity tó the two sources based on an analysis of the changing populations of ureolytic microorganisms and of urease activities of a glucose-amended soil. About 10% of the initial urease activity of the soil at or near steady state conditions, was attributed to the intact microbial populations. The proportion rose to about two thirds of the total activity when the microbial population was maximal, following incubation of the soil with glucose.

3.2.4. Soil pretreatment. Apart from the many studies of activity responses following amendment of soils with organic materials, the effects of various soil pretreatments have been tested, instigated in part by the necessity to establish satisfactory procedures for handling and storage of soils before assay. Pretreatments usually have included air drying, and storage of soils for different times at different temperatures in the air dried or moist (sometimes frozen) states. The range of pretreatments has extended to lyophilization, autoclaving, γ-irradiation, incubation with added proteinases, and fumigation or incubation of soils with organic solvents.

Effects on activity have varied with pretreatment, enzyme assayed and soil. Holding field moist soils at or near 5°C generally caused little change in activities over many months and appears to be a satisfactory storage procedure for many enzymes. Air drying frequently has decreased soil enzyme activities, but in some cases (arylsulphatase[143], acid phosphatase[39], urease[137]) activities have been enhanced. The extent to which air drying affects the activities of enzymes of a given group or subgroup has varied with assay substrate, eg. proteinases active against gelatine were less affected than those hydrolysing casein, those hydrolysing peptide or amide derivatives were less affected than those hydrolysing proteins[66]. More importantly perhaps, the effects of air drying and of other soil pretreatments have varied for a given enzyme, assayed for a range of soils in different laboratories. Soil properties obviously have had large effects on the responses obtained. Assay conditions, and the time of sampling in relation to season and vegetation (and hence the proportions of an activity due to different enzymes or to enzymes in different organisms or exocellular locations in soils) may also be important in determining activity responses to various pretreatments. Many of the effects of soil pretreatments on soil enzyme activities, as reported prior to 1978, have been summarized elsewhere[18]. More recent reports deal with effects on catalase[112,147]

polyphenol oxidase[147], peroxidase[147], proteinases[112,137], urease[40,42,100,
112,137], phosphatases[40,42,106,137], sulphatase[106,137], β-glucosidase[75][158],
invertase[112,137], amylase[137], cellulase[137], xylanase[137], rhodanese[126],
dehydrogenase[40,112], inorganic pyrophosphatase[146] and hydrogenase[32].

3.2.5. <u>Stabilization mechanism(s)</u>. The stability of many enzymes in
soils has been amply demonstrated. The mechanism(s) by which they are
stabilized may involve enzyme adsorption to clays or to resistant complex
organic heterocondensates and polymers in a manner analogous to those
proposed for the stabilization of organic nitrogenous compounds of soil
organic matter. In the case of soil organic N, only a minor proportion is
stabilized within microbial and faunal biomass. In the case of some enzymes
similar proportions may prevail in soil under "steady state"
conditions[104,105], but clearly proportions will vary with the enzyme (eg.
intracellular dehydrogenases vs. exocellular proteinases) and, temporarily at
least, with the availability of energy sources for microbial growth. Also,
stabilization of enzymes exocellularly must involve more than a simple
protection of nitrogenous substrates (in some cases fragments of large
molecules) from biological attack. With an enzyme, a protein must be
protected, and protected at the tertiary level to ensure that it is
maintained in a catalytically-active form, albeit perhaps with altered
kinetic properties.

Reactions of enzymes with clays or organic polymers usually decrease
enzymic activities, but may stabilize the bound enzymes against degradation
by proteinases or denaturating agents[3,4,50,62,67,68,71,118,132]. The
possible role of resistant organic colloids in stabilizing soil enzymes has
received more attention in recent years due in part to a perceived analogy
between humus-enzyme complexes and enzymes insolubilized by reaction with or
entrapment within a great variety of organic support materials. Such
reactions may (i) enhance the rigidity of the enzyme molecules, stabilizing
them against denaturation by heating or drying, (ii) render the enzymes
inaccessible to attack by free proteinases, and (iii) if the bound enzyme is
itself a proteinase, prevent destruction by autodigestion.

3.2.6. <u>Model enzyme complexes</u>. Early studies by Ladd and Butler[69]
showed that soil humic acids and trypsin reacted ionically to form insoluble
precipitates with initially decreased enzyme activity, compared with
unreacted trypsin, but with enhanced stability against autodigestion. In
this example the enzyme and humic acid could be easily separated. Later,
Rowell et al.[118] prepared complexes of trypsin and Pronase by reacting the

enzymes with polymerizing p-benzoquinone at pH8.0. Each proteinase yielded a soluble and an insoluble, enzymically-active product, differing from each other and the respective free enzyme in their N contents and kinetic properties. In these complexes bonding of the enzymes to the aromatic polyanionic moieties involved more than ionic linkages. The model proteinase-humic acid analogues had specific activities less than those of the free enzymes, irrespective whether or not proteins or low molecular weight compounds were used as assay substrates, but were more stable after incubation at 25°C or at elevated temperatures. The Pronase complexes were however less stable to lyophilization than free Pronase.

More recently Maignan[77] has described the preparation of a range of insoluble humic acid - invertase - Ca^{2+} complexes. By varying the concentration and order of addition of reactants, as well as the time of incubation of the organic components with Ca^{2+}, preparations were obtained which differed in their initial specific activities and their stabilities on repeated incubation with substrate. When complexes were formed by adding invertase to suspensions of insoluble Ca-humate, the invertase was bound by adsorption at the surface of the support. When the enzyme was mixed with the soluble neutralized humic acid before Ca^{2+} addition, the invertase was partly adsorbed to surfaces, possibly involving covalent bond formation, and partly entrapped within micelles.

These studies on model enzyme complexes imply that the organic compounds which stabilize exocellular enzymes in soil are brown-coloured humic materials, probably aromatic in character. However, knowledge of the nature of the organic ligands and of the manner in which they may complex active enzymes, rests in part on the extraction of enzymes from soil, preferably in high yield and without artefact formation, and their characterization after fractionation and purification.

4. ACTIVITY AND PROPERTIES OF EXTRACTED SOIL ENZYMES

Active enzymes have been extracted from soils by shaking suspensions for periods ranging from 5 min to 24 h, and at temperatures, sometimes ambient or below, but more usually at 37-40°C (Table 3). Generally conditions are chosen to avoid extensive damage to live cells and to favour extraction of abiontic enzymes. Mostly, soil extracts have been assayed for hydrolases, some have been fractionated and the organic matter associated with enzymic activity has been partly characterized. Kinetic constants, stabilities,

Table 3. Enzymes extracted from soils.

Enzyme	Major extractant, Conditions	Reference
o-Diphenol oxidase	0.2M phosphate, 0.1M EDTA, pH 7.0 1h	85,87
p-Diphenol oxidase	" " " " " " 0.05M citrate, pH 6.0, 30min, room temp.	85,87,88 140
Peroxidase	0.05M phosphate, pH 6.0, 5min	7,8
Carboxylesterase	0.1M phosphate, pH 7.5, 30min, 30°C 0.2M NaOH, 30min, room temp.	101 49,121
Arylesterase	0.25M phosphate, pH 7.0, 4h, 4°C	23
Phosphatase	0.1M phosphate, 1M KCl, 0.01M EDTA, 30min, 30°C 0.2M phosphate, 0.1M EDTA, pH 7.0, 1h 0.14M pyrophosphate, pH 7.1, 10min- 24h, 37°C	52 9,120 95
Cellulase	0.2M phosphate, 0.1M EDTA, pH 7.0, 1h	9,120
β-Glucosidase	" " " "	9,120
Invertase	" " " "	9,120
Proteinase/ Peptidase	0.1M Tris-borate, pH 8.1, 60min, 40°C 0.1M bicarbonate, pH 8.1, 30min, 40°C 0.2M phosphate, 0.1M EDTA, pH 7.0, 1h 0.14M pyrophosphate, pH 7.1, 10min-24h 37°C	9,65,120 73 83 95
Urease	" " " " 0.25M phosphate, 2M urea, 4M NaCl, 0.013M EDTA, 0.06M mercaptoethanol, 4.5h, 4°C distilled water, 20min, then 0.95M citrate, 0.05M phosphate, 0.05M glycine, 2.0M NaCl, pH 6.3, 2h, 10°C. 0.2M pyrophosphate, 30min 0.1M pyrophosphate, pH 7.1, 1-24h, 37°C phosphate, pH 6.0, 5°C	95 21 22 76 96 14
Amino acid decarboxylase	0.2M phosphate, 0.1M EDTA, pH 7, 1h	84

responses to added inhibitors, etc., of extracted enzymes have been compared with those respectively of unextracted soils.

o and p-Diphenol oxidases have been extracted aseptically from fresh soils and partially purified using a salmine precipitation technique, followed by Sephadex G-25 gel chromatography which removed some associated humic compounds[85]. The enzyme preparation was heterogeneous; acid hydrolysates contained sugars and amino sugars (36%) as well as NH_4^+ and inorganic compounds (10%), and further fractionation of the enzyme preparation by polyacrylamide gel electrophoresis revealed four peaks of o-diphenol oxidase activity.

The partially-purified extract oxidised a range of phenolic substrates, and also contained proteinases and amino acid decarboxylases[83]. Preincubation of a toluene-treated soil enzyme preparation for 12h at 37°C did not affect diphenol oxidase activities, ie. the oxidases appeared to be resistant to attack by the coextracted soil proteinases. Addition of hyaluronidase before preincubation also was without effect. Preincubation with the microbial proteinase, Pronase for 18h at 37°C decreased diphenol oxidase activities by 30%, and by 100% when both Pronase and hyaluronidase were added. The results suggested that the polysaccharides associated with the extracted soil oxidases protected the enzymes from proteolysis and may play a role in stabilizing exocellular enzymes in soils.

The extract after the Sephadex G-25 treatment was further separated by Sephadex G100 chromatography into three fractions, each of which contained o- and p-diphenol oxidases[87]. The first of these fractions was again fractionated by DEAE cellulose chromatography. Of four components, two were inactive, the third contained o-diphenol oxidases (with traces of p-diphenol oxidase activity) and the fourth solely p-diphenol oxidases. Another fractionation procedure, using DEAE DE 52 cellulose chromatography, yielded from the original soil extract, three enzymically-active fractions, one of which was free of humic compounds and appeared to be homogeneous on polyacrylamide gels. However, o- and p-diphenol oxidases in this fraction were readily separated by Sephadex G-100 chromatography, each subfraction having an enhanced specific activity compared with those of the crude soil extracts.

As with soil, diphenol oxidases extracted from forest litter also can be fractionated to yield components essentially free of co-extracted humic compounds[86]. A comparison of humic-free laccases (p-diphenol oxidases) purified from soil and litter extracts with those from Polyporus versicolor indicated that the former preparations were strongly electronegative and were only weakly adsorbed to humic colloids[88]. By contrast other enzymically-active fractions from soil extracts were associated with humic compounds in complexes which were not separable by chromatography or electrophoresis[87].

Suflita and Bollag[140] have extracted soil oxidases which catalysed the oxidative coupling of 2,6-dimethoxyphenol to a quinone dimer, $3,3^1,5,5^1$ -tetramethoxydiphenoquinone, a reaction which formed the basis of assay. The extracts were active against a range of substrates commonly used in phenoloxidase assays. Losses of activity on storage increased with temperature (5-50°C). About one half of the activity was lost after 24h at 30°C; complete losses occurred after 15 min at 100°C. The reaction was completely inhibited by H_2O_2 (10mM), dithiothreitol (0.1mM), KCN (10mM) and 2,3- dimercaptopropanol (0.1mM), the last two reagents, presumably active by complexing copper in the enzyme's prosthetic group. The inhibition by H_2O_2 indicated that the reaction was not catalysed by soil peroxidases. Soils previously incubated with sucrose yielded extracts with enhanced oxidative coupling activities.

Bartha and Bordeleau[8] found that *peroxidases* in sterile soil extracts catalysed the oxidation of o-dianisidine in the presence of H_2O_2. Unfiltered, non-sterile extracts were respectively about 5-15% more active. Extracts heated at 100°C for 15 min were inactive.

Carboxylesterase activity of soil extracts was demonstrated in a sensitive assay based on the formation of a fluorescent product, 7-hydroxy 4-methyl umbellipherone from the hydrolysis of the butyl ester of 7-hydroxy 4-methylcoumarin[101]. EDTA, Cu^{2+} and S^{2-} competitively inhibited esterase activities, Fe^{3+} inhibited non-competitively. Activities increased after addition of Ca^{2+}, Mg^{2+}, Na^+ or K^+.

Getzin and Rosefield[49] have extracted from soil and partially purified an *arylesterase* which hydrolysed the insecticide malathion to its monoacid. Under the conditions of assay the reaction proceeded with zero order kinetics. Purification of the extracted enzymes included precipitation of co-extracted humic compounds, resulting in a 9.3 fold increase in specific activity. Following further treatment with ammonium sulphate and separation by ion-exchange chromatography, the specific activity of the partially-purified enzyme had increased to 22 times that of the crude soil extract, with an overall recovery of activity of 32%. The enzyme was optimally active at pH 7.0. Measurements of initial velocities of reaction for different substrate concentrations showed that the reaction conformed to Michaelis-Menten kinetics; a Km value for the malathion esterase was calculated to be $2.12 \times 10^{-4}M$ for the two enzyme levels tested.

The esterase was unusually stable, retaining most of its activity in the 0.2M NaOH extracts of soil for several hours, but was irreversibly denatured when exposed for 24h at pHs <2.0 or >10.0. No losses of activity occurred when solutions of the purified enzyme were held in prolonged storage at 4°C or -10°C, or for 15 min at 60°C. The enzyme was partially inactivated at 70°C and completely inactivated at 90°C. The temperature inactivation curves for solutions of the purified, extracted enzyme and for unextracted soil suspensions were similar.

The malathion esterase activity of soil was destroyed by autoclaving. When the purified enzyme was added to an inoculated, autoclaved soil or to a soil of naturally-low esterase activity, and then immediately extracted with alkali, 50-65% of the added activity was recovered in neutral buffer extracts (Tris-HCl, pH 7.0) of the soils, indicative of a rapid adsorption of the added esterase to soil colloids. Further, the adsorbed enzyme remained in a biologically - stable state since no losses of activity occurred during an eight-week incubation of the amended non-sterile soils.

A modified purification procedure[121], involving precipitation of contaminants with protamine sulphate and fractionation of the esterase on QAE Sephadex A-50 gel, resulted in a 560 fold increase in specific activity towards malathion compared with that of the crude soil extract. The purified enzyme also hydrolysed phenyl and p-nitrophenyl esters but not aliphatic esters. The adjusted Km for malathion hydrolysis was 6.06×10^{-5}M. Esterase activities were not destroyed when the purified enzyme was incubated for 24h with a range of proteinases, including Pronase, papain and thermolysin. Incubation with various carbohydrate-hydrolysing enzymes were also ineffective, with one exception. Hyaluronidase action increased arylesterase activity 1.7 fold at the pH optimum (pH 7.5) and increased the sensitivity of the enzyme to changes in pH. Further, after hyaluronidase treatment, the esterase exhibited the characteristic absorbance at 280nm of proteins due to tyrosine residues, whereas in the untreated enzyme such residues are masked. The soil arylesterase is considered to be a glycoprotein; the effect of hyaluronidase was interpreted as due to the removal of a carbohydrate shield which allowed the enzyme easier access to substrate. The effect on absorbance suggested that the carbohydrate and protein moieties were linked through N-acetylhexosamine-tyrosine bonds. The presence of carbohydrate in the enzyme complex may protect the esterase from attack by proteinases as has been demonstrated for other glucoproteins.

Cacco and Maggioni[23] have described the extraction and partial purification of arylesterases active against α-naphthyl acetate. Fractionation by electrophoresis in polyacrylamide gels yielded several esterase-active bands, each of which contained coloured humic compounds. The results were inconclusive regarding the presence of different esterases in the crude extracts, or to the influence of bonding of an esterase to different humic compounds on enzyme electrophoretic mobility.

Extracts after dialysis, concentration, and sterilization by filtration, retained full activity when stored for several months at -18°C, but were completely inactivated by heating at 100°C for 20min. The properties of the esterase(s) extracted from soil differed from those of a naphthyl acetate esterase prepared from wheat kernels. The soil enzyme exhibited a higher optimal temperature for activity, indicative of a greater heat stability, but a lower affinity for substrate as shown by higher Km values, 2.95×10^{-3}M cf. 8.45×10^{-4}M. (Presumably such differences may also have resulted from different purities of the enzymes from the sources). In particular, the soil-derived arylesterase completely retained activity after incubation for 29h with the proteinase, Pronase, whereas the plant-derived arylesterase was completely destroyed or inactivated. · Bonding of soil arylesterases in humic-enzyme complexes may possibly have increased enzyme thermal stabilities and protection from enzymic attack, and decreased their affinity for substrate.

Hayano[52] has used a phosphate buffer containing KCl and EDTA to extract *phosphodiesterases* from a forest soil. After dialysis and concentration of the extracts, and precipitation of the enzymes and coextracted coloured humic compounds by ammonium sulphate, solutions of the phosphodiesterases were further purified by addition of protamine sulphate and a second dialysis treatment. Removal of flocculated humic material was accomplished with only small losses of phosphodiesterase activity. Per gram of soil, extracted enzymic activity was curvilinearly related to extractant volume, indicative perhaps of desorption of enzyme ionically bound to soil surfaces. The partially-purified enzyme retained activity after heating solutions for 10min at 50°C, but was completely inactivated at 80°C.

Batistic *et al.*[9] have used similar extraction and purification procedures to obtain from soil active *acid* and *alkaline phosphomonoesterases* (and other hydrolases, *cellulase, B-glucosidase, invertase* and *proteinase*). Salmine substituted for protamine as a precipitant of humic material from dialysed extracts. In this case complete flocculation of humic compounds was accompanied by extensive (about 60%) losses of enzymic activities from solution. The activities of the precipitated material were low and could not be restored. Enzymes remaining in dialysed supernatants were of enhanced (about ten fold) specific activities, and were further fractionated by chromatography in Sephadex (G-100, G-75) gels. Using Sephadex G-100 gel, three fractions were obtained, all of which were active in assays for each hydrolase. The fractions varied in their molecular weight range, specific activities for a given hydrolase, and carbohydrate contents (14-31%). Further separation using Sephadex G-75 gel yielded subfractions, some of which were of increased specific activities. Some separation of hydrolase activities was achieved.

A second approach was to fractionate extracts before salmine treatment by using anion-exchange (DEAE cellulose) chromatography. This technique yielded two fractions, both of which were active in all hydrolase assays and contained carbohydrates. One fraction also contained coloured humic material which could not be separated from enzymic activities. It was concluded that the extracted enzymes occurred exocellularly in soils as complexes with carbohydrate and with humic-carbohydrate associations.

In a later study[120] one of the humic-free, carbohydrate-enzyme complexes obtained by fractionation in Sephadex G-75 gel was further separated into five components by anion-exchange chromatography. All five subfractions contained the five hydrolases assayed, most showing enhanced specific activities. Relative activities varied between fractions as did carbohydrate contents, which ranged from 12 to 57%. Incubation of the carbohydrate-enzyme complex with lysozyme or hyaluronidase for 24h prior to assay did not affect hydrolase activities. Treatment with dodecylsulphate to dissociate enzyme and carbohydrate caused large activity losses. Analyses following acid hydrolysis of the carbohydrate-enzyme complex showed a wide range of sugars, galactose, glucose, mannose, arabinose, xylose, ribose and rhamnose. The carbohydrate-enzyme complexes are thought not to be glycoproteins. Rather, various carbohydrates, by physical adsorption of the enzymes, are considered to stabilize the enzyme protein against denaturation or proteolysis in soils. The presence of carbohydrates complexed with enzymes probably also affects the kinetics of enzyme action, perhaps decreasing the affinity of an enzyme for its substrate.

Nannipieri et al.[95] have used neutral pyrophosphate buffer to extract hydrolases from soils. Maximal acid phosphomonoesterase activities of extracts were achieved after 2-6 hours extraction of soils and were equivalent to 31-66% of the activities of the respective unextracted soils. Phosphatase activities for extracts were directly correlated with C and N contents and were relatively thermostable in crude extracts losing little activity after exposure for 2h at 80°C, whereas under the same conditions a

purified, commercial acid phosphatase was almost completely inactivated[98].
Kinetic analyses of phosphatase activities indicated that soil extracts
contained at least two enzymes (or two forms of the same enzyme) catalysing
the same reaction. Calculated kinetic constants (Km, V_{max}) of extracted
phosphatases differed not only for enzymes in the same extract but also for
enzymes in extracts from different soils[94].

Proteinases in soil extracts have been assayed using casein[73,83,95] and
substituted amides[95] and peptides[65] as assay substrates. Mayaudon et al.[83]
found that the casein hydrolysing activities of soil extracts were directly
related to soil organic matter contents and inversely related to clay
contents. Specific activities doubled when associated coloured humic
compounds were removed from the extracts. The partially-purified proteinases
appeared to be of a serine protease type as judged from their responses to
inhibitors. Optimal activity occurred at pH 8.5 and at 50°C. The estimated
Km value (22mg casein ml^{-1}) was considerably higher than those (0.15 and
0.54mg casein ml^{-1}) reported by Nannipieri et al.[94] for proteinases of crude
extracts from two soils.

Mayaudon et al.[83] also showed that the casein-hydrolysing activities of
soil extracts were poorly correlated to microbial numbers in soils. Ladd and
Paul[73] demonstrated with a glucose-amended soil that the highest activities
of proteinases towards casein were achieved during a rapid phase of turnover
of microbial cells and of net decline in microbial cell numbers, at which
time equivalent to 14% of the net gain in soil caseinase activity was
extractable with a dilute bicarbonate solution. Nannipieri et al.[95] reported
that the casein activities of pyrophosphate extracts of two soils were
equivalent to about 60% of the activities of the respective unextracted
soils; and, in the case of a third soil, to about 140%, indicative of higher
active enzyme concentrations or higher specific activities due perhaps to
facilitation of enzyme-substrate reactions in extracts.

Enzymes active against casein and BAA varied in the efficiency with
which they were extracted from soils[95], and differed in their responses to
changing pHs or preincubation temperatures[98]. ZPL-hydrolysing activities of
soil extracts were poorly correlated with the amounts of extracted humic
compounds, the majority of which were readily removed by precipitation with
$CaCl_2$ without loss of enzyme activity[65]. The sum of the activities of an
extract and an extracted soil invariably exceeded that of the unextracted
soil, suggesting increased specific activities due to enzyme extraction.
Extracts of a stored moist soil were less active than extracts from the same

soil stored dry, whereas activities of the extracted soil were similar whether stored moist or dry. Whereas the non-extractable ZPL-hydrolysing enzymes of the soil appeared to be stable for both soil storage conditions, the extractable enzymes were present in the soil in a form more amenable to microbial decomposition under adequate moisture conditions.

Soil differed in their relative activities towards ZPL and BAA[70]; the enzymes involved were extracted from a soil with different efficiencies[65]. Extracted ZPL-hydrolysing enzymes retained complete activity when freeze-dried, or dried at 30°C (90 min), or when incubated under bacteriostatic conditions for 10 days at 25°C. Incubation of the extracts for 1 day at 30°C with or without the addition of microbial proteinases, thermolysin or subtilisin decreased activities towards ZPL by about 10% only[65]. The extracted enzyme, in regard to its specificity of hydrolysis of Z-dipeptides and its response to inhibitors, exhibited some of the properties of carboxypeptidase.

Some of the most detailed studies of the properties of extracted soil enzymes has been devoted to soil ureases. Early studies by Briggs and Segal[14] reported the cyrstallization of urease-active material (N content, 8.8%) by acetone precipitation from a phosphate buffer extract of a surface forest soil. Ultracentrifugation separated the preparation into three components. Urease activity was maximal at pH 7.1, and was inhibited by p-chloromercuribenzoate and by heavy metal ions (Ag^+, Hg^{2+}, Cu^{2+}).

Burns et al.[21,22] extracted from soil, clay-free material with urease activity equivalent to about 20% of that of unextracted soil. Partial separation of brown-coloured humic compounds from the extracted ureases could be achieved with little loss of enzymic activity. The extracted soil urease retained activity when incubated for 24h with Pronase, whereas this proteinase was sufficiently stable in the presence of soil to demonstrate the destruction of added jack bean urease. Pronase action also markedly decreased the activities of complexes of bentonite and jack bean urease but was ineffective against a bentonite-urease-lignin complex. The persistence of urease activity in soils was attributed to the formation of exocellular urease-organic colloidal complexes, which permitted reaction between enzyme and substrate and diffusion of products, but which protected ureases from attack by soil proteinases. Further studies by Pettit et al.[107] compared the properties of extracted ureases, soil ureases and jack bean urease respectively. Km values were influenced by the type of buffer used. Ureases of soil extract had the highest Km value and were the most stable to thermal denaturation, and to lyophilization. Jack bean urease gave the lowest Km value and was least stable.

Extracted ureases were not uniformly associated with soil organic matter, as revealed by fractionation by ultracentrifugation, and by agarose gel chromatography[90]. Some enzymically-active fractions were nearly colorless and were of specific activities about an order of magnitude higher than those of coloured humus-enzyme fractions. Nannipieri et al.[96] used pyrophosphate buffer, pH 7.1, containing toluene, to extract ureases from soils under conditions which apparently did not affect the populations of ureolytic microflora, and which therefore were presumed to have solubilized exocellular soil ureases only. The specific activities of the extracted ureases increased with extraction time up to 18h, due to the early coextraction of humic materials of relatively low or of no urease activity[95,96]. Sequential application of ultrafiltration and gel chromatography techniques to the extracts successively increased both the total activity of extracted ureases and the specific activities of individual fractions.[25] Precipitation of humic compounds from extracts by addition of protamine sulphate failed to separate ureases from extracted organic matter suggesting that the enzymes and humic compounds may be bonded more strongly than by ionic linkages.

Ureases of crude soil extracts and of fractions of different molecular size ranges were more resistant to thermal denaturation, and to proteolysis by Pronase, than was free urease purified from jack bean[25,47,98]. Of the fractions the most stable were those of molecular weight nominally $>10^5$. Km values for soil ureases differed for extracts from different soils[94,97] and for extract fractions[97] (Table 4). Km values for fractions ranged from $8 \times 10^{-3} M$ to $40 \times 10^{-3} M$, reflecting perhaps differences in the kinetic properties of ureases of different origin (Km range, 20 to $40 \times 10^{-3} M$) or of ureases in different states of association or dissociation. Alternatively, bonding of ureases in complexes with soil organic matter, (which may differ in its chemical nature, molecular size and shape, etc.) may also induce differences in the structure and charge distribution at the enzyme's active site and/or differences in the rates of diffusion of urea substrate to the active site, any of which would affect measured Km values.

It is well established that bonding of enzymes to charged organic supports alters the H^+ ion concentrations within the vicinity of the enzyme compared with that in the bulk solution, and thus changes the enzyme's pH-activity profile. Extracted urease-humus complexes of different molecular

Table 4. Kinetic properties of ureases in soil extracts.

Urease source	Km*	Optimal pH**
Soil extract (crude)	100	8.3
Fraction AI (mol wt >10^5)	28	7.6, 8.8
Fraction AII (mol wt <10^5)	35	7.0, 7.8
Subfraction SI (mol wt >5×10^5)	11	8.2
Subfraction SII (mol wt $1.3-1.5 \times 10^5$)	25	7.5, 8.8
Subfraction SIII (mol wt $6-8 \times 10^4$)	8	8.7
Subfraction SIV (mol wt $7-9 \times 10^4$)	40	7.0, 7.8
Subfraction SV (mol wt $4-5 \times 10^4$)	11	7.8, 8.8

* assayed in phosphate buffer; ** assayed in phosphate or borate
buffers

Data taken from Nannipieri et al.[97]

weight differed in their pH optima as well as Km values (Table 4)[97]. All
fractions exhibited maximal specific activities in the pH range 8.0 to 8.8,
but some fractions showed a second pH optimum in phosphate buffer at pH 7.0
to 7.5 (which was close to that (6.5-7.6) of purified jack bean urease, also
assayed in phosphate buffer). On the assumption that soil ureases of varied
origin had, if free, properties similar to those of jack bean urease, then
the results of Nannipieri et al.[97], are consistent with the notion that the
extracted ureases are bonded in polyanionic humic matrixes, which increased
the H^+ ion concentrations near the bound enzymes and displaced their
pH-activity profiles to more alkaline regions. However, neither Km values
nor optimal pHs varied consistently with the molecular weight ranges of the
extracted enzyme complexes. Just as caution is needed in interpreting
differences in Km values of the soil fraction ureases, so too care is
required in explaining differences in their pH-activity profiles.

Mayaudon et al.[84,85] have demonstrated the presence of aromatic amino
acid decarboxylases in soil extracts by measuring the release of $^{14}CO_2$ from
carboxyl-labelled DL-3,4-dihydroxyphenylalanine (DOPA), DL-tyrosine and

DL-tryptophane. The respective rates of decarboxylation of these substrates by an extract, partially purified to remove humic compounds, were in the ratio 100:35:7. Decarboxylation of DOPA followed Michaelis-Menten kinetics, with a calculated Km value of 8.3 x 10^{-4}M; activation energy for decarboxylation was estimated to be 7.9 kcal mole^{-1} °C^{-1}. The extracts also contained o- and p-diphenol oxidases, and carbohydrate (36% w/w).

5. CONCLUSIONS

During the past 10-15 years the range of enzymes assayed in soil has continued to widen and more importantly, assay conditions have been described which allow measurements of activities on a sound kinetic basis. Thus potential enzymic activities of different soils or soil fractions, or of soils subjected to various pretreatments or field management practices, can be reliably compared. Nevertheless the failure to achieve widespread acceptance of standard assay conditions for a given enzyme has restricted comparisons of results from different laboratories.

Provided that satisfactory storage and pretreatment conditions are employed, assays can be expected to give measures of potential activities of soils at the time of their sampling. Such activities may vary substantially with the season according to the synthesis, release into soil, and persistence of plant, animal and microbial enzymes. Whether the potential activities of soils, sampled once or perhaps several times in a season, truly reflect natural soil activities integrated over the season, or longer, in unresolved. Such a relationship is implied in the many environmental studies based on the activities of selected enzymes, and on their responses to pesticide additions, pollutants etc. Nevertheless, it is a sobering thought that the rates of hydrolysis of added organic N, P and S substrates in assays devised to measure the potential activities respectively of soil proteinases, phosphatases and sulphatases, do not consistently correlate with the capacities of soils to release mineral N, P and S from natural sources in longer-term incubation tests. Studies of urea-hydrolysing activities of soils and of the use of inhibitors to control NH_4^+ production from fertilizer urea (and hence N losses and toxicities to germinating plants) would appear to be more soundly based. In these cases the assay substrate, urea, is also the 'natural' substrate, and at least during the brief period following fertilizer addition when control of hydrolysis rates is being sought, field concentrations of urea are temporarily high (perhaps insufficiently high to ensure zero order kinetics), and soil urease levels are likely to be more constant than over a year or the total growing season.

Considering that the net releases of plant nutrients from soil organic sources do not always reflect the potential activities of even the classes of enzymes likely to be involved, it is unsurprising that soil enzyme activities do not consistently reflect soil fertility levels. In some studies the chosen enzymes catalyse reactions which have no obvious connection with plant growth. The basis for their use as a soil fertility index may possibly lie with their frequently observed positive correlations with soil organic matter levels, which affect both chemical and physical properties of soils, and plant growth. The difficulties that arise in seeking to establish soil enzyme activities as broad indices of soil fertility is exemplified by the studies of Zhou and his colleagues[110,161]. They found that the fertility of black soils of Northeast China with long and different histories of cultivation and exploitation was related to soil organic matter status and to the potential activities of a group of enzymes (catalase, polyphenol oxidase, invertase, urease and neutral and alkaline phosphatases). However, for fertilized black soils the relationship between fertility and enzyme activities held only for catalase, invertase and urease : polyphenol oxidase and phosphatase activities were lower in fertilized than in unfertilized soils.

Enzymic activities of crude soil suspensions have been demonstrated to follow Michaelis-Menten kinetics. Calculated K_m values have varied for different soils and for active fractions of soil extracts. The extent to which kinetic constants of soil enzymes are influenced by the state in which the enzymes occur in soils, is unknown. For some activities, two K_m values have been distinguished for the one crude soil extract. Fractionation has revealed that enzymes may exist in tightly- and loosely- bound complexes with soil coloured humic compounds. Enzymes freed of coloured materials may nevertheless still be bound in complexes by their association with carbohydrates. These may not only influence enzyme kinetic properties but evidence suggests that they may also confer some degree of stability on the enzymes in soil. Whereas early speculation on the mechanism(s) by which enzymes are protected in soil tended, in the case of abiontic enzymes, to focus on the role of clay or humic colloids, fractionation studies have drawn attention to the potentially important role of soil carbohydrates. The manner in which carbohydrates are bonded to the enzymes in soil has not as yet been established and may be a fruitful line of enquiry.

Differences in the manner in which enzyme complexes are formed, in the concentrations of reactants, in the nature of the ligands, and in the positions and types of bonding of the ligand to enzyme will have important consequences for enzyme kinetics and stability. The use of model humus-enzyme complexes has limitations in the context of soil enzyme properties. Although the properties of model complexes have been as generally anticipated for enzymes bonded to polyanionic support materials, it has not been established that the models chosen are representative of the enzyme complexes in soils.

Comparisons of the properties (Km, pH-activity profiles) of soil enzymes with those of "free" enzymes, ie. enzymes purified from known plant or microbial sources, have served to reinforce concepts of the mechanisms by which abiontic enzymes exist in soil, but otherwise they also have limited value. Not only may the soil enzymes differ from each other and from the selected reference enzymes, both in their origins and in their intrinsic properties, but the reference enzyme itself may exhibit different kinetic properties according to its state of purification.

As yet no generalized picture has emerged regarding the relationship between activities and turnover of enzymes in soils. Aspects of special interest include (a) the extent to which activities against substrates of ranging molecular size are dependent upon the availability of energy supplies, ie. on de novo enzyme synthesis, and (b) the possibility that, under given environmental conditions, soils have characteristic capacities to stabilize or protect microbial biomass (and hence intracellular enzymes) and abiontic enzymes. Some evidence based on urea hydrolysis rates[160] suggests not only that soils have such capacities but that the stabilization process(es) may be specific.

6. REFERENCES

1. AHMED M., OADES J.M. and LADD J.N. 1982. Determination of ATP in soils: effect of soil treatments. Soil Biology and Biochemistry, 14, 273-279.
2. AL-KHAFAJI A.A. and TABATABAI M.A. 1979. Effects of trace elements on arylsulfatase activity in soils. Soil Science, 127, 129-133.
3. AMBROZ Z. 1966. Adsorption of two proteases occurring frequently in soil. Sbornik Vysoke Skoly Zemedelske Y Brne, A2, 161-167.
4. AOMINE S. and KOBAYASHI Y. 1964. Effects of allophane on the enzymatic activity of a protease. Soil, Plant and Food, Tokyo, 10, 28-32.
5. ASHWORTH J., AKERBOOM H.M. and CREPIN J.M. 1980. Inhibition by xanthates of nitrification and urea hydrolysis in soil. Soil Science Society of America Journal, 44, 1247-1249.
6. BARMAN T.E. 1969. Enzyme Handbook, Vol II. Springer-Verlag, Berlin.
7. BARTHA R. and BORDELEAU L.M. 1969. Transformation of herbicide-derived chloroanilines by cell free peroxidases in soil. Bacteriological Proceedings, 4, A26.

8. BARTHA R. and BORDELEAU L.M. 1969. Cell-free peroxidases in soil. Soil Biology and Biochemistry, 1, 139-143.
9. BATISTIC L., SARKAR J.M. and MAYAUDON J. 1980. Extraction, purification and properties of soil hydrolases. Soil Biology and Biochemistry, 12, 59-63.
10. BERI V. and BRAR S.S. 1978. Urease activity in subtropical, alkaline soils of India. Soil Science, 126, 330-335.
11. BERI V., GOSWAMI K.P. and BRAR S.S. 1978. Urease activity and its Michaelis constant for soil systems. Plant and Soil, 49, 105-115.
12. BREMNER J.M. and DOUGLAS L.A. 1971. Inhibition of urease activity in soils. Soil Biology and Biochemistry, 3, 297-307.
13. BREMNER J.M. and MULVANEY R.L. 1978. Urease activity in soils Chapter 5 in Soil Enzymes. Ed. Burns R.G. Academic Press, London.
14. BRIGGS M.H. and SEGAL L. 1963. Preparation and properties of a free soil enzyme. Life Sciences, 1, 69-72.
15. BROWMAN M.G. and TABATABAI M.A. 1978. Phosphodiesterase activity of soils. Soil Science Society of America Journal, 42, 284-290.
16. BUNDY L.G. and BREMNER J.M. 1973. Effects of substituted p-benzoquinones on urease activity in soils. Soil Biology and Biochemistry, 5, 847-853.
17. BURNS R.G. 1977. Soil enzymology. Science Progress (Oxford), 64, 275-285.
18. BURNS R.G. (Ed.) 1978. Soil Enzymes. Academic Press, London.
19. BURNS R.G. 1978. Enzyme activity in soil: some theoretical and practical considerations. Chapter 8 in Soil Enzymes. Ed. Burns R.G. Academic Press, London.
20. BURNS R.G. 1981. Enzyme activity in soil: location and a possible role in microbial ecology. Soil Biology and Biochemistry, 14, 423-427.
21. BURNS R.G., EL-SAYED M.H. and McLAREN A.D. 1972. Extraction of a urease-active organo-complex from soil. Soil Biology and Biochemistry, 4, 107-108.
22. BURNS R.G., PUKITE A.H. and McLAREN A.D. 1972. Concerning the location and persistence of soil urease. Soil Science Society of America Proceedings, 36, 308-311.
23. CACCO G. and MAGGIONI A. 1976. Multiple forms of acetyl-naphthyl-esterase activity in soil organic matter. Soil Biology and Biochemistry, 8, 321-325.
24. CAPLAN J.A. and FAHEY J.W. 1980. A rapid assay for the examination of protein degradation in soils. Bulletin of Environmental Contamination and Toxicology, 25, 424-426.
25. CECCANTI B., NANNIPIERI P., CERVELLI S. and SEQUI P. 1978. Fractionation of humus-urease complexes. Soil Biology and Biochemistry, 10, 39-45.
26. CERVELLI S., NANNIPIERI P., CECCANTI B. and SEQUI P. 1973. Michaelis constants of soil acid phosphatase. Soil Biology and Biochemistry, 5, 841-845.
27. CERVELLI S., NANNIPIERI P., GIOVANNINI G. and PERNA A. 1976. Relationships between substituted urea herbicides and soil urease activity. Weed Research, 16 365-368.
28. CERVELLI S., NANNIPIERI P., GIOVANNINI G. and PERNA A. 1977. Effect of soil on urease inhibition by substituted urea herbicides. Soil Biology and Biochemistry, 9, 393-396.
29. CERVELLI S., NANNIPIERI P. and SEQUI P. 1978. Interactions between agrochemicals and soil enzymes. Chapter 7 in Soil Enzymes. Ed. Burns R.G. Academic Press, London.

30. CHENDRAYAN K., ADHYA T.K. and SETHUNANTHAN N. 1980. Dehydrogenase and invertase activities of flooded soils. Soil Biology and Biochemistry, 12, 271-273.
31. COLE, M.A. 1977. Lead inhibition of enzyme synthesis in soil. Applied and Environmental Microbiology, 33, 262-268.
32. CONRAD R. and SEILER W. 1981. Decomposition of atmospheric hydrogen by soil microorganisms and soil enzymes. Soil Biology and Biochemistry, 13, 43-49.
33. COOPER P.J.M. 1972. Arylsulphatase activity in Northern Nigerian soils. Soil Biology and Biochemistry, 4, 333-337.
34. DICK W.A. and TABATABAI M.A. 1978. Inorganic pyrophosphatase activity of soils. Soil Biology and Biochemistry, 10, 59-65.
35. DOELMAN P. and HAANSTRA L. 1979. Effect of lead on soil respiration and dehydrogenase activity. Soil Biology and Biochemistry 11, 475-479.
36. DORAN J.W. 1980. Soil microbial and biochemical changes associated with reduced tillage. Soil Science Society of America Journal, 44, 765-771.
37. DURAND G. 1965. Enzymes in soil. Revue d'Ecologie et de Biologie du Sol, 2, 141-205.
38. DUTZLER-FRANZ G. 1977. Influence of soil chemical and physical properties on the enzyme activity of various soil types. Zeitschrift für Pflanzenernahrung und Bodenkunde, 140, 329-350.
39. EIVAZI F. and TABATABAI M.A. 1977. Phosphatases in soils. Soil Biology and Biochemistry, 9, 167-172.
40. EL-SHINNAWI M.M., SHEHATA S.M. and EL-SHIMI S.A. 1982. Enzymatic activities during incubation of remoistened soil samples stored for different periods. Zentralblatt fur Mikrobiologie, 137, 86-90.
41. FOSTER R.C. and MARTIN J.K. 1981. Chapter 2 in Soil Biochemistry, Vol.5. Eds. Paul E.A. and Ladd J.N. Marcel Dekker, New York.
42. FRANKENBERGER W.T. and JOHANSON J.B. 1982. Effect of pH on enzyme stability in soils. Soil Biology and Biochemistry, 14, 433-437.
43. FRANKENBERGER W.T. and TABATABAI M.A. 1980. Amidase activity in soils: I. Method of assay. Soil Science Society of America Journal, 44, 282-287.
44. FRANKENBERGER W.T. and TABATABAI M.A. 1980. Amidase activity in soils: II. Kinetic parameters. Soil Science Society of America Journal, 44, 532-536.
45. FRANKENBERGER W.T. and TABATABAI M.A. 1981. Amidase activity in soils: III. Stability and distribution. Soil Science Society of America Journal, 45, 333-338.
46. FRANKENBERGER W.T., TROEH F.R. and DUMENIL L.C. 1979. Bacterial effects of hydraulic conductivity of soils. Soil Science Society of America Journal, 43, 333-338.
47. GALSTAYAN A.Sh. 1978. Standardization of methods for determining the activity of soil enzymes. Pochvovedenie, 2, 107-114.
48. GALSTAYAN A.Sh. and BAZOYAN G.V. 1974. Activity of soil arylsulphatase. Doklady Akademii Nauk Armyanskoi S.S.R. 59, 184-187.
49. GETZIN L.W. and ROSEFIELD I. 1971. Partial purification and properties of a soil enzyme that degrades the insecticide malathion. Biochimica et Biophysica Acta, 235, 442-453.
50. GOLDSTEIN J.L. and SWAIN T. 1965. The inhibition of enzymes by tannins. Phytochemistry, 4, 185-192.
51. GOULD W.D., COOK F.D. and BULAT J.A. 1978. Inhibition of urease activity by heterocyclic sulfur compounds. Soil Science Society of America Journal 42, 66-72.
52. HAYANO K. 1977. Extraction and properties of phosphodiesterase from a forest soil. Soil Biology and Biochemistry, 9, 221-223.

216

53. HODSON R.C., WILLIAMS S.K. and DAVIDSON W.R. 1975. Metabolic control of urea catabolism in *Chlamydomonas reinhardi* and *Chlorella pyrenoidosa*. Journal of Bacteriology, 121, 1022-1035.

54. ISHII T. and HAYANO K. 1974. A method for the estimation of phosphodiesterase activity in soil. Journal of the Science of Soil and Manure, Japan, 45, 505-508.

55. JENKINSON D.S. and LADD J.N. 1981. Microbial biomass in soil: measurement and turnover. Chapter 10 in Soil Biochemistry, Vol.5. Eds. Paul E.A. and Ladd J.N. Marcel Dekker, New York.

56. KAPLAN D.L. and HARTENSTEIN R. 1979. Problems with toluene and the determination of extracellular enzyme activity in soils. Soil Biology and Biochemistry, 11, 335-338.

57. KISS S., DRAGAN-BULARDA M. and RADULESCU D. 1972. Biological significance of the enzymes accumulated in soil. 3rd Symposium of Soil Biology, Bucharest, 19-78.

58. KISS S., DRAGAN-BULARDA M. and RADULESCU D. 1975. Biological significance of enzymes accumulated in soil. Advances in Agronomy, 27, 25-87.

59. KISS S., DRAGAN-BULARDA M. and RADULESCU D. 1978. Soil polysaccharidases: activity and agricultural importance. Chapter 4 in Soil Enzymes. Ed. Burns R.G. Academic Press, London.

60. KLEIN T.M. and KOTHS J.S. 1980. Urease, protease and acid phosphatase in soil continuously cropped to corn by conventional or no-tillage methods. Soil Biology and Biochemistry, 12, 293-294.

61. KNOWLES R. and DENIKE D. 1974. Effect of ammonium-, nitrite-, and nitrate-nitrogen on anaerobic nitrogenase activity in soil. Soil Biology and Biochemistry, 6, 353-358.

62. KOBAYASHI Y. and AOMINE S. 1967. Mechanism of the inhibitory effect of allophane and montmorillonite on some enzymes. Soil Science and Plant Nutrition, Tokyo, 13, 189-194.

63. KOWALENKO C.G. and LOWE L.E. 1975. Mineralization of sulphur from four soils and its relationship to soil carbon, nitrogen and phosphorus. Canadian Journal of Soil Science, 55, 9-14.

64. KUPREVICH V.F. and SHCHERBAKOVA T.A. 1971. Comparative enzymatic activity in diverse types of soil. Chapter 7 in Soil Biochemistry, Vol.2. Eds.McLaren A.D. and Skujins J.J. Marcel Dekker, New York.

65. LADD J.N. 1972. Properties of proteolytic enzymes extracted from soil. Soil Biology and Biochemistry, 4, 227-237.

66. LADD J.N. 1978. Origin and range of enzymes in soil. Chapter 2 in Soil Enzymes. Ed Burns R.G. Academic Press, London.

67. LADD J.N. and BUTLER J.H.A. 1969. Inhibitory effect of soil humic compounds on the proteolytic enzyme, Pronase. Australian Journal of Soil Research, 7, 241-251.

68. LADD J.N. and BUTLER J.H.A. 1969. Inhibition and stimulation of proteolytic enzyme activities by soil humic acids. Australian Journal of Soil Research, 7, 253-261.

69. LADD J.N. and BUTLER J.H.A. 1970. The effect of inorganic cations on the inhibition and stimulation of protease activity by soil humic acids. Soil Biology and Biochemistry, 2, 33-40.

70. LADD J.N. and BUTLER J.H.A. 1972. Short-term assays of soil proteolytic enzyme activities using proteins and dipeptide derivatives as substrates. Soil Biology and Biochemistry, 4, 19-30.

71. LADD J.N. and BUTLER J.H.A. 1975. Humus-enzyme systems and synthetic organic polymer-enzyme analogs. Chapter 5 in Soil Biochemistry, Vol.4. Eds.Paul E.A. and McLaren A.D. Marcel Dekker, New York.

72. LADD J.N. and JACKSON R.B. 1982. Biochemistry of ammonification. Chapter 5 in Nitrogen in Agricultural Soils. Ed.Stevenson F.J. Agronomy Monograph No. 22, American Society of Agronomy, Madison, Wisconsin.

73. LADD J.N. and PAUL E.A. 1973. Changes in enzymic activity and distribution of acid-soluble, amino acid-nitrogen in soil during nitrogen immobilization and mineralization. Soil Biology and Biochemistry, 5, 825-840.

74. LEFTLEY J.W. and SYRETT P.J. 1973. Urease and ATP: Urea amidolyase activity in unicellular algae. Journal of General Microbiology, 77 109-115.

75. LETHBRIDGE G., BULL A.T. and BURNS R.G. 1978. 1,3-β-Glucanase activity in soil and its response to soil amendment. Revue d'Ecologie et de Biologie du Sol, 17, 479-489.

76. LLOYD A.B. 1975. Extraction of urease from soil. Soil Biology and Biochemistry, 7, 357-358.

77. MAIGNAN C. 1982. Activite des complexes acides humiques-invertase: influence du mode de preparation. Soil Biology and Biochemistry, 14, 439-445.

78. MATHUR S.P., MACDOUGALL J.I. and McGRATH M. 1980. Levels of activities of some carbohydrases, protease, lipase, and phosphatase in organic soils of differing copper content. Soil Science, 129 376-385.

79. MATHUR S.P. and RAYMENT A.F. 1977. Influence of trace element fertilization on the decomposition rate and phosphatase activity of a mesic fibrisol. Canadian Journal of Soil Science, 57, 397-408.

80. MATHUR S.P. and SANDERSON R.B. 1980. The partial inactivation of degradative soil enzymes by residual fertilizer copper in Histosols. Soil Science Society of America Journal, 44, 750-755.

81. MATZEL W., HEBER R., ACKERMANN W. and TESKE W. 1978. Ammonia losses in urea fertilizing. 3. Effect of urease inhibitors on ammonia volatilization. Archiv für Acker- und Pflanzenbau und Bodenkunde, 22, 185-191.

82. MAY P.B. and DOUGLAS L.A. 1976. Assay for soil urease activity. Plant and Soil, 45 301-305.

83. MAYAUDON J., BATISTIC L. and SARKAR J.M. 1975. Proprietes des activites proteolytiques extraites des sols frais. Soil Biology and Biochemistry 7, 281-286.

84. MAYAUDON J., EL-HALFAWI M. and BELLINCK C. 1973. Decarboxylation des acides amines aromatiques-1-^{14}C par les extraits de sol. Soil Biology and Biochemistry, 5, 355-367.

85. MAYAUDON J., EL-HALFAWI M. and CHALVIGNAC M.-A. 1973. Proprietes des diphenol oxydases extraites des sols. Soil Biology and Biochemistry, 5, 369-383.

86. MAYAUDON J. and SARKAR J.M. 1974. Etude des diphenol oxydases extraites d'une litiere de foret. Soil Biology and Biochemistry, 6, 269-274.

87. MAYAUDON J. and SARKAR J.M. 1974. Chromatographie et purification des diphenol oxydases du sol. Soil Biology and Biochemistry, 6, 275-285.

88. MAYAUDON J. and SARKAR J.M. 1975. Laccases de *Polyporus versicolor* dans le sol et la litiere. Soil Biology and Biochemistry, 7, 31-34.

89. McGARITY J.W. and MYERS M.G. 1967. A survey of urease activity in soils of northern New South Wales. Plant and Soil, 27, 217-238.

90. McLAREN A.D., PUKITE A.H. and BARSHAD I. 1975. Isolation of humus with enzymatic activity from soil. Soil Science, 119, 178-180.

91. MISHRA M.M., FLAIG W. and SOECHTIG H. 1980. The effect of quinoid and phenolic compounds on urease and dehydrogenase activity and nitrification in soil. Plant and Soil, 55, 25-33.

92. MULVANEY R.L. and BREMNER J.M. 1978. Use of *p*-benzoquinone and hydroquinone for retardation of urea hydrolysis in soils. Soil Biology and Biochemistry, 10, 297-302.

93. MULVANEY R.L. and BREMNER J.M. 1981. Control of urea transformations in soils. Chapter 4 in Soil Biochemistry, Vol.5. Eds. Paul E.A. and Ladd J.N. Marcel Dekker, New York.

94. NANNIPIERI P., CECCANTI B., CERVELLI S. and CONTI C. 1982. Hydrolases extracted from soil: Kinetic parameters of several enzymes catalysing the same reaction. Soil Biology and Biochemistry, 14, 429-432.

95. NANNIPIERI P., CECCANTI B., CERVELLI S. and MATARESE E. 1980. Extraction of phosphatase, urease, proteases, organic carbon, and nitrogen from soil. Soil Science Society of America Journal, 44, 1011-1016.

96. NANNIPIERI P., CECCANTI B., CERVELLI S. and SEQUI P. 1974. Use of 0.1M pyrophosphate to extract urease from a podzol. Soil Biology and Biochemistry, 6, 359-362.

97. NANNIPIERI P., CECCANTI B., CERVELLI S. and SEQUI P. 1978. Stability and kinetic propertis of humus-urease complexes. Soil Biology and Biochemistry, 10, 143-147.

98. NANNIPIERI P., CECCANTI B., CONTI C. and BIANCHI D. 1982. Hydrolases extracted from soil: their properties and activities. Soil Biology and Biochemistry, 14, 257-263.

99. NANNIPIERI P., PEDRAZZINI F., ARCARA P.G. and PIOVANELLI C. 1979. Changes in amino acids, enzyme activities, and biomasses during soil microbial growth. Soil Science, 127, 26-34.

100. PAL S. and CHHONKAR P.K. 1979. Thermal sensitivity and kinetic properties of soil urease. Journal of the Indian Society of Soil Science, 27, 43-47.

101. PANCHOLY S.K. and LYND J.Q. 1972. Quantitative fluorescence analysis of soil lipase activity. Soil Biology and Biochemistry, 4, 257-259.

102. PANCHOLY S.K. and RICE E.L. 1973. Soil enzymes in relation to old field succession : amylase, cellulase, invertase, dehydrogenase and urease. Soil Science Society of America Proceedings, 37, 47-50.

103. PANCHOLY S.K. and RICE E.L. 1973. Carbohydrases in soil as affected by successional stages of revegetation. Soil Science Society of America Proceedings, 37, 227-229.

104. PAULSON K.N. and KURTZ L.T. 1969. Locus of urease activity in soil. Soil Science Society of America Proceedings, 33, 897-901.

105. PAULSON K.N. and KURTZ L.T. 1970. Michaelis constant of soil urease. Soil Science Society of America Proceedings, 34, 70-72.

106. PETTIT N.M., GREGORY L.J., FREEDMAN R.B. and BURNS R.G. 1977. Differential stabilities of soil enzymes. Assay and properties of phosphatase and arylsulphatase. Biochimica et Biophysica Acta, 485, 357-366.

107. PETTIT N.M., SMITH A.R.J., FREEDMAN R.B. and BURNS R.G. 1976. Soil urease : activity, stability and kinetic properties. Soil Biology and Biochemistry, 8, 479-484.

108. PUGH K.B. and WAID J.S. 1969. The influence of hydroxamates on ammonia loss from an acid loamy sand treated with urea. Soil Biology and Biochemistry, 1, 195-206.

109. PUGH K.B. and WAID J.S. 1969. The influence of hydroxamates on ammonia loss from various soils treated with urea. Soil Biology and Biochemistry, 1, 207-217.

110. QIU F.Q., ZHOU L.K., CHEN E.F., DING Q.T., ZHANG Z.M. and DANG L.C. 1981. Relationships between organic matter and enzymatic activities and soil fertility in black soils. Acta Pedologica Sinica, 18, 244-254.

111. RAMIREZ-MARTINEZ J.R. 1968. Organic phosphorus mineralization and phosphatase activity in soil. Folia Microbiologica, 13, 161-174.
112. RAS'KOVA N.V. and ZVYAGINTSEV D.G. 1977. Influence of storage of soil specimens on enzyme activity and thermostability. Moscow University Soil Science Bulletin, 32, 47-51.
113. ROBERGE M.R. 1978. Methodology of soil enzyme measurement and extraction. Appendix in Soil Enzymes. Ed. Burns R.G. Academic Press, London.
114. ROON R.J., HAMPSHIRE J. and LEVENBERG B. 1972. The involvement of biotin in urea cleavage. The Journal of Biological Chemistry, 247, 7539-7545.
115. ROSS D.J. 1976. Invertase and amylase activities in ryegrass and white clover plants and their relationships with activities in soils under pasture. Soil Biology and Biochemistry, 8, 351-356.
116. ROSS D.J. 1976. Distribution of invertase and amylase activities in pasture topsoil fractions isolated by ultrasonic dispersion in Nemagon and a surfactant. Soil Biology and Biochemistry, 8, 485-490.
117. ROSS D.J. and McNEILLY B.A. 1975. Studies of a climosequence of soils in tussock grasslands. 3. Nitrogen mineralization and protease activity. New Zealand Journal of Science, 18, 361-375.
118. ROWELL M.J., LADD J.N. and PAUL E.A. 1973. Enzymatically active complexes of proteases and humic acid analogues. Soil Biology and Biochemistry, 5, 699-703.
119. SAHRAWAT K.L. 1979. Evaluation of some chelating compounds for retardation of urea hydrolysis in soil. Fertilizer Technology 16, 244-245.
120. SARKAR, J.M., BATISTIC L. and MAYAUDON J. 1980. Les hydrolases du sol et leur association avec les hydrates de carbone. Soil Biology and Biochemistry, 12, 325-328.
121. SATYANARAYANA T. and GETZIN L.W. 1973. Properties of a stable cell-free esterase from soil. Biochemistry, 12, 1566-1572.
122. SCHNURER J. and ROSSWALL T. 1982. Fluorescein diacetate hydrolysis as a measure of total microbial activity in soil and litter. Applied and Environmental Microbiology, 43, 1256-1261.
123. SEMLER B.L., HODSON R.C., WILLIAMS S.K. and HOWELL S.H. 1975. The induction of allophanate lyase during the vegetative cell cycle in light-synchronized cultures of Chlamydomonas reinhardi. Biochimica et Biophysica Acta, 399, 71-78.
124. SEQUI P. 1974. Enzymes in soil. Italia Agricola, 111, 91-109.
125. SHORER J., ZELMANOWICZ I. and BARASH I. 1972. Utilization and metabolism of urea during spore germination by Geotrichum candidum. Phytochemistry, 11, 595-605.
126. SINGH B.B. and TABATABAI M.A. 1978. Factors affecting rhodanese activity in soils. Soil Science, 125, 337-342.
127. SKUJINS J. 1976. Extracellular enzymes in soil. CRC Critical Reviews in Microbiology, 4, 383-421.
128. SKUJINS J. 1967. Enzymes in soil. Chapter 15 in Soil Biochemistry Vol.1., Eds. McLaren A.D. and Peterson G.H. Marcel Dekker, New York.
129. SKUJINS J.J. and McLAREN A.D. 1969. Assay of urease activity using ^{14}C-urea in stored, geologically preserved, and in irradiated soils. Soil Biology and Biochemistry, 1, 89-99.
130. SMITH M.S., FIRESTONE M.K. and TIEDJE J.M. 1978. The acetylene inhibition method for short-term measurement of soil denitrification and its evaluation using nitrogen-13. Soil Science Society of America Journal, 42, 611-615.

220

131. SMITH M.S. and TIEDJE J.M. 1979. Phases of denitrification following oxygen depletion in soil. Soil Biology and Biochemistry, 11, 261-267.
132. SØRENSEN L.H. 1969. Fixation of enzyme protein in soil by the clay mineral montmorillonite. Experientia, 25, 20-21.
133. SPEIR T.W. 1976. Studies on a climosequence of soils in tussock grasslands. 8. Urease, phosphatase, and sulphatase activities of tussock plant materials and of soil. New Zealand Journal of Science, 19, 383-387.
134. SPEIR T.W. 1977. Studies on a climosequence of soils in tussock grasslands. 10. Distribution of urease, phosphatase, and sulphatase activities in soil fractions. New Zealand Journal of Science, 20, 151-157.
135. SPEIR T.W., LEE R., PANSIER E.A. and CAIRNS A. 1980. A comparison of sulphatase, urease and protease activities in planted and in fallow soils. Soil Biology and Biochemistry, 12, 281-291.
136. SPEIR T.W. and ROSS D.J. 1978. Soil phosphatase and sulphatase. Chapter 6 in Soil Enzymes. Ed. Burns R.G. Academic Press, London.
137. SPEIR T.W. and ROSS D.J. 1981. A comparison of the effects of air-drying and acetone dehydration on soil enzyme activities. Soil Biology and Biochemistry, 13, 225-229.
138. SPIERS G.A. and McGILL W.B. 1979. Effects of phosphorus addition and energy supply on acid phosphatase production and activity in soils. Soil Biology and Biochemistry, 11, 3-8.
139. STEVENSON F.J. 1982. Origin and distribution of nitrogen in soil. Chapter 1 in Nitrogen in Agricultural Soils. Agronomy Monograph, no.22. Ed. Stevenson F.J. American Society of Agronomy, Madison, Wisconsin.
140. SUFLITA J.M. and BOLLAG J.-M. 1980. Oxidative coupling activity in soil extracts. Soil Biology and Biochemistry, 12, 177-183.
141. TABATABAI M.A. 1973. Michaelis constants of urease in soils and soil fractions. Soil Science Society of America Proceedings, 37, 707-710.
142. TABATABAI M.A. 1977. Effect of trace elements on urease activity in soils. Soil Biology and Biochemistry, 9, 9-13.
143. TABATABAI M.A. and BREMNER J.M. 1970. Factors affecting soil arylsulfatase activity. Soil Science Society of America Proceedings, 34, 427-429.
144. TABATABAI M.A. and BREMNER J.M. 1971. Michaelis constants of soil enzymes. Soil Biology and Biochemistry, 3, 317-323.
145. TABATABAI M.A. and BREMNER J.M. 1972. Assay of urease activity in soil. Soil Biology and Biochemistry, 4, 479-487.
146. TABATABAI M.A. and DICK W.A. 1979. Distribution and stability of pyrophosphatase in soils. Soil Biology and Biochemistry, 11, 655-659.
147. TARARINA L.F., VOINOVA V.N. and EMTSEV V.T. 1981. Effect of γ-radiation and high temperatures on enzymatic activity of grey forest soil. Izvestiya Timiryazevskoi Sel'skokhozyaistvennoi Akademii, 1, 93-101.
148. TATE R.L. 1979. Effect of flooding on microbial activities in organic soils: carbon metabolism. Soil Science, 128, 267-273.
149. THENTE B. 1970. Effects of toluene and high energy radiation on urease activity in soil. Lantbrukshogskolans Annaler, 36, 401-418.
150. TYLER G. 1974. Heavy metal pollution and soil enzymatic activity. Plant and Soil, 41, 303-311.
151. TYLER G. 1981. Heavy metals in soil biology and biochemistry. Chapter 9 in Soil Biochemistry Vol.5. Eds. Paul E.A. and Ladd J.N. Marcel Dekker, New York.
152. VOETS J.P. and DEDEKEN M. 1966. Soil enzymes. Mededelingen Rijksafaculteit Landbouwettenschappen, Ghent, 31, 177-190.

153. WAINWRIGHT M. 1978. A review of the effects of pesticides on microbial activity in soils. Journal of Soil Science, 29, 287-298.
154. WAINWRIGHT M. 1979. Assay of phenol o-hydroxylase activity in soil. Soil Biology and Biochemistry, 11, 549-551.
155. WHITNEY P.A. and COOPER T.G. 1972. Urea carboxylase and allophanate hydrolase: Two components of a multienzyme complex in *Saccharomyces cerevisiae*. Biochemical and Biophysical Research Communications, 49, 45-51.
156. WHITNEY P.A. and COOPER T.G. 1972. Urea carboxylase and allophanate hydrolase. The Journal of Biological Chemistry, 247, 1349-1353.
157. WILKE B.M. 1982. Lead sorption and the effect of lead pollution on biological activity of different types of humus forms. Zeitschrift fur Pflanzenernahrung und Bodenkunde, 145, 52-65.
158. YOSHIKURA J., HAYANO K. and TSURU S. 1980. Effects of drying and preservation on β-glucosidases in soil. Soil Science and Plant Nutrition, 26, 37-42.
159. ZANTUA M.I. and BREMNER J.M. 1976. Production and persistence of urease activity in soils. Soil Biology and Biochemistry, 8, 369-374.
160. ZANTUA M.I. and BREMNER J.M. 1977. Stability of urease in soils. Soil Biology and Biochemistry, 9, 135-140.
161. ZHOU L.K., ZHANG Z.M. and CHEN E.F. 1981. Enzyme activities in black soils. Acta Pedologica Sinica, 18, 158-166.

CHAPTER 6

THE SOIL BIOMASS

G.P. SPARLING

The Macaulay Institute for Soil Research, Aberdeen, Scotland.

CONTENTS

1. INTRODUCTION

The role of the microbial biomass in the transformation of organic matter in soil is a crucial one and the rates of turnover and mineralization of organic substrates are largely governed by the activity of the soil biomass. Inhibition of microbial activity by a low or high temperature, drought, waterlogging, extremes of pH or xenobiotic substances may result in the persistence in soil of potentially decomposable and mineralizable compounds with subsequent effects on fertility, nutrient cycling and soil structure. The microbial biomass is itself part of the soil organic matter, typically about two percent of the total organic C [110] and is defined as the living microbial component of the soil and includes bacteria, actinomycetes, fungi, protozoa, algae and microfauna. Usually, plant roots and fauna larger than 5×10^3 μm^3, such as earthworms, are not included. For the purpose of this chapter, the biomass will be considered as an undifferentiated whole, usually expressed quantitatively as biomass-C, cell biomass or biovolume (μm^3). However, the activities of various components of the biomass - protozoa, nematodes, etc. - have been included where this seems appropriate.

The diverse activity of the biomass in soil organic matter studies integrates with many topics covered in other parts of this book. The reader is, therefore, referred to the appropriate chapters for further information on rates of decomposition and turnover, microbial polysaccharide and soil structure, phytotoxic compounds, enzyme activities, and mineral cycling.

2. METHODS

2.1. Measurement of the soil biomass

A quantitative assessment of the biomass requires that the method used must be appropriate to all the diverse components of the biomass; an alternative approach is to use several different methods to build up a composite picture. The former method is currently the most popular although no single technique has, as yet, been found to be satisfactory for all soils. Many of the methods used to estimate biomass in pure cultures, tissues, aquatic environments and sediments have been applied to soils [120, 132, 136]. Examples are cell counts, rates of respiration, ATP content, enzyme activities and analyses for particular cellular components, such as n-glucosamine. A major advance in the method of estimating the soil biomass has been the introduction of the chloroform-fumigation technique. A full description of this method and its development has been published [105, 106, 110, 113, 114] and a critical review of its use and relationship with other methods of biomass assessment, such as direct microscopy or analysis for biomass constituents, is given by Jenkinson and Ladd [110]. Therefore, only a brief resumé is given here, with emphasis on recent developments, anomalies that have occurred, and some of the more recent studies where the relationships between various biomass indices have been measured.

2.2. The decomposition of the fumigated biomass

A value of 0.5 was initially suggested by Jenkinson and Powlson [114] for the k-factor, which is the fraction of the biomass mineralised to CO_2, and is used to calculate the biomass from the flush of CO_2 over 10 days at 25°C. Anderson and Domsch [10] found that bacteria were decomposed more rapidly than fungi, and, making allowances for a bacterial to fungal ratio of the biomass of 1:3, they suggested that for incubation at 22°C a factor of 0.41 was more applicable. A factor of 0.45 seems a reasonable compromise for 25°C incubations. The flush of CO_2 from some fumigated acid soils is anomalously low [50, 109, 230] and may not be detectable in soils that have had recent amendments of available substrates [187, 236]. As yet there seems no satisfactory way of measuring biomass-C on acid (<pH 5) soils using the fumigation method, and adding lime or nutrients to neutralise the soil, and to stimulate the flush, results in complex patterns of CO_2 release. Paul and Voronay [187] suggested that the biomass of a recently amended soil could be estimated without taking into account the respiration of the unfumigated soil, but provided meagre evidence in support of this modification. Lynch and Panting [147] also showed that the biomass could be estimated using the CO_2-C respired solely from the fumigated soil, but in contrast to Paul and Voronay derived a regression curve linking the soil biomass (determined by

the usual method) with the CO_2 respired from the fumigated soil, where B_k = 0.673x - 3.53, x = CO_2-C produced in 10 days and k was typically 0.41 - 0.50. This modification removes the requirement to measure the CO_2 level on the non-fumigated (control) soils, but it is not clear whether this would apply to soils recently amended with substrates.

2.3. Relationship between biomass-C determined by fumigation and other methods of estimating the biomass

2.3.1 Biovolume estimated by microscopy.

There is general agreement between biovolume and fumigation biomass-C in some soils [116] and in compost [234]. However, biovolume estimates involve a number of assumptions about extraction, shrinkage, cell density, moisture content and composition. Some of these assumptions have been challenged [175, 253] and it seems that the cell density of the soil biomass is higher than commonly assumed. To convert biovolume to biomass a value of 0.33 g cm^{-3} is suggested for fungi [253], while 0.28 g cm^{-3} is suggested for the whole biomass [110]. A large proportion of organisms viewed by direct methods may be non-viable or inactive, but the use of fluorescent vital stains can help to reveal the active cells [14, 104, 225, 226]. Direct counts of bacteria and estimates of hyphal length and biovolume show that the fungi are the dominant biomass component in the majority of soils. Baath and Soderstrom [23] emphasise the importance of the method used to calculate the mean hyphal cross section, showing that taking the arithmetic mean can lead to an underestimation of the biovolume. On balance, there are probably fewer errors associated with the determination of biomass-C by the fumigation technique than with biovolume estimates. However, when trying to assess the relative merits of various methods it should be borne in mind that the justification for them may rely on a good agreement with biomass as determined from the biovolume.

2.3.2. Biomass-C and the rate of respiration.

There is rarely any consistent relationship between the basal rate of respiration of a soil and the fumigation biomass, this being because a large proportion of the biomass is inactive [147]. However, the addition of saturating quantities of a readily available substrate such as glucose generally causes a large immediate increase in the rate of CO_2 respiration.

The patterns of CO_2 respiration may differ between amended soils, but can be related to the biomass, determined by fumigation, by the equation: $y = 40.04x + 0.37$ where y = mg biomass-C $100g^{-1}$ soil and x = rate of respiration, ml CO_2 $100g^{-1}$ soil at 22°C [11]. Specific inhibitors may allow differentiation between the bacterial and fungal components of the biomass [11] although inhibitors do not appear satisfactory on some soils [208]. The inhibitor studies also substantiate the evidence for dominance of fungi over the bacteria at a ratio of about 3:1. The rates of respiration from the glucose-amended soil biomass (about 25 μlCO$_2$ mgC^{-1} h^{-1}) are similar to those of glucose-amended fungi in late linear and stationary phases of growth (18 - 42 μlCO$_2$ mgC^{-1} h^{-1}) [11]. Poor agreement between the biomass-C in leaf litter using the Anderson and Domsch technique and the biomass of active cells using microscopy and fluorescent staining was reported by Ineson and Anderson [102], who suggested that either dead hyphae were being stained, or that the respiration of some active hyphae had not been stimulated by glucose.

2.3.3. The ATP content of the biomass.

The adenosine triphosphate (ATP) content of organisms has long been used as a measure of their metabolic activity [132], but problems in applying the method to soil have centred around optimizing extraction from the organisms, prevention of hydrolysis, and minimising sorption by soil colloids. Various acidic, neutral, alkaline and organic reagents have been advocated, but the three main ones currently in use are cold 0.5 M H_2SO_4 [67, 83], sodium bicarbonate-chloroform [185] and trichloracetic acid-phosphate- paraquat [112]. In comparisons of these three main extractants, Jenkinson et al. [109] showed that trichloracetic acid-phosphate-paraquat (TCA-P-P) was better for native soil ATP than bicarbonate-chloroform, while Sparling and Eiland [232] showed that TCA-P-P was better than H_2SO_4. However, TCA-P-P is a highly corrosive and toxic reagent and these characteristics have probably inhibited its widespread use. In fact, H_2SO_4 may prove to be a satisfactory extractant, because with the increasing sensitivity of photometers, lower concentrations of ATP can be measured accurately, and recoveries improved by using higher extractant-to-soil ratios [69]. Where ATP is measured by photometry either immediately or shortly after the addition of luciferin-luciferase, the sensitivity of the assay can be increased by replacing arsenate with Tris buffer. Arsenate buffer is appropriate for light emmission measured up to 1h later on a scintillation spectrophotometer [112] because the light emmission, although initially lower, decays less rapidly. The specificity of the assay can be improved by using purified enzymes [244]; for reliable measurement of the microbial ATP, soil treatments such as air drying, freeze-drying, waterlogging or prolonged storage should be avoided [5, 243].

Measurement of the ATP content of the soil and the biomass-C by fumigation has enabled the ATP content of the biomass to be calculated. Oades and Jenkinson [182] estimated this to be ca. 8.33 mg ATP g^{-1} biomass-C and the biomass-C:ATP ratio to be about 120. They suggested that this ratio was sufficiently constant for it to be used to estimate the biomass-C from the ATP content of soil. Subsequent work [109] revised this ratio to 138. Modification of the soil storage conditions, use of purified enzymes and calculation of the biomass using the CO_2 flush between 0-10 days rather than 10-20 days have revised these estimates. Powlson and Jenkinson [192] found the biomass-C:ATP ratio in cultivated soils to be 152 or 193 depending on how the biomass was calculated, whereas Tate and Jenkinson [243] suggest 171 as the revised figure after allowing for GTP interference. This ratio was changed from 234 to 168 by the comparatively mild treatment of soil storage at 25°C for 7 days [243]. Ross et al. [207] also found that storage and temperature affected the ATP content of soil, and recommended storage at -20°C for ATP and biomass-C determinations. This suggests that ATP is such a labile biomass constituent that laboratory measurement of ATP may not relate very closely to levels in the field. Various effects of soil P level on the biomass ATP content have been noted [109, 170, 232], but again the effect does not appear consistent. Ahmed et al. [5] did not consider the use of a single factor to convert ATP content to biomass-C to be justified, and Ross et al. [208] have found large seasonal variations in the biomass-C:ATP ratio. Until further data becomes available on how the ATP level of the biomass may fluctuate, it seems best to be cautious when estimating biomass-C from the ATP content.

2.4. Estimation of biomass-N, biomass-P and biomass-S by the fumigation technique

The apparent specificity of $CHCl_3$ for biomass-C as opposed to the native organic C also seems to apply to the biomass-N, P and S. Ayanaba et al.[18] showed that following fumigation there was increased mineralization of N during the 10 day incubation. The increased N (N-flush) was assumed to have been derived from the killed biomass, the ratio of biomass-C to N-flush being roughly 8:1. This fairly high N content of the biomass probably refers more accurately to the more easily decomposable cytoplasmic components that are mineralized during the 10 day incubation[110]. Marumoto et al.[158, 159] studied the decomposition of fumigated ^{15}N-labelled cells in soil and concluded that the fraction of microbial-N mineralized over 28 days was 38%, i.e. a k_N factor of 0.38. Other k_N values have been derived indirectly: Ladd et al.[131] proposed a provisional value of 0.625 for a 10 day incubation, and Voronay et al.[256] used the much lower value of 0.28 whilst Carter and Rennie[49] used a provisional value of 0.4. The use of ^{15}N-labelled materials and the consequent detection of ^{15}N in the N-flush following fumigation has allowed the incorporation and conversion of ^{15}N-substrates into biomass to be estimated[7, 131, 158, 159].

A similar approach has been adopted to measure biomass-P, the main difference being that it is possible to detect a P-flush immediately after fumigation, without the need for any further incubation and mineralization phase. Presumably much of the P is released from the chloroform-lysed organisms in organic forms (P_o) but is converted to inorganic forms (P_i) by the action of soil phosphatases. The phosphatases seem unaffected by the chloroform treatment, and in the organic soils we have tested, have sufficient activity to account for the levels of P_i detected (Table 1). Brookes et al.[45] suggest that the level of P_o is sufficiently low for it to be ignored and that biomass-P can be determined from the P_i in a bicarbonate extract. An allowance is made for incomplete extraction of P_i by calculating the recovery of P_i from a "spiked" sample. An alternative scheme is proposed by Hedley and Stewart[99] to fractionate soil P (Chapter 9) and to estimate biomass-P, again using chloroform lysis, but because the method uses soil that has been previously dried and ground, the biomass is likely to have been drastically altered, and the measurements may not be representative of samples from the field. Both methods allow for only partial mineralization of microbial P to P_i, and an overall k_p factor of 0.4 is suggested, which is similar to the k-factors currently used for biomass-C and biomass-N.

Table 1. Biomass, extractable P_i and phosphatase activity on Scots Pine F-H layers amended with glucose or glucose and NPK salts. (From Sparling and Williams, unpublished).

Amendment (% w/w)	Biomass[1] ($\mu gC\ g^{-1}$)	Extractable P_i[2] Control	Fumigated	Phosphatase activity[3] μmoles PNP $g^{-1}h^{-1}$
1 None	851	4.5	96.1	5.59
2 1% glucose	2084	18.2	80.0	4.97
3 5% glucose	4440	13.8	56.3	7.98
4 1% glucose+NPK	3829	939.5	952.2	3.77
5 5% glucose+NPK	4771	4050	4236	29.12

[1] Estimated by the Anderson and Domsch method.
[2] Extracted in 0.01 M $CaCl_2$, not corrected for % recovery ($\geq$90%).
[3] Estimated on fumigated samples as μmoles p-nitrophenol $g^{-1}h^{-1}$ at 25°C at pH 4.8.

A similar scheme has been proposed for the estimation of biomass-S. Again the basis of the method is the selective action of $CHCl_3$ on the release of S from the biomass rather than the native organic matter. The S is detected in $CaCl_2$ or $NaHCO_3$ extracts [216,217] and biomass-S estimated from the difference between fumigated and non-fumigated soil, using an assumed mineralization of S in 10 days of 34.6% or 41.2% depending upon the extractant used. This factor is of a similar order to those previously mentioned for the estimates of C, N and P.

2.5. Trace elements

Soil microorganisms are known to accumulate certain trace elements, while they may exclude others [250, 251]. Little work has been reported on the trace element composition of the biomass, but trace elements have been estimated in ammonium acetate extracts from fumigated and non-fumigated soils (Sparling and Berrow, unpublished data). While there were generally higher levels in fumigated than non-fumigated soils, the concentrations were very different from those that would be expected if the trace element concentration in the biomass bore any resemblance to that of fungal sporophores (Table 2) [250, 251].

Table 2. Trace elements extracted by 1 M ammonium acetate and CO_2-C respired from fumigated and non-fumigated soil (μg g^{-1} dry soil).

Element	Immediately after fumigation			After 10 day incubation at 25°C		
	Control	Fumigated	Flush	Control	Fumigated	Flush
Mg	108	113	5	96	99	3
Cu	0.214	0.253	0.039	0.184	0.192	0.008
Fe	1.616	1.560	<0	1.470	1.284	<0
Mn	0.468	1.054	0.586	0.602	0.660	0.058
Zn	0.782	0.750	<0	0.668	0.554	<0
K	84.16	87.04	2.88	91.21	96.96	5.75
CO_2-C	0	0	0	81.80	195.1	113.3

2.6. Microcalorimetry

The heat output from organisms reflects their catabolic activity [32,74] and is largely independent of species and reaction pathway. The use of microcalorimetry for measuring microbial activity in soil was suggested by Mortensen et al. [166] and Ljungholm et al. [137, 138] who recorded changes in catabolic activity in various soils. Sparling [231] related the rate of heat output from soils to the biomass-C and the rate of respiration. The rate of CO_2 respiration and heat output were closely related, but, in unamended soils, showed little relationship to the biomass-C determined by fumigation. However, in an analogous way to the Anderson and Domsch respiration technique discussed previously, the rate of heat output 30-90 min after the addition of saturating quantities of glucose was linearly related to the biomass-C, estimated by the fumigation method, the relationship being roughly 1 g biomass-C = 180 mW.

2.7. Analyses for other constituents of the biomass
2.7.1 Hexosamines.

The two main hexosamines of interest for estimating the biomass are muramic acid and n-acetyl glucosamine. Unfortunately, neither of these compounds appears to be specific for the living soil biomass. Thus, although muramic acid is a constituent of the cell walls of bacteria and blue-green algae, measurement of the muramic acid content of soil [162] gave biomass figures that were unrealistically large, assuming that the muramic acid content of the living biomass is about 19.4 mg g^{-1} biomass-C [110]. It is, therefore, likely that a large proportion of muramic acid extracted from soils is from dead cells.

N-acetyl glucosamine is a component of those fungal cell walls which contain chitin, but is almost absent from prokaryotes and higher plants [262]. Fungal biomass has been estimated by measuring the glucosamine content [77] but the glucosamine content of the hyphae changes with age and the nutrient content of the substrate, making conversion of glucosamine content to biomass-C difficult. Frankland et al. [77] suggested that for leaf litter measurement of hyphal biovolume by microscopy was both more accurate and rapid than the n-glucosamine assay. Chitin is also present in soil as non-fungal components, such as insect exo-skeletons, which is a further source of error.

2.7.2. Nucleic acids.

All organisms contain nucleic acids and thus analysis for the nucleic acid content of soil should give a measure of the biomass. However, although the initial decomposition of free RNA and DNA added to soil appears to be rapid [84, 96, 97] and the DNA in soil seems predominantly of microbial rather than plant origin [8], the levels extracted from soil seem too high for it to have originated solely from the living biomass [110]. Some microbial DNA does seem to survive in soil even after decomposition of the cell structure [8] making the method unsuitable for use with soil samples.

2.8. Relationship between various methods of measuring the biomass

Mention has been made already of the relationships between biovolume, biomass-C from the C-flush or N-flush following fumigation, the ATP content of the biomass, and the heat output of the biomass. Few studies have attempted to measure many more paramenters, but Domsch et al. [65] measured the rates of O_2 and CO_2 respiration, the ATP content, biomass by the respiration/amendment method, enzyme activities and microbial numbers by plate counts and direct observation. Good agreement between the relative values was obtained for biomass-C, the ATP content and the activity of the soil enzymes dehydrogenase and amylase. In contrast, Sparling [230] found very few significant correlations between seven methods used to estimate biomass and activity in 11 soils, although good agreement between the biomass estimated from the C-flush, rate of respiration after substrate addition, biovolume and rate of heat output was obtained with a mushroom compost [234]. Ross et al. [209] found the relationship of the C-flush, N-flush and ATP content to be correlated, but the ratios varied in a series of nine soils, while in a further study [208], they noted seasonal changes in the relationships.

Nannipieri et al. [170, 171] investigated biomass, ATP content, respiration and enzyme activities in soils, but found that most of the parameters varied independently, especially after nutrient amendments; a similar observation was made by Sparling et al. [235]. It is unlikely that any one parameter can satisfactorily measure the biomass and activity in all soils and much will depend on the particular aspect being investigated. Powlson and Jenkinson [192] suggested that the biomass-C and biomass-N determined by fumigation were sensitive indicators of changing soil processes, while Ross et al. [206] have suggested that the invertase activity of soil may be a reasonable indicator of the soil fertility status.

3. SOIL ORGANIC MATTER AND THE BIOMASS

3.1. Organic matter as a source of nutrient

The bulk of the soil biomass is inactive, normally because of nutrient limitation, and the addition of available organic substrate to soil generally results in a large increase in microbial activity and the biomass. The main source of organic input to the biomass is from plant material, in the form of roots, leaf and stem litter, and what may be broadly termed root exudates. Overall, those soil systems that have the highest organic matter inputs also tend to have the highest biomass with the greatest microbial activity. However, this tenet is modified by the availability of the organic substrate to the biomass, and environmental factors may place a restraint on the activity of the biomass. Only a small proportion of the total soil organic matter input is readily utilized by the organisms, such that the biomass-C generally comprises only 1-3% of the soil organic-C in mineral soils, and a much smaller proportion in organic soils and litter layers [12, 37, 113, 145, 146, 254].

3.2. Cultivation and the soil biomass

Cultivation of natural vegetation such as forest or prairie decreases both the soil organic C and the soil biomass [1, 18, 256]. Those crops that have copious root systems produce a larger soil biomass than those with less extensive root systems. For example, an arable field in England contained 374 µg biomass-C (1.8% of the organic-C) while a nearby grassland on similar soil contained 1061 µg biomass-C (4.6% of the organic-C) [143]. Direct drilling of crops increases the amount of root and biomass in the top 5 cm of soil [49, 145, 146] although over the whole depth profile there may be little difference between direct-drilled and ploughed soils [192]. Burning of straw stubble decreased the biomass from 278 to 149 µgC g^{-1} in the surface soil [146]. The deeper soil horizons have a lower biomass than the surface horizons. Cropping increases the proportion of soil C and N comprised by the biomass in the A soil horizon, but decreases the proportion in the B horizon [256].

Fertilisers generally increase the biomass; organic fertilisers cause a much greater increase than inorganic fertilisers. Part of the increase in the biomass caused by organic fertilisers results from a direct stimulation to the biomass, rather than from increased crop growth, because considerable increases in biomass occur on fallow soils receiving regular fertiliser additions (Table 3) [114]. Periodic addition of farmyard manures or slurries to soils causes an initial small increase in the biomass and the FDA-active fungi over 20 days, but microbial activity then declines almost to the pretreatment level [68]. Field soils in Denmark receiving regular additions of farmyard manure behaved very similarly to the Broadbalk soils (Table 3) which had higher biomass in the manured soils than those given NPK fertiliser [68].

Table 3.   The soil biomass (µgC g^{-1}) in soils from long-term cultivation experiments (Broadbalk).

Fertiliser treatment	Wheat	Fallow
None	128	96
Farmyard manure	387	344
N, P, K, Mg, Na	189	176

Recalculated from Jenkinson and Powlson [114] using the CO_2-C flush between 0-10 days and a k-factor of 0.45.

3.3. <u>The rhizosphere</u>

The rhizosphere is that zone of the soil which is influenced by the presence of a root. The rhizosphere has been the subject of several reviews 44, 176, 210, 211, 266 which note that the rhizosphere is a zone of increased microbial activity, largely resulting from the increased levels of organic compounds in that region, originating from the exudation of soluble-C compounds, the secretion of mucilages and the lysis of root cap cells, root hairs and epidermal and cortical cells.

It has long been known that bacterial numbers are much higher in the rhizosphere than the bulk soil, and that the organisms differ in their species and physiological requirements. It is only recently, however, that attempts have been made to quantify the rhizosphere <u>biomass</u> rather than <u>numbers</u> and to relate this to the organic matter availability and root exudation. Roots can lose a considerable proportion of their photosynthate to their growth medium, and for soil-grown plants losses of up to 23% of the fixed photosynthate are recorded 155, 176 the losses being even greater if root and microbial respiration over the whole growth period are included 261.

Despite the apparently rich supply of organic substrates and the large increases in microbial numbers detected by plate counts, direct observation of the root surface shows that only a small proportion of the root surface is colonised by microorganisms 213. The root tips are free of organisms, but typically 0.3 - 9% of the older root surface is colonised by bacteria and 0.2 - 4% by fungi 177. Bacteria appear to be the major component of the rhizosphere biomass 168 and Vancura and Kunc 252, using "specific" inhibitors, estimated the rhizosphere biomass to be 70-80% bacteria and 10-35% fungi, compared to 20-30% bacteria and 70-80% fungi in non-rhizosphere soil. It is not clear whether mycorrhizal hyphae would be estimated by this method. However, the weight of extramatrical mycelium of the endo-mycorrhizal fungi is known to be related to the amount of internal infection of the roots 129, 219, and infected plants would be expected to have a higher fungal component in the rhizosphere biomass.

Newman and Watson 177 brought together much available data into a mathematical model which related the amounts of available substrate to microbial growth in the rhizosphere. The model revealed that in terms of increased microbial biomass the rhizosphere was very narrow, typically only fractions of a millimeter, and that the gradients were very steep. Subsequent work suggests that the rhizosphere is even more restricted than they suggested. Newman and Watson used a microbial concentration of 200 μg cm^{-3} or about 100 μgC g^{-1} soil for their maximum biomass, whereas the biomass-C estimated from fumigation can be up to 2000 μgC g^{-1} soil; hence in terms of microbial growth the rhizosphere effect is probably even more restricted than the model predicts [177]. Direct microscopy of wheat rhizoplanes revealed that microbial (mainly bacteria) proliferation was less than predicted and that growth was related to the lysis of root epidermal cells rather than exudation of soluble substrates [255].

The relatively small difference between the size of the biomass in rhizosphere and in non-rhizosphere soil is not necessarily in conflict with earlier reports of large increases in microbial numbers in the root zone. The dry weight of a single bacterium is about $0.2x10^{-6}$ μg, i.e. $0.1x10^{-6}$ μgC. Thus 1 μg biomass-C is equivalent to some $10x10^{6}$ bacteria, and a small (10%) change in the biomass from (say) 500 to 550 μgC could represent $5x10^{8}$ bacteria which would appear as a dramatic increase in numbers determined by direct or plate counts. Old grasslands, in which the soil may be considered to be almost all rhizosphere, typically have biomass levels 3-4 times higher than fallow soils. The rhizosphere effect is, therefore, much less dramatic when expressed as biomass rather than microbial numbers.

4. GROWTH AND ACTIVITY OF THE BIOMASS

4.1. Activity

There is general agreement among authors, using a variety of techniques, that only a small proportion of the total biomass is active, values ranging from 2.4 - 27.2% (Table 4). The main restriction on activity is a lack of organic substrate [80, 144, 186, 187, 257, 268] and the additions of readily available substrates cause a virtually instantaneous increase in respiration [11] and heat output [231]. Soil organisms maintain a high adenylate energy charge (AEC) which partly explains their ability to make a rapid response to added exogenous substrates [46]. Generally, however, exogenous substrates are in very low concentrations so that growth and activity of the biomass is restricted, and the small amounts of substrate available are used for maintenance.

Table 4. Estimates of the proportion (%) of active organisms in soil by various methods.

Soil	Organisms	% active	Method	Reference
Bare fallow	All	10.9	Respiration rates	unpublished
Grassland	All	27.2	Respiration rates	unpublished
Pine litter	Fungi	2.4-4.3	FDA staining	226
Peat	Bacteria	15-30	Direct observation	56
Arable	Bacteria	15	Histochemistry	148
Garden soil	Bacteria	11	Histochemistry	148
Compost	Bacteria	23	Histochemistry	148

4.2. Maintenance and growth

Despite the apparent need to maintain a high AEC, the biomass is well adapted to survive long periods of starvation in soil and the maintenace requirements are extremely low [80, 81, 110, 187, 240]. Typically, the biomass shows a slow decline when deprived of organic substrates, e.g. during periods of soil storage or under bare fallow, [50, 114, 131, 159, 192, 256].

Table 5. Conversion of substrate-C to biomass-C[*].

Substrate	Incubation (°C)	Duration (days)	% Substrate C in biomass	Reference
Glucose	25	51	11	113
Medic	25	62	9.9	7
Medic	Field	56	8-18	131
Rye grass	25	6	9.3	50
Rye grass	25	62	10.8	50
Glucose	22	10	4-11	154
Acetate	22	84	3.9-12.5	121
Pyruvate	22	84	0.9-7.5	121
Cellulose	22	84	6.1-7.2	121
Microbial cells	22	84	0.8-4.9	121
Melanoid cells	22	84	0.2-1.3	121
Phenolic acids	22	84	0.3-6.7	122

[*] calculated using the fumigation method.

The amount of substrate incorporated into new microbial biomass is much lower than expected from theoretical considerations and pure culture experiments. Payne[188] suggested that a conversion efficiency of about 60%

of substrate-C to cell-C was possible under aerobic conditions, and a conversion effeciency of 50% was assumed by Van Veen and Paul [254]. Table 5 shows the amount of substrate-C incorporated into the biomass from a variety of C sources. All were much lower than the theoretical level and lower than previous estimates where the biomass was not estimated by fumigation [33], [257]. The efficiency of conversion is higher at lower substrate concentrations [154] and varies from 0.2-18% depending on the substrate, soil and incubation conditions. Some substrates remain as relatively undecomposed residues [122] but other material may accumulate as non-biomass components. Martens [154] noted that a high proportion of labelled non-biomass material accumulates at high rates of glucose addition, this material appeared to be a microbial metabolite and had the characteristics of a polysaccharide. The accumulation of C in the organic matter rather than cells may be influenced by nitrogen availability [101].

The growth and turnover of the "native" soil biomass, in contrast to that newly formed by adding substrates, is very slow [50, 187, 228]. Data were presented by Jenkinson and Ladd [110] showing that on an unmanured wheat field in England which had reached a "steady state" the biomass turned over in 2.5 years. Various model systems have been proposed to relate biomass growth to C input and decomposition [43, 110, 117, 177, 187, 256, 257]. It is recognised that only a proportion of the incoming substrate is available to the biomass, and the substrate may be utilised for respiration, growth and metabolites, and that conversion efficiencies are generally low. For example, Cerri and Jenkinson [50] suggested that a suitable overall factor for the conversion of ryegrass to biomass over 62 days was 0.15.

The generally low rates of biomass growth make some of the apparently large fluctuations difficult to understand. Thus Ross et al. [208] reported annual fluctuations in the biomass of up to 480 μgC g^{-1} and Lynch and Panting [147] recorded apparent increases in the biomass of 275 μgC seemingly caused by a sieving procedure. Bearing in mind the comparatively poor efficiency of substrate utilization, it is doubtful if there would be sufficient C input to the soils to achieve this increase. Ross et al. [206] showed that the biomass was slow to re-establish on stripped soils planted to grass and Jenkinson and Powlson [115] found little effect of soil sieving on the size of the flush. Further information on C input to soil, microbial growth and the uniformity of response following the fumigation treatment is required.

Biomass newly formed following substrate addition declines much more quickly than the "native" biomass [50, 121, 122, 187, 222, 228, 256]. the reason for the rapid decline is uncertain, but presumably the rate of decline reflects the "protective capacity" of a soil. It has been known for many years that cells added to soil die very rapidly, but that the proportion surviving is influenced by the organic matter, $CaCO_3$, clay, predators, pH and competition for available substrates [123, 124, 139, 260]. Clay and silt can stabilise the biomass on some soils [131, 152, 154, 229] and may affect the response of the biomass to "stress" conditions such as soil drying [227, 228]. Many of the parameters affecting the survival of pathogens in soil [25, 221] also appear applicable to the survival and stability of the soil biomass.

4.3. Environmental limitations

Various factors that decrease the biomass or the ATP content have been mentioned in section 2 of this chapter. These treatments such as soil drying, freezing and waterlogging are natural occurrences in many environments, and may severely restrict the size and activity of the biomass. Certain groups of organisms are very tolerant of extreme conditions [130] and the relationship between environment, microbial activity and rates of decomposition has been the subject of several reviews [9, 107, 126, 186, 242, 256, 257]. Briefly, however, it may be mentioned that soil moisture is of particular importance [24, 72, 73, 128, 168, 180, 264] and that even in mire and forest soils increased microbial activity was noted following rain although moisture was not apparently limiting [56].

Other restrictions on microbial activity may result in organic matter accumulation. Such examples are: arctic and alpine tundra where the main limitation on microbial activity is low temperature, low available moisture and acidity, and cool/temperate peats where a combination of low temperature, acidity and poor aeration inhibit decomposition [242]. Many of these organic soils can be made productive by appropriate treatments such as liming and drainage and the biochemistry and microbiology of such soils is reviewed elsewhere [79]. Tropical soils may show marked annual fluctuations in rates of decomposition where drying of the soil during arid seasons limits the microbial activity [179, 256]. Despite the generally rapid rates of decomposition under tropical conditions [108], in some environments, particularly estuaries and deltas, the rate of plant productivity can be so high that organic matter (as peat) does accumulate [79].

Other soil factors can also affect to the rates of decomposition [72, 73]. Examples are: organic-matter metal complexing [156], phenolic polymers [173], enzyme inactivation [34, 35, 36], nutrient availability [60, 238], clays [152, 153, 237], oxygenation [87] and soil physical conditions [2, 3, 157, 269]. Some of these factors are discussed in more detail elsewhere in this volume. Herbicides and pesticides affect various soil microbial processes [118, 203, 258], inhibit decomposition [89, 193] and, depending on type and rate of application, can alter the biomass quantitatively and qualitatively in both the short and long term [13].

5. THE SOIL BIOMASS AND PLANT NUTRITION

5.1. Plant-microbial interactions

There are numerous plant-microbial interactions, and it is beyond the scope of this review to include them all. Some of the more important, in terms of plant nutrition and soil conditions, are mentioned here briefly. Various general reviews are available dealing with other topics such as plant pathogens, nitrification, denitrification, legumes, mycorrhizae, soil structure, root growth, biological control, and the physiology of both plants and soil microorganisms [25, 26, 39, 62, 63, 64, 70, 78, 79, 80, 81, 93, 215, 239, 247].

5.2. Soil organisms and root uptake of nutrients

Organisms or their by-products can have a direct effect on the uptake properties of roots [26, 211] and both stimulation and inhibition of root uptake and translocation of various ions, including Na, K, P, Zn, Fe and Mn, are reported [27, 28, 29, 151, 210, 211, 220]. Some of these effects can be explained in terms of microbial uptake of nutrients and direct competition with the plant. In the case of Mn at least, a water soluble factor was isolated from the microbial medium that increased uptake of Mn by plants [29]. The mechanism of the stimulation is uncertain, but microorganisms are known to produce substances that can alter the availability of some elements by a chelating action, such as the calcium and iron siderophores [127, 249].

Rhizosphere organisms produce organic acids such as 2-keto-gluconic acid which was suggested to have a chelating action on calcium, thereby improving P availabity [66]. The acid is also an efficient extractant of trace elements [38] and is common in soils [164], but it has been suggested that its effect on P solubility may be related more to pH changes rather than to a chelating action [163]. Soil organisms can solubilize "insoluble" phosphates [4] and organisms able to decompose organic phosphates are more common in the rhizosphere than the bulk soil [82].

5.3. Mycorrhizas

Uptake of P was improved when plants were inoculated with both P-solubilizing bacteria and mycorrhizal fungi [19]. The beneficial effects of both ecto- and endomycorrhizal fungi on plant growth and particularly P-uptake are well documented [78, 93, 161, 165, 167, 219]. The mechanism for the improved P-nutrition and greater root uptake of mycorrhizal plants is related to the relative immobility of P in soil [178]. The mycorrhizal hyphae have been shown to take up P from the soil beyond the depleted rhizosphere zone [40] and can translocate P, Zn or S to the root [59, 200, 201, 202]. The fungi generally appear to utilise the same inorganic sources as the root, rather than cause increased breakdown of organic compounds [98].

5.4. Associative N-fixation

There are many reports of plant growth and N-content being improved by inoculation with non-symbiotic nitrogen-fixing bacteria such as Azospirillum [183]. It is generally assumed that the growth responses are caused by the improved nitrogen nutrition of the plants, which is in turn assumed to have been derived from atmospheric-N_2 fixation by the rhizosphere organisms [62]. However, critical examination of the acetylene-reduction technique, often used to estimate nitrogenase activity and N-fixation, has shown the very real possibility of overestimation [133, 265] and the efficiency of N_2-conversion is low, such that there seems to be insufficient substrate in the rhizosphere to account for the amounts of N_2 apparently fixed [30]. Further investigations into alleged specificity of the association [172, 197] and investigations into crop responses using ^{15}N-dilution assay under temperate conditions in Britain have shown no evidence of N-fixation in associations between various cereals and bacteria [134]. Notwithstanding these recent negative observations, numerous reports of beneficial effects following inoculation have been published and reviewed [183] although the responses are not always attributed to N-fixation [199, 245].

Most of the growth responses following inoculation of plants with N-fixing organisms have been obtained under tropical or continental climates. It has been suggested that this is a result of greater amounts of C-substrates in the rhizosphere from plants with C_4-metabolism grown under high light conditions. The organisms are also reported to form a closer association within the root cortex rather than in the soil or rhizoplane [62, 184]. However, in many cases the numbers of N-fixing organisms still seem very low, the mechanism of transfer of N to the plant is not defined, and the energetics of the process are obscure.

5.5. Influence on soil conditions

Microorganisms have been shown to affect soil aggregration and thus to influence indirectly the root environment and plant growth. Polysaccharides have an important role in soil aggregate formation and stability [54, 85, 86, 195, 196] and these polysaccharides can be of microbial origin [53, 88, 119]. Microorganisms added to soil promote aggregate formation [142] and much of the improvement in soil texture in pastures has been attributed to the binding effect of mycorrhizal hyphae [246, 247, 248] as well as to the materials released from the roots [181, 194, 195]. The stabilising effect of fungal and mycorrhizal hyphae is particularly noticeable on sandy soils [75, 76] and helps to stabilise mobile dunes both by a binding and adhesive action [57, 241]. The polysaccharides in soil, as well as stabilising aggregates, can affect water retention (Chapter 7). Stable aggregates have an apparently protective effect on the material within them and disruption of soil crumbs generally results in a flush of C and N mineralization [190, 212].

5.6. Microbial products affecting plant growth

Microorganisms are known to produce various products that can affect plant growth. Two possible mechanisms are by their toxicity to plants and by acting as hormone analogues [47, 48, 95, 125, 141, 143, 174, 189, 259]. Some of these factors are reviewed in Chapter 4.

6. THE SOIL BIOMASS AS A SOURCE AND SINK OF PLANT NUTRIENTS

6.1 The nutrient content of the soil biomass

The biomass is a small but comparatively labile part of the organic fraction in soil, and is important as a source and sink of nutrients. Estimates of the amounts of nutrient involved have been made by measuring the biomass-C in soil by various methods, and then assuming the mineral composition of the biomass to be similar to that of laboratory-grown cells. The nutrient content of such laboratory-grown organisms seems rather high, Eiland [68] used values of 50% C, 7.5% N, 5.8% P, 4.9% K. The biomass in Danish soils was then estimated to contain 30-120 kgN ha^{-1}, 30-120 kgP ha^{-1} and 20-78 kgK ha^{-1}. Mean values for 29 soils in Germany were: 108, 83, 70 and 11 kg ha^{-1} of N, P, K and Ca respectively [12]. Equivalent values for the fungal biomass of a forest soil were 27, 5 and 8 kg ha^{-1} for the N, P and K content respectively [22]. This method of calculation is open to the criticism that, in natural soils, the composition of cells is altered by such factors as moisture stress, nutrient balance and species [12, 175, 253].

Estimation of the nutrient content of the biomass by the fumigation technique is a more direct method, but still assumes that the 'flushes' of nutrients from the killed biomass behave in a similar way to those from the organisms used to estimate the k-factors. At the time of writing there have been comparatively few estimates using the fumigation technique and the majority of these have been on biomass-N.

The biomass-N comprises roughly the same proportion of total N as the biomass-C does of the total C. Broadbalk unmanured soil was estimated to contain 95 kgN ha^{-1} (3.5% of total N) [110] while New Zealand grassland soil contained 220-265 kgN ha^{-1} (2.4 - 3.9% of total N) [204]. Biomass N in Irish soils was 36-1100 kgN ha^{-1} [1]. The biomass-N is closely linked to the biomass-C, so that on soils with a low biomass-C the biomass-N is also low. Thus on stockpiled soils biomass-N was only a small proportion of the total N [204]. On stripped soils replanted to pasture the proportion of N as biomass increased as the pasture became established, although the process took several years [206]. Prairie soils have a higher biomass-N in all soil horizons (77-605 kgN ha^{-1}) than soils which have been under cultivation (65-360 kgN ha^{-1}) [256].

The C:N ratio of the biomass ranges between 5 and 15 [187]. The ratio has at various times been calculated to be 10.9, 8, 7, 6.7 and 5 [1, 12, 49, 68,

131, 256. Clearly there is a need to establish how the C and N content of the biomass varies under different conditions.

The rate of decomposition of organic matter is related to the C:N ratio [43], and the soil biomass, with a comparatively high N content, is a relatively labile fraction. For an unmanured soil continuously cropped to wheat, the turnover of the biomass was 2.5 years and the biomass-N was 95 kgN ha^{-1}, so that the annual flux of N through the biomass was estimated to be 38 kg [110]. This amount was similar to that removed as N in the straw and grain. Depending on the rate of turnover, the nutrients in the biomass could make a significant contribution to the plants' requirements [12, 223]. Much of the increase in the level of available nutrients following a period of soil drying may originate from the killed biomass [41, 42, 158, 159, 160, 191], and N from the killed organisms is readily utilised by plants [111, 135].

Biomass-N is a slightly more enduring fraction than biomass-C, both as added cells or as part of the native biomass. During a 28 day incubation of killed biomass, 49% of the biomass-C was mineralized compared to 33% mineralization of the biomass-N [158]. In a soil labelled with $^{14}C^{15}N$ medic, biomass-^{14}C and biomass-^{15}N accounted for 14 and 22% of the labelled organic residues after 4 yrs, at earlier sampling times (32 wks) the proportions were 8 and 12% respectively [131].

The annual flux of P through the biomass was estimated to be 5 kgP ha^{-1}, similar to that taken up by the plant [110]. As for N, much of the increase in available P caused by soil drying is derived from the killed biomass [45].

There seems little doubt that in future the fumigation technique will be widely used to estimate fluxes of nutrients through the biomass. While current interest has centred on N and P fluxes, there seems no reason in principle why the method could not be applied to other elements.

6.2. Microbial immobilization of nutrients

The biomass comprises a substantial pool of nutrients, but the extent to which the organisms compete with the plant for soil nutrients is not clearly established. The organisms are generally not in active growth and so their requirements are minimal. However, the addition of an available substrate, particularly carbohydrate, can stimulate microbial growth and the soil can become depleted of available nutrients.

In _Liriodendron_ forest, P was preferentially accumulated over C in the O1 litter layer, and N was preferentially accumulated in all the soil horizons. At some times of year, almost all of the N in the O1 litter was as biomass [17]. This was not considered a disadvantage to the plants because leaching of soil nutrients was thereby much reduced, and the periods of maximum microbial immobilization coincided with periods of minimum root activity and low plant demand. The effects of organisms were not always beneficial, and in a laboratory study microbial immobilization of N decreased pine seedling growth in glucose-amended humus [21], the main organisms responsible for the immobilization were fungi and yeasts.

Laboratory studies have also shown microbial immobilization of P and S in cellulose-amended soils. There was considerable fluctuation in the C:P and C:S ratios [51, 52, 217], suggesting that a C-rich biomass is formed following substrate addition. The greater the level of solution P the greater the immobilization, but the reduction in solution P levels was partially compensated by increased mineralization of organic P sources [52]. Increased mineralization following amendments to forest soils has also been noted (Table 1). The biomass was increased by the amendments, but the subsequent biomass-P recovery was not much greater than the controls, again suggesting variability in the C:P ratio. Addition of P salts did not affect this overall response (Table 1) and an essentially similar pattern was obtained for N on other forest soils and an acid mineral soil.

The mobility of P in a calcareous soil was increased by sucrose amendment. However, the mobile P was largely colloidal and probably unavailable to plants [92]. The colloidal P was almost certainly biomass-P because both microbial numbers and P mobility were lower when formaldehyde was added along with the glucose.

6.3. Effect of decomposer and predator organisms on nutrient fluxes

It is the flux through the various nutrient pools, as well as the absolute size of the latter, that determines the amount of plant-available nutrient at a particular time. Estimates of the biomass nutrients were mentioned earlier, as well as some factors, such as soil temperature or drying, that may influence the fluxes. It has become apparent, however, that the decomposer and predator organisms of the biomass have a crucial role in determining the rates of nutrient turnover [58].

Bacterial-feeding nematodes increased the rates of N, P and C mineralization in a soil-bacteria-nematode microcosm, the bacteria alone taking considerably longer to achieve the same amount of mineralization [15]. Similarly, P mineralization by a Pseudomonas from an oil shale was faster in the presence of the bacterial-feeding nematode Rhabditis [16]. Nematodes were also found to affect the amount of N leached from forest humus [20, 21]. Amoebae, rather than flagellates or ciliates, were considered to be the main protozoan predators controlling bacterial numbers in forest soils [55]. Amoebae increased the N mineralization from Pseudomonas in soil [224], but in the presence of bacterial-feeding nematodes as well as amoebae, no further increase in N mineralization was detected until the nematode population declined. The effect of nematodes on N-availability may be through direct excretion of N compounds by the nematode rather than an effect on the physiology and growth of the bacteria [267].

Greater amounts of NH_4^+, NO_3^- and Ca^{2+} were leached from oak litter in the presence of collembola, but the leaching of K^+ and Na^+ was not affected by these invertebrates. The collembola feed on fungi, but in the presence of small numbers of the animals, fungal biomass was greater than in their absence. In contrast, large numbers of collembola decreased the fungal biomass [103].

The soil invertebrates may form a small pool of nutrients in their own right, but groups such as the Enchytraeidae, Dorylaimidae and Elateridae probably have greater importance in increasing the rates of transformation of nutrients, particularly P [149].

The beneficial effects of earthworms in increasing the rates of organic matter turnover are well known [150, 215] but they also affect the biomass, enzyme activities [205]. Other invertebrates also increase organic matter turnover by ingestion and excretion of plant materials [9, 100, 198]. However, in some cases invertebrates or fertilizers have little effect on the rate of decay, which seems an intrinsic characteristic of the plant material [60, 61].

The factors controlling the size and activity of the various soil populations have only been investigated for a limited number of organisms [90, 91]. There have been many more studies on simpler aquatic systems [9] or in laboratory cultures [31, 71, 263] and it would be desirable to extend this knowledge to the soil habitat.

7. SUMMARY AND CONCLUSIONS

The introduction of the chloroform fumigation technique represents a significant advance in the assessment of the soil biomass. In conjunction with other methods, the technique has shown that the biomass is large, but greatly restricted in its activity because of the low levels of available nutrients. The fungi form a major component of the biomass. The biomass comprises only a small proportion of the total soil organic matter, but it is a comparatively labile fraction.

The ability to differentiate biomass from non-living soil organic matter has enabled the distribution of C, N, P and S in the soil to be measured, and the use of isotopes can allow the rate of turnover to be estimated.

The biomass has a multiple role in soil, affecting the decomposition and turnover of organic matter, nutrient immobilization and cycling, root physiology and soil structure. Future studies will no doubt extend the range of parameters being currently measured, and enable quantitative assessments to be made of the influence of the biomass on soil fertility and plant growth.

8. REFERENCES

1. ADAMS T.McM. and LAUGHLIN R.J. 1981. The effects of agronomy on the carbon and nitrogen contained in the soil biomass. Journal of Agricultural Science, Cambridge, 97, 319-327.
2. ADU J.K. and OADES J.M. 1978. Physical factors influencing decomposition of organic materials in soil aggregates. Soil Biology and Biochemistry, 10, 109-115.
3. ADU J.K. and OADES J.M. 1978. Utilization of organic materials in soil aggregates by bacteria and fungi. Soil Biology and Biochemistry, 10, 117-122.
4. AGNIHOTRI V.P. 1970. Solubilization of insoluble phosphates by some soil fungi isolated from nursery seedbeds. Canadian Journal of Microbiology, 16, 877-880.
5. AHMED M., OADES J.M. and LADD J.N. 1982. Determination of ATP in soils: effect of soil treatments. Soil Biology and Biochemistry, 14, 273-279.
6. ALLEN E.B. and ALLEN M.F. 1980. Natural re-establishment of vesicular-arbuscular mycorrhizae following stripmine reclamation in Wyoming. Journal of Applied Ecology, 17, 139-147.
7. AMATO M. and LADD J.N. 1980. Studies of nitrogen immobilization and mineralization in calcareous soils - V. Formation and distribution of isotope-labelled biomass during decomposition of ^{14}C- and ^{15}N-labelled plant material. Soil Biology and Biochemistry, 12, 405-411.
8. ANDERSON G. 1979. Bacterial DNA in soil. Soil Biology and Biochemistry, 11, 213.

248

9. ANDERSON J.M. and MACFADYEN A. 1976 (Eds). The role of terrestrial and aquatic organisms in decomposition processes. The 17th Symposium of the British Ecological Society, 15-18th April 1975. Blackwell Scientific Publications.

10. ANDERSON J.P.E. and DOMSCH K.H. 1978. Mineralization of bacteria and fungi in chloroform fumigated soils. Soil Biology and Biochemistry, 10, 207-213.

11. ANDERSON J.P.E. and DOMSCH K.H. 1978. A physiological method for the quantitative measurement of microbial biomass in soils. Soil Biology and Biochemistry, 10, 215-221.

12. ANDERSON J.P.E. and DOMSCH K.H. 1980. Quantities of plant nutrients in the microbial biomass of selected soils. Soil Science, 130, 211-216.

13. ANDERSON J.P.E., ARMSTRONG R.A. and SMITH S.N. 1981. Methods to evaluate pesticide damage to the biomass of the soil microflora. Soil Biology and Biochemistry, 13, 149-153.

14. ANDERSON J.R. and SLINGER J.M. 1975. Europium chelate and fluorescent brightener staining of soil propagules and their photomicrographic counting - 1. Methods. Soil Biology and Biochemistry, 7, 205-209.

15. ANDERSON R.V., COLEMAN D.C., COLE C.V. and ELLIOTT E.T. 1981. Effect of the nematodes Acrobeloides sp. and Mesodiplogaster cheritieri on substrate utilization and nitrogen and phosphorus mineralization in soil. Ecology, 62, 549-555.

16. ANDERSON R.V., TROFYMOW J.A., COLEMAN D.C. and REID C.P.P. 1982. Phosphorus mineralization by a soil pseudomonad in spent oil shale as affected by a rhabditid nematode. Soil Biology and Biochemistry, 14, 365-371.

17. AUSMUS B.S., EDWARDS N.T. and WITKAMP M. 1976. Microbial immobilization of carbon, nitrogen, phosphorus and potassium, implication for forest ecosystem processes. In: The role of terrestrial and aquatic organisms in decomposition processes pp. 397-416. Eds. Anderson J.M. and Madfadyen A. Blackwell Scientific Publications.

18. AYANABA A., TUCKWELL S.B. and JENKINSON D.S. 1976. The effects of clearing and cropping on the organic reserves and biomass of tropical forest soils. Soil Biology and Biochemistry, 8, 519-525.

19. AZCON R., BAREA J.M. and HAYMAN D.S. 1975. Utilization of rock phosphate in alkaline soils by plants inoculated with mycorrhizal fungi and phosphate-solubilizing bacteria. Soil Biology and Biochemistry, 8, 135-138.

20. BÅÅTH E., LOHM V., LUNDGREN B., ROSSWALL T., SÖDERSTRÖM B and SOHLENIUS B. 1981. Impact of microbial-feeding animals on total soil activity and nitrogen dynamics: a soil microcosm experiment. Oikos, 37, 257-264.

21. BÅÅTH E., LOHM V., LUNDGRÉN B., ROSSWALL T., SÖDERSTRÖM B., SOHLENIUS B. and WIREN A. 1978. The effect of nitrogen and carbon supply on the development of soil organism populations and pine seedlings. Oikos, 31, 153-163.

22. BÅÅTH E. and SÖDERSTRÖM B. 1979. Fungal biomass and fungal immobilization of plant nutrients in Swedish coniferous forest soils. Revue d'Ecologie et de Biologie du Sol, 16, 477-489.

23. BÅÅTH E. and SODERSTROM B. 1979. The significance of hyphal diameter in calculation of fungal biovolume. Oikos, 33, 11-14.

24. BÅÅTH E. and SÖDERSTRÖM B. 1982. Seasonal and spatial variation in fungal biomass in a forest soil. Soil Biology and Biochemistry, 14, 353-358.

25. BAKER K.F. and SNYDER W.C. 1965 (Eds). Ecology of soil-borne plant pathogens. John Murray, London.

26. BARBER D.A. 1968. Microoorganisms and the Inorganic nutrition of Higher plants. Annual Review of Plant Physiology, 19, 71-88.

27. BARBER D.A. 1974. The absorption of ions by microorganisms and excised roots. New Phytologist, 73, 91-96.

28. BARBER D.A., BOWEN G.D. and ROVIRA A.D. 1976. Effects of micro-organisms on absorption and distribution of phosphate in barley. Australian Journal of Plant Physiology, 3, 801-808.

29. BARBER D.A. and LEE R.B. 1974. The effect of microorganisms on the absorption of manganese by plants. New Phytologist, 73, 97-106.

30. BARBER D.A. and LYNCH J.M. 1977. Microbial growth in the rhizosphere. Soil Biology and Biochemistry, 9, 305-308.

31. BARSDATE R.J., PRENTKI R.T. and FENCHEL T. 1974. Phosphorus cycle of model ecosystems: significance for decomposer food chains and effect of bacterial grazers. Oikos, 25, 239-251.

32. BEEZER A.E. 1980. Biological Microcalorimetry. Academic Press.

33. BEHERA, B and WAGNER G.H. 1974. Microbial growth rate in glucose-amended soil. Proceedings of the Soil Science Society of America, 38, 591-594.

34. BENOIT R.E. and STARKEY R.L. 1968. Enzyme inactivation as a factor in the inhibition of decomposition of organic matter by tannins. Soil Science, 105, 203-208.

35. BENOIT R.E. and STARKEY R.L. 1968. Inhibition of decomposition of cellulose and some other carbohydrates by tannin. Soil Science, 105, 291-296.

36. BENOIT R.E., STARKEY R.L. and BASARABA J. 1968. Effect of purified plant tannin on decomposition of some organic compounds and plant materials. Soil Science, 105, 153-158.

37. BERG B. and SÖDERSTRÖM B. 1979. Fungal biomass and nitrogen in decomposing Scots pine needle litter. Soil Biology and Biochemistry, 11, 339-341.

38. BERROW M.L., DAVIDSON M.S. and BURRIDGE J.C. 1982. Trace elements extractable by 2-ketogluconic acid from soils and their relationship to plant contents. Plant and Soil, 66, 161-171.

39. BERKELEY R.C.W., LYNCH J.M., MELLING J., RUTTER P.R. and VINCENT B. 1980 (Eds). Microbial adhesion to surfaces. Ellis Horwood Ltd., Chichester.

40. BHAT K.K.S and NYE P.H. 1974. Diffusion of phosphate to plant roots in soil. III. Depletion around onion roots without root hairs. Plant and Soil, 41, 383-394.

41. BIRCH H.F. 1958. The effect of soil drying on humus decomposition and nitrogen availability. Plant and Soil, 10, 9-31.

42. BIRCH H.F. and FRIEND M.T. 1961. Resistance of humus to decomposition. Nature (London), 191, 731-732.

43. BOSATTA E. and STAAF H. 1982. The control of nitrogen turnover in forest litter. Oikos, 39, 143-151.

44. BOWEN G.D. 1980. Misconceptions, concepts and approaches in rhizosphere biology. In: Contemporary Microbial Ecology, pp. 283-304. Eds. Ellwood D.C., Hedger J.N., Latham M.J., Lynch J.M. and Slater J.H. Academic Press.

45. BROOKES P.C., POWLSON D.S. and JENKINSON D.S. 1982. Measurement of microbial biomass phosphorus in soil. Soil Biology and Biochemistry, 14, 319-329.

46. BROOKES P.C., TATE K.R. and JENKINSON D.S. 1983. The adenylate energy charge of the soil microbial biomass. Soil Biology and Biochemistry, 15, 9-16.

47. BROWN M.E. 1972. Plant growth substances produced by microorganisms of soil and rhizosphere. Journal of Applied Bacteriology, 35, 443-451.

48. BROWN M.E. and BURLINGHAM S.K. 1968. Production of plant growth substances by Azotobacter chroococcum. Journal of General Microbiology, 53, 135-144.

49. CARTER M.R. and RENNIE D.A. 1982. Changes in soil quality under zero tillage farming systems: distribution of microbial biomass and mineralizable C and N potentials. Canadian Journal of Soil Science, 62, 587-597.

50. CERRI C.C and JENKINSON D.S. 1981. Formation of microbial biomass during the decomposition of ^{14}C labelled rye grass in soil. Journal of Soil Science, 32, 619-626.

51. CHAUHAN B.S., STEWART J.W.B. and PAUL E.A. 1979. Effect of carbon additions on soil labile inorganic, organic and microbially held phosphate. Canadian Journal of Soil Science, 59, 387-396.

52. CHAUHAN B.S., STEWART J.W.B. and PAUL E.A. 1981. Effect of labile inorganic phosphate status and organic carbon additions on the microbial uptake of phosphorus in soils. Canadian Journal of Soil Science, 61, 373-385.

53. CHESHIRE M.V. 1977. Origins and stability of soil polysaccharide. Journal of Soil Science, 28, 1-10.

54. CHESHIRE M.V., SPARLING G.P. and MUNDIE C.M. 1983. Effect of periodate treatment of soil on carbohydrate constituents and soil aggregation. Journal of Soil Science, 34, 105-112.

55. CLARHOLM M. 1981. Protozoan grazing of bacteria in soil-impact and importance. Microbial Ecology, 7, 343-350.

56. CLARHOLM M. and ROSSWALL T. 1980. Biomass and turnover of bacteria in a forest soil and tundra peat. Soil Biology and Biochemistry, 12, 49-51.

57. CLOUGH K.S. and SUTTON J.C. 1978. Direct observation of fungal aggregates in sand dune soil. Canadian Journal of Microbiology, 24, 333-335.

58. COLE C.V., ELLIOTT E.T., HUNT H.W. and COLEMAN D.C. 1978. Trophic interactions in soils as they affect energy and nutrient dynamics. V. Phosphorus transformations. Microbial Ecology, 4, 381-387.

59. COOPER K.M. and TINKER P.B. 1978. Translocation and transfer of nutrients in vesicular-arbuscular mycorrhizas. II Uptake and translocation of phosphorus, zinc and sulphur. New Phytologist, 81, 43-52.

60. COULSON J.C. and BUTTERFIELD J. 1978. An investigation of the biotic factors determining the rates of plant decomposition on blanket bog. Journal of Ecology, 66, 631-650.

61. CURRY J.P. 1969. The decomposition of organic matter in soil. Part 1. The role of the fauna in decaying grassland herbage. Soil Biology and Biochemistry, 1, 253-258.

62. DOBEREINER J. 1974. Nitrogen-fixing bacteria in the rhizosphere. In: The Biology of Nitrogen Fixation, pp. 86-120. Ed. Quispel A. North Holland Publishing Co.

63. DOMMERGUES Y.R., BELSER L.W. and SCHMIDT E.L. 1977. Limiting factors for microbial growth and activity in soil. In: Advances in Microbial Ecology Volume 2, pp. 49-104. Ed. Alexander M. Plenum Press.

64. DOMMERGUES Y.R. and KRUPA S.V. 1978. Interactions between non-pathogenic soil microorganisms and plants. Elsevier Scientific Publishing Co.

65. DOMSCH K.H., BECK T., ANDERSON J.P.E., SÖDERSTRÖM B., PARKINSON D. and TROLLDENIER G. 1979. A comparison of methods for soil microbial population and biomass studies. Zeitschrift für Pflanzenernahrung und Bodenkunde, 142, 520-533.

66. DUFF R.B., WEBLEY D.M. and SCOTT R.O. 1963. Solubilization of minerals and related materials by 2-ketogluconic acid-producing bacteria. Soil Science, 95, 105-114.

67. EILAND F. 1979. An improved method for determination of adenosine triphosphate (ATP) in soil. Soil Biology and Biochemistry, 11, 31-35.

68. EILAND F. 1981. Organic manure in relation to microbiological activity in soil. In: Agricultural Yield Potentials in Continental Climates, pp 147-156. Proceedings of 16th Colloquium International Potash Institute, Bern (1981).

69. EILAND F. 1983. A simple method for quantitative determination of ATP in soil. Soil Biology and Biochemistry. 15 .

70. ELLWOOD D.C., HEDGER J.N., LATHAM M.J., LYNCH J.M. and SLATER J.H. 1980 (Eds). Contemporary Microbial Ecology. Academic Press.

71. FENCHEL T. and HARRISON P. 1976. The significance of bacterial grazing and mineral cycling for the decomposition of particulate detritus. In: The role of terrestrial and aquatic organisms in decomposition processes, pp. 285-299. Eds. Anderson J.M. and MacFadyen A. Blackwell Scientific Publications.

72. FLOATE M.J.S. 1970. Decomposition of organic materials from hill soils and pastures. III. The effect of temperature on the mineralisation of carbon, nitrogen and phosphorus from plant materials and sheep faeces. Soil Biology and Biochemistry, 2, 187-196.

73. FLOATE M.J.S. 1970. Decomposition of organic materials from hill soils and pastures. IV. The effects of moisture content on the mineralisation of carbon, nitrogen and phosphorus from plant materials and sheep faeces. Soil Biology and Biochemistry, 2, 275-283.

74. FORREST W.W. 1972. Microcalorimetry. In: Methods in Microbiology, pp. 285-318. Eds. Norris J.R. and Ribbons D.W. Academic Press.

75. FORSTER S.M. and NICOLSON T.H. 1981. Aggregation of sand from a maritime embryo sand dune by microorganisms and higher plants. Soil Biology and Biochemistry, 13, 199-203.

76. FORSTER S.M. and NICOLSON T.H. 1981. Microbial aggregation of sand in a maritime dune succession. Soil Biology and Biochemistry, 13, 205-208.

77. FRANKLAND J.C., LINDLEY D.K. and SWIFT M.J. 1978. A comparison of two methods for the estimation of mycelial biomass in leaf litter. Soil Biology and Biochemistry, 10, 323-333.

78. GERDEMANN J.W. 1968. Vesicular-arbuscular mycorrhiza and plant growth. Annual Review of Phytopathology, 6, 397-418.

79. GIVEN P.H. and DICKINSON C.H. 1975. Biochemistry and microbiology of peats. In: Soil Biochemistry, volume 3, pp. 123-212. Eds. Paul E.A. and McLaren A.D. Marcel Dekker.

80. GRAY T.R.G. 1976. Survival of vegetative microbes in soil. Symposium of the Society of General Microbiology, 26, 327-364.

81. GRAY T.R.G. and WILLIAMS S.T. 1971. Microbial productivity in soil. Symposium of the Society of General Microbiology, 21, 255-286.

82. GREAVES M.P. and WEBLEY D.M. 1965. A study of the breakdown of organic phosphates by microorganisms from the root region of certain pasture grasses. Journal of Applied Bacteriology, 28, 454-465.

83. GREAVES M.P., WHEATLEY R.E., SHEPHERD H. and KNIGHT A.H. 1973. Relationship between microbial populations and adenosine triphosphate in a basin peat. Soil Biology and Biochemistry, 5, 685-687.

84. GREAVES M.P. and WILSON M.J. 1970. The degradation of nucleic acids and montmorillonite-nucleic acid complexes by soil microorganisms. Soil Biology and Biochemistry, 2, 257-268.

85. GREENLAND D.J., LINDSTROM G.R. and QUIRK J.P. 1962. Organic materials which stabilize natural soil aggregates. Proceedings of the Soil Science Society of America, 26,366-371.

86. GREENLAND D.J., LINDSTROM G.R. and QUIRK J.P. 1961. Role of polysaccharides in stabilization of natural soil aggregates. Nature (London), 191, 1283-1284.

87. GREENWOOD D.J. 1961. The effect of oxygen concentration on the decomposition of organic materials in soil. Plant and Soil, 14, 360-376.

88. GRIFFITHS E. and JONES D. 1965. Microbiological aspects of soil structure. I. Relationships between organic amendments, microbial colonisation and changes in aggregate stability. Plant and Soil, 23, 17-33.

89. GROSSBARD E. and WINGFIELD G.I. 1978. Effects of paraquat, aminotriazole and glyphosate on cellulose decomposition. Weed Research, 18, 347-353.

90. HABTE M. and ALEXANDER M. 1977. Further Evidence for the Regulation of Bacterial Populations in Soil by Protozoa. Archives of Microbiology, 113, 181-183.

91. HABTE M. and ALEXANDER M. 1978. Mechanisms of persistence of low numbers of bacteria preyed upon by protozoa. Soil Biology and Biochemistry, 10, 1-6.

92. HANNAPEL R.J., FULLER W.H. and FOX R.H. 1964. Phosphorus movement in a calcareous soil. II. Soil microbial activity and organic phosphorus movement. Soil Science, 97, 421-427.

93. HARLEY J.L. 1972. The Biology of Mycorrhiza. 2nd Edition. Plant Science Monographs. Leonard Hill.

94. HARLEY J.L. and RUSSELL R.S. 1979. The Soil-Root Interface. Proceedings of an International Symposium held in Oxford, England, March 28-31 1978. Academic Press.

95. HARPER S.H.T. and LYNCH J.M. 1981. Effect of fungi on barley seed germination. The Journal of General Microbiology, 122, 55-60.

96. HARRISON A.F. 1982. ^{32}P-method to compare rates of mineralization of labile organic phosphorus in woodland soils. Soil Biology and Biochemistry, 14, 337-341.

97. HARRISON A.F. 1982. Labile organic phosphorus mineralization in relationship to soil properties. Soil Biology and Biochemistry, 14, 343-351.

98. HAYMAN D.S. 1982. Practical aspects of vesicular-arbuscular mycorrhiza. In: Advances in Agricultural Microbiology, pp. 325-373. Ed. Subba Rao N.S. Butterworth Scientific.

99. HEDLEY M.J. and STEWART J.W.B. 1982. Method to measure microbial phosphate in soils. Soil Biology and Biochemistry, 14, 377-385.

100. HOLTER P. 1979. Effect of dung-beetles (Aphodius spp.) and earthworms on the disappearance of cattle dung. Oikos, 32, 393-402.
101. HUNTJENS J.L.M. and ALBERS R.A.J.M. 1978. A model experiment to study the influence of living plants on the accumulation of soil organic matter in pastures. Plant and Soil, 50, 411-418.
102. INESON P. and ANDERSON J.M. 1982. Microbial biomass determination in deciduous leaf litter. Soil Biology and Biochemistry, 14, 607-608.
103. INESON P., LEONARD M.A. and ANDERSON J.M. 1982. Effect of collembolan grazing upon nitrogen and cation leaching from decomposing leaf litter. Soil Biology and Biochemistry, 14, 601-605.
104. INGHAM E.R. and KLEIN D.A. 1982. Relationship between fluorescein diacetate-stained hyphae and oxygen utilization, glucose utilization and biomass of submerged fungal batch cultures. Applied and Environmental Microbiology, 44, 363-370.
105. JENKINSON D.S. 1966. Studies on the decomposition of plant material in soil. II. Partial sterilization of soil and the soil biomass. Journal of Soil Science, 17, 280-302.
106. JENKINSON D.S. 1976. The effect of biocidal treatments on metabolism in soil. IV. The decomposition of fumigated organisms in soil. Soil Biology and Biochemistry, 8, 203-208.
107. JENKINSON D.S. 1981. The fate of plant and animal residues in soil. Chapter 9 in: The Chemistry of Soil Processes. Eds. Greenland D.J. and Hayes M.H.B. John Wiley and Sons Ltd.
108. JENKINSON D.S. and AYANABA A. 1977. Decomposition of carbon-14 labelled plant material under tropical conditions. Journal of the Soil Science Society of America, 41, 912-915.
109. JENKINSON D.S., DAVIDSON S.A. and POWLSON D.S. 1979. Adenosine triphosphate and microbial biomass in soil. Soil Biology and Biochemistry, 11, 521-527.
110. JENKINSON D.S. and LADD J.N. 1981. Microbial biomass in soil: measuremant and turnover. Chapter 10 in: Soil Biochemistry, volume 5. Eds. Paul E.A. and Ladd J.N. pp. 415-471. Marcel Dekker.
111. JENKINSON D.S., NOWAKOWSKI T.Z. and MITCHELL J.D.D. 1972. Growth and uptake of nitrogen by wheat and ryegrass in fumigated and irradiated soil. Plant and Soil, 36, 149-158.
112. JENKINSON D.S. and OADES J.M. 1979. A method for measuring adenosine triphosphate in soil. Soil Biology and Biochemistry, 11, 193-199.
113. JENKINSON D.S. and POWLSON D.S. 1976. The effects of biocidal treatments on metabolism in soil. I. Fumigation with chloroform. Soil Biology and Biochemistry, 8, 167-177.
114. JENKINSON D.S. and POWLSON D.S. 1976. The effects of biocidal treatments on metabolism in soil. V. A method for measuring soil biomass. Soil Biology and Biochemistry, 8, 209-213.
115. JENKINSON D.S. and POWLSON D.S. 1980. Measurement of microbial biomass in intact soil cores and in sieved soil. Soil Biology and Biochemistry, 12, 579-581.
116. JENKINSON D.S., POWLSON D.S. and WEDDERBURN R.W.M. 1976. The effects of biocidal treatments on metabolism in soil. III. The relationship between soil biovolume, measured by optical microscopy, and the flush of decomposition caused by fumigation. Soil Biology and Biochemistry, 8, 189-202.
117. JENKINSON D.S. and RAYNER J.H. 1977. The turnover of soil organic matter in some of the Rothamsted Classical Experiments. Soil Science, 123, 298-305.

118. JOHNEN B.G. and DREW E.A. 1977. Ecological effects of pesticides on soil microorganisms. Soil Science, 123, 319-324.
119. JONES D. and GRIFFITHS E. 1967. Microbial aspects of soil structure. Plant and Soil, 27, 187-199.
120. JONES J.G. 1979. A guide to methods for estimating microbial numbers and biomass in fresh water. Freshwater Biological Association: Scientific Publications 39.
121. KASSIM G., MARTIN J.P. and HAIDER K. 1981. Incorporation of a wide variety of organic substrate carbons into soil biomass as estimated by the fumigation procedure. Journal of the Soil Science Society of America, 45, 1106-1112.
122. KASSIM G., STOTT D.E., MARTIN J.P. and HAIDER K. 1982. Stabilization and incorporation into biomass of phenolic and benzenoid carbons during biodegradation in soil. Journal of the Soil Science Society of America, 46, 305-309.
123. KATZNELSON H. 1940. Survival of microorganisms inoculated into sterilized soil. Soil Science, 49, 211-217.
124. KATZNELSON H. 1940. Survival of microorganisms introduced into soil. Soil Science, 49, 283-293.
125. KATZNELSON H. and COLE S.E. 1965. Production of gibberellin-like substances by bacteria and actinomycetes. Canadian Journal of Microbiology, 11, 733-741.
126. KILBERTUS G, REISINGER O., MOUREY A. and CANCELA DA FONSECA J.A. 1974. Biodegradation et Humification. Rapport du 1°er Colloque International, Nancy.
127. KLOEPPER J.W., LEONG J., TEINTZE M. and SCHROTH M.N. 1980. Enhanced plant growth by siderophores produced by plant growth promoting rhizobacteria. Nature (London), 286, 885-886.
128. KNIGHT W.G. and SKUJINS J. 1981. ATP concentration and soil respiration at reduced water potentials in arid soils. Journal of the Soil Science Society of America, 45, 657-660.
129. KUCEY R.M.N. and PAUL E.A. 1982. Biomass of mycorrhizal fungi associated with bean roots. Soil Biology and Biochemistry, 14, 413-414.
130. KUSHNER D.J. 1978. Microbial life in extreme environments. Academic Press.
131. LADD J.N., OADES J.M. and AMATO M. 1981. Microbial biomass formed from ^{14}C, ^{15}N-labelled plant material decomposing in soils in the field. Soil Biology and Biochemistry, 13, 119-126.
132. LEE C.C., HARRIS R.F., WILLIAMS J.D.H., SYERS J.K. and ARMSTRONG D.E. 1971. Adenosine triphosphate in lake sediments. II. Origins and significance. Proceedings of the Soil Science Society of America, 35, 86-91.
133. LETHBRIDGE G., DAVIDSON M.S. and SPARLING G.P. 1982. Critical evaluation of the acetylene reduction test for estimating the activity of nitrogen-fixing bacteria associated with the roots of wheat and barley. Soil Biology and Biochemistry, 14, 27-35.
134. LETHBRIDGE G. and DAVIDSON M.S. 1983. The role of root-associated nitrogen-fixing bacteria in the nitrogen nutrition of wheat estimated by ^{15}N isotope dilution. Soil Biology and Biochemistry. In press.
135. LETHBRIDGE G. and DAVIDSON M.S. 1983. Microbial biomass as a source of nitrogen for cereals. Soil Biology and Biochemistry, 15 365.
136. LITCHFIELD C.D. and SEYFRIED P.L. 1979. Methodology for biomass determinations and microbial activities in sediments. A.S.T.M. Special

Technical Publication 673. American Society for Testing and Materials, Philadelphia.

137. LJUNGHOLM K., NORÉN B., SKÖLD R. and WADSÖ I. 1979. Use of microcalorimetry for the characterisation of microbial activity in soil. Oikos, 33, 15-23.

138. LJUNGHOLM K., NORÉN B. and WADSÖ I. 1979. Microcalorimetric observation of microbial activity in normal and acidified soils. Oikos, 33, 24-30.

139. LOCHHEAD A.G. and THEXTON R.H. 1947. Growth and survival of bacteria in peat. II. Peat pellets. Canadian Journal of Research, 25, 14-19.

140. LYNCH J.M. 1976. Products of soil micro-organisms in relation to plant growth. C.R.C. Critical Reviews in Microbiology, 5, 67-107.

141. LYNCH J.M. 1978. Production and phytotoxicity of acetic acid in anaerobic soils containing plant residues. Soil Biology and Biochemistry, 10, 131-135.

142. LYNCH J.M. 1981. Promotion and inhibition of soil aggregate stabilization by micro-organisms. Journal of General Microbiology, 126, 371-375.

143. LYNCH J.M. 1981. Crop rotation and plant residues in relation to biological activity in soil. In: Agricultural Yield Potientials in Continental Climates, pp. 147-157. Proceedings of 16th Colloquium of the International Potash Institute, Bern (1981).

144. LYNCH J.M. 1982. Limits to microbial growth in soil. Journal of General Microbiology, 128, 404-410.

145. LYNCH J.M. and PANTING L.M. 1980. Cultivation and the soil biomass. Soil Biology and Biochemistry, 12, 29-33.

146. LYNCH J.M. and PANTING L.M. 1980. Variations in the size of the soil biomass. Soil Biology and Biochemistry, 12, 547-550.

147. LYNCH J.M. and PANTING L.M. 1981. Measurement of the microbial biomass in intact cores of soil. Microbial Ecology, 7, 229-234.

148. MACDONALD R.M. 1980. Cytochemical demonstration of catabolism in soil micro-organisms. Soil Biology and Biochemistry, 12, 419-423.

149. McKERCHER R.B., TOLLEFSON T.S. and WILLARD J.R. 1979. Biomass and phosphorus contents of some soil invertebrates. Soil Biology and Biochemistry, 11, 387-391.

150. MANSELL G.P., SYERS J.K. and GREGG P.E.H. 1981. Plant availability of phosphorus in dead herbage ingested by surface-casting earthworms. Soil Biology and Biochemistry, 13, 163-167.

151. MARSCHNER H. and BARBER D.A. 1975. Iron uptake by sunflower plants under sterile and non-sterile conditions. Plant and Soil, 43, 515-518.

152. MARSHALL K.C. 1975. Clay mineralogy in relation to survival of soil bacteria. Annual Review of Phytopathology, 13, 357-373.

153. MARSHALL K.C. 1976. Nonspecific Interfacial Interactions in Microbial Ecology: terrestrial ecosystems. Chapter 5 in: Interfaces in Microbial Ecology. Harvard University Press.

154. MARTENS R. 1982. Apparatus to study the quantitative relationships between root exudates and microbial populations in the rhizosphere. Soil Biology and Biochemistry, 14, 315-317.

155. MARTIN J.K. and KEMP J.R. 1980. Carbon loss from roots of wheat cultivars. Soil Biology and Biochemistry, 12, 551-554.

156. MARTIN J.P., ERVIN J.O. and SHEPHERD R.A. 1966. Decomposition of Iron, Aluminium, Zinc and Copper Salts or Complexes of some Microbial and Plant Polysaccharides in Soil. Proceedings of the Soil Science Society of America, 30, 196-200.

157. MARTIN J.P., ZUNINO H., PIERANO P., CALOZZI M. and HAIDER K. 1982. Decomposition of ^{14}C-labelled lignins, model humic acid polymers, and fungal melanins in allophanic soils. Soil Biology and Biochemistry, 14, 289-293.

158. MARUMOTO T., ANDERSON J.P.E. and DOMSCH K.H. 1982. Decomposition of ^{14}C and ^{15}N-labelled microbial cells in soil. Soil Biology and Biochemistry, 14, 461-467.

159. MARUMOTO T., ANDERSON J.P.E. and DOMSCH K.H. 1982. Mineralization of nutrients from soil microbial biomass. Soil Biology and Biochemistry, 14, 469-475.

160. MARUMOTO T., KAI H., YOSHIDA T. and HARADA T. 1977. Relationship between an accumulation of soil organic matter becoming decomposable due to drying of soil and microbial cells. Soil Science and Plant Nutrition, 23, 1-18.

161. MARX D.H. 1972. Ectomycorrhizae as biological deterrents to pathogenic root infections. Annual Review of Phytopathology, 10, 429-454.

162. MILLAR W.N. and CASIDA L.E. 1970. Evidence for muramic acid in soil. Canadian Journal of Microbiology, 16, 299-304.

163. MOGHIMI A. and TATE M.E. 1978. Does 2-ketogluconate chelate calcium in the pH range 2.4 - 6.4? Soil Biology and Biochemistry, 10, 289-292.

164. MOGHIMI A., TATE M.E. and OADES J.M. 1978. Characterisation of rhizosphere products especially 2-ketogluconic acid. Soil Biology and Biochemistry, 10, 283-287.

165. MOLINA R. and TRAPPE J.M. 1982. Applied Aspects of Ectomycorrhizae. In: Advances in Agricultural Microbiology, pp 305-324. Ed. Subba Rao N.S. Butterworth Scientific.

166. MORTENSÉN U., NORÉN B. and WADSÖ I. 1973. Microcalorimetry in the study of the activity of microorganisms. Bulletins from the Ecological Research Committee (Stockholm), 17, 187-197.

167. MOSSE B. 1973. Advances in the study of vesicular-arbuscular mycorrhizas. Annual Review of Phytopathology, 11, 171-196.

168. NAKAS J.P. and KLEIN D.A. 1979. Decomposition of microbial cell components in a semi-arid grassland soil. Applied and Environmental Microbiology, 38, 454-460.

169. NAKAS J.P. and KLEIN D.A. 1980. Mineralization capacity of bacteria and fungi from the rhizosphere-rhizoplane of a semi-arid grassland. Applied and Environmental Microbiology, 39, 113-117.

170. NANNIPIERI P., JOHNSON R.L. and PAUL E.A. 1978. Criteria for measurement of microbial growth and activity in soil. Soil Biology and Biochemistry, 10, 223-229.

171. NANNIPIERI P., PEDRAZZINI E., ARCARA P.G. and PIOVANELLI C. 1979. Changes in amino acids, enzyme activities and biomasses during soil microbial growth. Soil Science, 127, 26-34.

172. NEAL J.L. and LARSEN R.I. 1976. Acetylene reduction by bacteria isolated from the rhizosphere of wheat. Soil Biology and Biochemistry, 8, 151-155.

173. NELSON D.W., MARTIN J.P. and ERVIN J.O. 1979. Decomposition of Microbial Cells and Components in Soil and Their Stabilization through complexing with Model Humic Acid-type Phenolic Polymers. Journal of the Soil Science Society of America, 43, 84-88.

174. NEWBERG D.McC. 1979. The effects of decomposing roots on the growth of grassland plants. Journal of Applied Ecology, 16, 613-622.

175. NEWELL S.Y. and STATZELL-TALLMAN A. 1982. Factors for conversion of fungal biovolume values to biomass, carbon and nitrogen: variation with

mycelial ages, growth conditions and strains of fungi from a salt marsh. Oikos, 39, 261-268.

176. NEWMAN E.I. 1978. Root Microorganisms: their significance in the ecosystem. Biological Reviews (of the Cambridge Philosophical Society), 53, 511-554.

177. NEWMAN E.I. and WATSON A. 1977. Microbial · abundance in the rhizosphere: a computer model. Plant and Soil, 48, 17-56.

178. NYE P.H. and TINKER P.B. 1977. Solute movement in the root-soil system. Blackwell Scientific Publications.

179. NYHAN J.W. 1975. Decomposition of carbon-14 labeled plant materials in a grassland soil under field conditions. Proceedings of the Soil Science Society of America, 39, 643-648.

180. NYHAN J.W. 1976. Influence of Soil Temperature and Water Tension on the decomposition rate of carbon-14 labeled Herbage. Soil Science, 121, 288-293..

181. OADES J.M. 1978. Mucilages at the root surface. Journal of Soil Science, 29, 1-16.

182. OADES J.M. and JENKINSON D.S. 1979. The adenosine triphosphate content of the soil microbial biomass. Soil Biology and Biochemistry, 11, 201-204.

183. PATRIQUIN D.G. 1982. New developments in grass-bacteria associations. In: Advances in Agricultural Microbiology, pp 139-190. Ed. Subba Rao N.S. Butterworth Scientific.

184. PATRIQUIN D.G. and DOBEREINER J. 1978. Light microscopy observations of tetrazolium-reducing bacteria in the endorhizosphere of maize and other grasses in Brazil. Canadian Journal of Microbiology, 24, 734-742.

185. PAUL E.A. and JOHNSON R.L. 1977. Microscopic counting and adenosine 5'-triphosphate measurement in determining microbial growth in soils. Applied and Environmental Microbiology, 34, 263-269.

186. PAUL E.A. and McLAREN A.D. 1975. Biochemistry of the soil subsystem. Chapter 1 in: Soil Biochemistry, volume 3, pp. 1-36. Eds. Paul E.A. and McLaren A.D.. Marcel Dekker.

187. PAUL E.A. and VORONEY R.P. 1980. Nutrient and energy flows through soil microbial biomass. In: Contemporary Microbial Ecology, pp 216-227. Eds Ellwood D.C., Hedger J.N., Latham M.J., Lynch J.M. and Slater J.H. Academic Press.

188. PAYNE W.J. 1970. Energy yields and growth of heterotrophs. Annual Review of Microbiology, 24, 17-52.

189. PENN D.J. and LYNCH J.M. 1981. Effect of decaying couch grass rhizomes on the growth of barley. Journal of Applied Ecology, 18, 669-674.

190. POWLSON D.S. 1980. The effects of grinding on microbial and non-microbial organic matter in soil. Journal of Soil Science, 31, 77-85.

191. POWLSON D.S. and JENKINSON D.S. 1976. The effects of biocidal treatments on metabolism in soil - II. Gamma irradiation, autoclaving, air drying and fumigation with chloroform or methylbromide. Soil Biology and Biochemistry, 8, 179-188.

192. POWLSON D.S. and JENKINSON D.S. 1981. A comparison of the organic matter, biomass, adenosine triphosphate and mineralizable nitrogen contents of ploughed and direct-drilled soils. Journal of Agricultural Science, Cambridge, 97, 713-721.

193. PUGH G.J.F. and WILLIAMS J.I. 1971. Effect of an organo-mercury fungicide on saprophytic fungi and on litter decomposition. Transactions of the British Mycological Society, 57, 164-166.

194. REID J.B. and GOSS M.J. 1980. Changes in the aggregate stability of a sandy loam effected by growing roots of perennial ryegrass (Lolium perenne). Journal of the Science of Food and Agriculture, 31, 325-328.

195. REID J.B. and GOSS M.J. 1981. Effect of living roots of different plant species on the aggregate stability of two arable soils. Journal of Soil Science, 32, 521-541.

196. REID J.B. and GOSS M.J. 1982. Interactions between soil drying due to plant water use and decreases in aggregate stability caused by maize roots. Journal of Soil Science, 33, 47-53.

197. RENNIE R.J. and LARSON R.I. 1979. Dinitrogen fixation associated with disomic chromosome substitution lines of spring wheat. Canadian Journal of Botany, 57, 2771-2775.

198. REYES V.G. and TIEDJE J.M. 1976. Metabolism of ^{14}C-labeled plant materials by woodlice (Tracheoniscus rathkei BRANT) and soil microorganisms. Soil Biology and Biochemistry, 8, 103-108.

199. REYNDERS L. and VLASSAK K. 1982. Use of Azospirillum brasilense as biofertilizer in intensive wheat cropping. Plant and Soil, 66, 217-273.

200. RHODES L.H. and GERDEMANN J.W. 1975. Phosphate uptake zones of mycorrhizal and non-mycorrhizal onions. New Phytologist, 75, 555-561.

201. RHODES L.H. and GERDEMANN J.W. 1978. Hyphal translocation and uptake of sulfur by vesicular-arbuscular mycorrhizae of onion. Soil Biology and Biochemistry, 10, 355-360.

202. RHODES L.H. and GERDEMANN J.W. 1978. Influence of phosphorus nutrition on sulfur uptake by vesicular-arbuscular mycorrhizae of onion. Soil Biology and Biochemistry, 10, 361-364.

203. ROSS D.J. 1974. Influence of four pesticide formulations on microbial processes in a New Zealand pasture soil. II. Nitrogen mineralization. New Zealand Journal of Agricultural Research, 17, 9-17.

204. ROSS D.J. and CAIRNS A. 1981. Nitrogen availability and microbial biomass in stockpiled topsoils in Southland. New Zealand Journal of Science, 24, 137-143.

205. ROSS D.J. and CAIRNS A. 1982. Effects of earthworms and ryegrass on respiratory and enzyme activities of soil. Soil Biology and Biochemistry, 14, 583-587.

206. ROSS D.J., SPIER T.W., TATE K.R., CAIRNS A., MEYRICK K.F. and PANSIER E.A. 1982. Restoration of pasture after topsoil removal: effects of soil carbon and nitrogen mineralization, microbial biomass and enzyme activities. Soil Biology and Biochemistry, 14, 575-581.

207. ROSS D.J., TATE K.R., CAIRNS A. and MEYRICK K.F. 1980. Influence of storage on soil microbial biomass estimated by three biochemical procedures. Soil Biology and Biochemistry, 12, 369-374.

208. ROSS D.J., TATE K.R., CAIRNS A. and MEYRICK K.F. 1981. Fluctuations in microbial biomass indices at different sampling times in soils from tussock grasslands. Soil Biology and Biochemistry, 13, 109-114.

209. ROSS D.J., TATE K.R., CAIRNS A. and PANSIER E.A. 1980. Microbial biomass estimations in soils from tussock grasslands by three biochemical procedures. Soil Biology and Biochemistry, 12, 375-383.

210. ROVIRA A.D. 1965. Interactions between plant roots and soil microorganisms. Annual Review of Microbiology, 19, 241-266.

211. ROVIRA A.D. and DAVEY C.B. 1974. Biology of the rhizosphere. In: The Plant Root and its Environment, pp 153-204. Ed. Carson E.W. University Press of Virginia.

212. ROVIRA A.D. and GREACEN E.L. 1957. The effect of aggregate disruption on the activity of micro-organisms in the soil. Australian Journal of Agricultural Research, 8, 659-673.

213. ROVIRA A.D., NEWMAN E.I, BOWEN H.J. and CAMPBELL R. 1974. Quantitative assessment of the rhizoplane microflora by direct microscopy. Soil Biology and Biochemistry, 6, 211-216.

214. RUSSELL E.W. 1961. Soil conditions and plant growth. 9th Edition. Longman.

215. RUSSELL R.S. 1977. Plant root systems: their function and interaction with the soil. McGraw-Hill.

216. SAGGAR S., BETTANY J.R. and STEWART J.W.B. 1981. Measurement of microbial sulfur in soil. Soil Biology and Biochemistry, 13, 493-498.

217. SAGGAR S., BETTANY J.R. and STEWART J.W.B. 1981. Sulfur transformations in relation to carbon and nitrogen in incubated soils. Soil Biology and Biochemistry, 13, 499-511.

218. SANDERS F.E., MOSSE B. and TINKER P.B. 1975 (Eds). Endomycorrhizas. Proceedings of a Symposium held at University of Leeds, 22-25 July 1974. Academic Press.

219. SANDERS F.E., TINKER P.B., BLACK R.L.B and PALMERLEY S.M. 1977. The development of endomycorrhizal root systems. I. Spread of infection and growth-promoting effects with four species of vesicular-arbuscular endophyte. New Phytologist, 78, 257-268.

220. SARKER A.N. and WYN JONES R.G. 1982. Influence of rhizosphere on the nutrient status of dwarf french beans. Plant and Soil, 64,369-380.

221. SCHIPPERS B. and GAMS W. 1979 (Eds). Soil-borne plant pathogens. Academic Press.

222. SHIELDS J.A., PAUL E.A. and LOWE W.E. 1974. Factors influencing the stability of labelled microbial materials in soils. Soil Biology and Biochemistry, 6, 31-37.

223. SHIELDS J.A., PAUL E.A., LOWE W.E. and PARKINSON D. 1973. Turnover of microbial tissue in soil under field conditions. Soil Biology and Biochemistry, 5, 753-764.

224. SINCLAIR J.L., McCLELLAN J.F. and COLEMAN D.C. 1981. Nitrogen mineralization by Acanthamoeba polyphaga in grazed Pseudomonas paucimobilis populations. Applied and Environmental Microbiology, 42, 667-671.

225. SÖDERSTRÖM B.E. 1979. Some problems in assessing the fluorescein diacetate-active fungal biomass in the soil. Soil Biology and Biochemistry, 11, 147-148.

226. SÖDERSTRÖM B.E. 1979. Seasonal fluctuations of active fungal biomass in horizons of a podzolized pine-forest in central Sweden. Soil Biology and Biochemistry, 11, 149-154.

227. SØRENSEN L.H. 1975. The influence of clay on the rate of decay of amino acid metabolites synthesised in soils during decomposition of cellulose. Soil Biology and Biochemistry, 7, 171-177.

228. SØRENSEN L.H. 1979. Decomposition of straw in soil after stepwise repeated additions. Soil Biology and Biochemistry, 11, 23-29.

229. SØRENSEN L.H. 1981. Carbon-nitrogen relationships during the humification of cellulose in soils containing different amounts of clay. Soil Biology and Biochemistry, 13, 313-321.

230. SPARLING G.P. 1981. Microcalorimetry and other methods to assess biomass and activity in soil. Soil Biology and Biochemistry, 13, 93-98.

231. SPARLING G.P. 1983. Estimation of microbial biomass and activity in soil using microcalorimetry. Journal of Soil Science, 34, in press.

232. SPARLING G.P. and EILAND F. 1983. A comparison of methods for measuring ATP and microbial biomass in soils. Soil Biology and Biochemistry, 15, 227-229.

234. SPARLING G.P., FERMOR T.R. and WOOD D.A. 1982. Measurement of the microbial biomass in composted wheat straw, and the possible contribution of the biomass in the nutrition of Agraricus bisporus. Soil Biology and Biochemistry, 14, 609-611.

235. SPARLING G.P., ORD B.G. and VAUGHAN D. 1981. Microbial biomass and activity in soils amended with glucose. Soil Biology and Biochemistry, 13, 99-104.

236. SPARLING G.P., ORD B.G. and VAUGHAN D. 1981. Changes in microbial biomass and activity in soils amended with phenolic acids. Soil Biology and Biochemistry, 13, 455-460.

237. STOTZKY G. 1972. Activity, ecology and population dynamics of microorganisms in soil. C.R.C. Critical Rewiews in Microbiology, 2, 59-137.

238. STOTZKY G. and NORMAN A.G. 1961. Factors limiting microbial activities in soil. I. The level of substrate, nitrogen and phosphorus. Archives of Microbiology, 40, 341-369.

239. SUBBA RAO N.S. 1982. (Ed.) Advances in Agricultural Microbiology. Butterworth Scientific.

240. SUSSMAN A.S. 1965. Dormancy of soil microorganisms in relation to survival. In: Ecology of Soil-Borne Plant Pathogens, pp. 99-110. Eds. Baker K.F. and Snyder W.C. John Murray.

241. SUTTON J.C. and SHEPPARD B.R. 1976. Aggregation of sand-dune soil by endomycorrhizal fungi. Canadian Journal of Botany, 54, 326-333.

242. SWIFT M.J., HEAL O.W. and ANDERSON J.M. 1979. Decomposition in terrestrial ecosystems. (Studies in Ecology, volume 5). Blackwell Scientific Publications.

243. TATE K.R. and JENKINSON D.S. 1982. Adenosine Triphosphate (ATP) and microbial biomass in soil: Effects of storage at different temperatures and at different moisture levels. Communications in Soil Science and Plant Analysis, 13, 899-908.

244. TATE K.R. and JENKINSON D.S. 1982. Adenosine Triphosphate measurement in soil: an improved method. Soil Biology and Biochemistry, 14, 331-335.

245. TIEN T.M., GASKIN M.H. and HUBBELL D.H. 1979. Plant growth substances produced by Azospirillum brasilense and their effect on the growth of pearl millet (Pennisetum americanum L.). Applied and Environmental Microbiology, 37, 1016-1024.

246. TISDALL J.M. and OADES J.M. 1979. Stabilization of soil aggregates by the root systems of ryegrass. Australian Journal of Soil Research, 17, 429-441.

247. TISDALL J.M. and OADES J.M. 1982. Organic matter and water-stable aggregates in soils. Journal of Soil Science, 33, 141-163.

248. TISDALL J.M., COCKROFT B. and UREN N.C. 1978. The stability of soil aggregates as affected by organic materials, microbial activity and physical disruption. Australian Journal of Soil Research, 16, 9-17.

249. TRUTER M.R. 1976. Chemistry of the calcium ionophores. In: Calcium in Biological Systems. Symposium XXX of the Society for Experimental Biology, pp. 19-40. Ed. Duncan C.J. Cambridge University Press.

250. TYLER G. 1980. Metals in sporophores of basidiomycetes. Transactions of the British Mycological Society, 74, 41-49.

251. TYLER G. 1982. Accumulation and exclusion of metals in Collybia peronata and Amanita rubescens. Transactions of the British Mycological Society, 79, 239-245.

252. VANCURA V. and KUNC F. 1977. The effect of streptomycin and actidione on respiration in the rhizosphere and non-rhizosphere Soil. Zentralblatt für Bakteriologie Parasitenkunde, Infectionskrankheiten und Hygiene, 132, 472-478.

253. VAN VEEN J. and PAUL E.A. 1979. Conversion of biovolume measurements of soil organisms grown under various moisture tensions, to biomass and their nutrient content. Applied and Environmental Microbiology, 37, 686-692.

254. VAN VEEN J.A. and PAUL E.A. 1981. Organic carbon dynamics in grassland soils. 1. Background information and computer simulation. Canadian Journal of Soil Science, 61, 185-201.

255. VAN VUURDE J.W.L. and SCHIPPERS B..1980. Bacterial colonization of seminal wheat roots. Soil Biology and Biochemistry, 12, 559-565.

256. VORONEY R.P., VAN VEEN J.A. and PAUL E.A. 1981. Organic carbon dynamics in grassland soils. 2. Model validation and simulation of the long-term effects of cultivation and . rainfall erosion. Canadian Journal of Soil Science, 61, 211-224.

257. WAGNER G.H. 1975. Microbial growth and carbon turnover. In: Soil Biochemistry, volume 3, pp. 269-305. Eds. Paul E.A. and McLaren A.D.. Marcel Dekker.

258. WAINWRIGHT M. 1978. A review of the effects of pesticides on microbial activity in soils. Journal of Soil Science, 29, 287-298.

259. WALLACE J.M. and ELLIOTT L.F. 1979. Phytotoxins from anaerobically decomposing wheat straw. Soil Biology and Biochemistry, 11, 325-330.

260. WAKSMAN S.A. and WOODRUFF H.B. 1940. Survival of bacteria added to soil and the resultant modification of soil population. Soil Science, 50, 421-427.

261. WAREMBOURG F.R. and PAUL E.A. 1977. Seasonal transfers of assimilated ^{14}C in grassland: plant production and turnover, translocation and respiration. Soil Biology and Biochemistry, 9, 295-301.

262. WHIPPS J.M., HESELWANDTER K., McGEE E.E.M. and LEWIS D.H. 1982. Use of biochemical markers to determine growth, development and biomass of fungi in infected tissues, with particular reference to antagonistic and mutualistic biotrophs. Transactions of the British Mycological Society, 79, 385-400.

263. WILLIAMS F.M. 1980. On understanding predator-prey interactions. In: Contemporary Microbial Ecology, pp. 349-375. Eds. Ellwood D.C., Hedger J.N., Latham M.J., Lynch J.M. and Slater J.H. Academic Press.

264. WILSON J.M. and GRIFFIN D.M. 1975. Water potential and the respiration of microorganisms in the soil. Soil Biology and Biochemistry, 7, 199-204.

265. WITTY J.F. 1979. Overestimation of N_2-fixation in the rhizosphere by the acetylene reduction method. In: The Root-Soil Interface, pp. 137-144. Eds. Harley J.L. and Russell R.S. Academic Press.

266. WOLDENDORP J.W. 1981. Nutrients in the rhizosphere. In: Agricultural Yield Potentials in Continental Climates. Proceedings of 16th Colloquium of the International Potash Institute, Bern (1981).
267. WOODS L.E., COLE C.V., ELLIOTT E.T., ANDERSON R.V. and COLEMAN D.C. 1982. Nitrogen transformations in soil as affected by bacterial-microfaunal interactions. Soil Biology and Biochemistry, 14, 93-98.
268. YODER D.L. and LOCKWOOD J.L. 1973. Fungal spore germination on natural and sterile soil. Journal of Gereral Microbiology, 74, 107-117.
269. ZUNINO H., BORIE F., AGUILERA S., MARTIN J.P. and HAIDER K. 1982. Decomposition of ^{14}C-labelled glucose, plant and microbial products and phenols in volcanic ash derived soils of Chile. Soil Biology and Biochemistry, 14, 37-43.

CHAPTER 7

CARBOHYDRATES IN RELATION TO SOIL FERTILITY

M.V. CHESHIRE

The Macaulay Institute for Soil Research, Aberdeen, Scotland.

CONTENTS

1.   INTRODUCTION

The carbohydrate in soil, which, on average, accounts for about ten per cent of the organic carbon content, forms one of the largest of the soil organic matter fractions, in so far as these may be defined. Most of the carbohydrate is in the form of polysaccharide, but small amounts of monosaccharides (free sugars) and oligosaccharides are also present[21]. It plays a vital role in soil fertility in three areas; biological nutrition, soil structure and soil water relationships. More specific functions involve plant root growth, adhesion in microorganisms, and the protection of enzymes.

2.   BIOLOGICAL NUTRITION

2.1. A source of nutrient for microorganisms

In theory, carbohydrates in soil should provide a rich source of available energy for microorganisms. Because of the repetitive bonding in homogeneous polysaccharides, where one type of sugar is present with perhaps only one type of linkage, a single enzyme molecule should be able to uncouple all the monosaccharide units and so give a good return of metabolizable sugar for the energy used to synthesize the enzyme[2]. Even the more complex carbohydrate forms should be far more easily degraded enzymatically than the humified organic matter of soils. However, there are a number of reasons why polysaccharides may not be readily utilized.

2.1.1. Chemical structure of the substrate.

First, there is the linkage between monomers. Enzymes which hydrolyse polysaccharides are specific not only for the different types of sugars, but also for the different types of linkage between like or unlike sugars. Endopolymerases hydrolyse bonds in the interior of the molecule between particular monomers, whereas another class of enzyme, the exopolymerases, only hydrolyse bonds next to the non-reducing end group. Both types of enzymic reaction are specific for different kinds of sugar. Heteropolysaccharides, therefore, may require many different enzymes in a particular sequence for the complete hydrolysis to monosaccharides. Carbohydrates in soil include those which originate from decaying plant tissue as well as root exudates from living plants. Cellulose and starch are common homopolysaccharides from this source, but the ill-defined plant hemicellulose, though rich in β-(1-4) xylan, contains glucose, galactose, arabinose and xylose and glucuronic acid in various combinations. Pectin is predominantly an α-(1-4)-D galacturonan, but also contains arabinose, xylose, galactose, rhamnose and glucose. In microorganisms, hetero-polysaccharides usually form a greater proportion of the carbohydrate than in plants, although fewer different kinds of sugar are involved. Pentoses are relatively uncommon.

Second, another factor which could limit metabolism would be the isomeric form of the sugar. Arabinose, for example, usually occurs in plants in the L form and only rarely in the D form, whereas in microorganisms, the D form is prevalent. These forms would each require specific enzymes.

Third, substitution of the hydrogen of sugar hydroxyl groups is a common feature of some polysaccharides. For example, the xylan of grasses appears to be highly substituted with acetyl groups. When grass is eaten by ruminants, there is an increase in the proportion of acetyl groups in the grass residues in the faeces[3], indicative of a greater resistance of the acetylated part to the hydrolysing enzymes of the microflora in the gut. If there is a sufficient degree of substitution (more than every alternate sugar in a chain[96]), there will be no enzymic hydrolysis by glycanases. The substituting groups must first be removed by yet another enzyme.

Small quantities of methyl sugars occur in soil[14,37], and the methoxyl group shows the same protective effect as the acetyl group[96]. It has been queried whether these methyl groups are present in polysaccharides as a defence mechanism against attack on tissue structure by "foreign" enzymes[37].

In addition to carbon, hydrogen and oxygen, soil carbohydrates also contain nitrogen in the form of amino sugars. Amino groups occur in chitin in the form of N-acetyl glucosamine and soils possess strong hydrolytic activity for chitin. From the amounts of amino sugars in soil, it may be calculated that they account for 1-12% of the nitrogen. However, different forms of N differ in availability to microorganisms, the amino sugar nitrogen being one of the more readily available[94].

Soil polysaccharide also contains phosphorus and sulphur[69]. The latter ranges from one S atom per 40-250 hexose units and it may be in the form of ester sulphate groups[70], whereas the phosphorus is usually of the order of 1%. It is considered that neither the sulphur nor the phosphorus form would contribute in any special way nutritionally. In fact they may be more closely associated with the peptide which appears to be covalently linked to the polysaccharide and it becomes a matter of definition as to what is encompassed by the term polysaccharide.

Substances which are strictly carbohydrate are primarily used metabolically as sources of energy, but other organic molecules will also serve for this purpose. There are two major alternatives, amino acids from protein and fatty acids, but the availability of these substances in soil may govern their use as energy sources. Amino acids occur in soil in peptides and proteins, which may be partially tanned by reaction with polyphenols or with humic substances, which have tanning capability. This is more likely when the protein is derived from decaying plant matter, and there is an abundance of polyphenols. Tanning gives protection against attack by proteases which makes the return of energy to the organism much less worthwhile. Possibly the enzymes themselves may undergo tanning inactivation. Long chain fatty acids and waxes occur to the extent of 1-6% of the humus of soils[108] and would appear a better source of energy-rich food and yet there may be problems of insolubility to be overcome. Calcium is usually the dominant cation amongst the exchangeable cations in soil and so fatty acids will be found predominantly in the calcium form, which is insoluble. Under some conditions lipids account for a greater proportion of the humus of more acid soils such as podzols. Possibly the acidic forms of fats and waxes are more hydrophobic than the calcium form and this is the greatest constraint to their decomposition.

The availability of the carbohydrate may be directly affected by these characteristics either because of association or linkage with the substance as with peptide, or the presence of lipopolysaccharide, in which there is a covalent bond between the lipid and carbohydrate parts.

In isolation, lignin would be a far less rewarding source of metabolizable material than carbohydrate, but not in the context of the soil, where there is a considerable input from decaying plants, and there are constraints on polysaccharide decomposition, but in the initial stages of decay, at least, the carbohydrate is oxidized to CO_2 by the biomass much faster than is the lignin[54]. Furthermore, the decomposition of lignin by organisms such as the white rot fungus <u>Phanerochaete</u> <u>chrysosporium</u> needs the presence of a "growth" substrate such as glucose or cellulose[63].

2.1.2. <u>Other constraints on the rate of metabolism</u>.

The soil <u>milieu</u> also imposes restrictions which include both chemical and physical effects. Carbohydrates may form complexes with metal ions through reaction of the carboxyl groups of uronic acids or their hydroxyl groups. Vicinal <u>cis</u> hydroxyl, in the most advantageous conformational arrangement, may contribute three attachment points[1]. Whereas the most common ions available to form complexes in soils are Ca, Mg, K and Na, certain transition elements present in lower amounts form stronger complexes and may successfully compete for sites. Preformed complexes of microbial polysaccharides with copper, zinc, iron and aluminium are decomposed at a slower rate than the uncomplexed material [74,75].

Aluminium complexes were held to be responsible for the accumulation of organic matter that occurs in soils developed on granitic parent materials[19], the addition of further aluminium as a hydroxide gel decreasing the decomposition of both glucose and maize straw.

The effects of clay on the rate of decomposition of carbohydrates can be either positive or negative as has been observed with other substrates[110]. Carbohydrates are adsorbed by clays and their addition to a soil may serve to concentrate soluble carbohydrate materials on surface sites, which are the habitat of most soil microorganisms. The rate of decomposition of both glucose and starch was reported to increase when small additions of bentonite were made to soil[84]. Such effects, however, are usually outweighed by the negative effects resulting from the association of the organic matter with the clay. Thus in other experiments the addition of montmorillonite to sand or soil reduced the rate of decomposition of glucose[51].

Soils derived from volcanic ash (andosols) are rich in allophane. This alumino-silicate mineral appears to retard the decomposition of organic matter and carbohydrate[73,83,122]. The soils become rich in organic substances, with the carbohydrate fraction composition becoming dominated by sugars such as

mannose, fucose and ribose, characteristic of microbially synthesized
polymers. Similarly when andosols are buried by fresh volcanic activity
there appears to be a gradual change in sugar composition with time, so that
on hydrolysis by dilute acid mannose, and not glucose, becomes the
predominant sugar[119,120]. It is envisaged that slow chemical changes could
occur over the many thousands of years involved which would include the
epimerization of the glucose to mannose. Mannose, having vicinal <u>cis</u>
hydroxyl groups, could form complexes with aluminium. Such complexes
would prevent further epimerization of the mannose molecule and might
prevent it from entering a liquid phase as a result of its becoming part of
an insoluble solid. In this way the equilibrium between glucose and mannose
would be upset in favour of more mannose being formed from glucose. The
activity of allophane has been established by adding it to decomposing
substrates[73], but another feature of andosols is the presence of much
active alumina which must also contribute to the preserving effect.

In addition to a direct adsorptive effect on the carbohydrate, the
degrading enzymes may also be adsorbed and their action impaired. Thus both
cellulase and hemicellulase have been observed to show decreased activity in
mixtures with clay[55,95]. Humic substances probably also affect soil enzyme
activities[116]. This aspect is discussed in Chapter 2.

There is no evidence that the decomposition rates of carbohydrates are
more greatly affected by clays than are those of other organic substances.

Finally, the physical disposition of the carbohydrate in soil must
influence its availability as a nutrient. Whether it acts as the binding
agent, or is present incidentally, carbohydrate occurs in soil aggregates.
This must make it less accessible both to organisms and enzymes. The
enhanced breakdown of soil organic matter following disruption of soil
aggregates has been partially attributed to the exposure of previously
inaccessible sites[100].

2.2 Influence on biodegradation

Biodegradation is a most important parameter for any potential agro-
chemical. It is essential that residues of chemicals which are applied to
soil, or reach it from other sources, are decomposed at a rate fast enough
to prevent their build up in soil. It has been observed that the addition
to soil of a readily metabolizable substrate, such as a carbohydrate, may
enable an organism to degrade recalcitrant compounds. Examples of such

compounds are the phenylurea carbamate, phenoxy acid and triazene herbicides[77]. With the insecticide mexacarbate (4-dimethylamino-3, 5-xylyl methylcarbamate) the decomposition pathway is altered[6]. It is suggested that the additional nutrient provides energy or reductant required for enzyme synthesis[16]. This implies that the rate of decomposition of such chemicals in an unamended soil will be governed by the levels of readily available nutrients which are present.

3. THE FORMATION OF A STABLE SOIL STRUCTURE BY THE AGGREGATION OF SOIL PARTICLES

The clay, silt and sand materials in soil are rarely present in discrete particles, but are bonded together in aggregates which may range in size from <.2μm to >5000μm, although in many soils more than 70% of the aggregates are in the range 20-250μm[115]. The larger aggregates appear to be made up by the grouping together of numbers of smaller ones.

Because of the range of particle and aggregate sizes, a number of different aggregating agents are operative. For example, large particles are held together by a network of fungal hyphae[13] or fine plant roots[28,29,46]. These fungal hyphae are sticky as a result of a surface layer of poly-saccharide, and plant roots are similarly coated with mucigel. On the other hand, micro-aggregates in the size range 20-50μm appear to depend for cohesion on amorphous soil organic matter components. This aggregation is very largely dependent on soil polysaccharide[57].

Aggregation of soils is necessary to obtain optimum growth conditions for plants, and to enable soil cultivation to be carried out. Aggregation also helps to prevent most forms of soil erosion and reduce soil crusting. In addition aggregation is very important in allowing free movement of water and air in soil.

There are three main lines of evidence to support the involvement of polysaccharide in soil aggregation; firstly, the strong adhesive qualities possessed by many polysaccharides; secondly, the effect of the addition of polysaccharide to soil particles on the formation of stable synthetic aggregates, and thirdly, the effect of periodate and tetraborate, which remove carbohydrate, in disrupting natural aggregates.

The adhesiveness of polysaccharides is a well-known characteristic. Over a long period polysaccharides have been used as gums, for example modified-starch wallpaper pastes. The stickiness of a polysaccharide is governed by the size and shape of the molecule, the most effective being those which are 'spherical' rather than linear and which possess a high degree of branching[117]. Stickiness depends on weak physical forces such as van der Waals forces, and on hydrogen bonding. Covalent bonding is apparently not involved. In fact adsorption on the large surface area offered by soil clay is probably the predominant cohesive mechanism.

Both neutral and acidic polysaccharides have a binding action on soil particles[72]. The effectiveness of the acidic polymers increases with uronic acid content, but is dependent on the numbers of exchangeable cations present, increasing base saturation resulting in reduced binding capability. The ability of neutral polysaccharides to form stable synthetic aggregates, tested using dextrans produced by the blue-green alga Leuconostoc meserentoides, was found to increase with increasing molecular weight[87]. The effect was accounted to the greater number of linkages which could be formed per polymer molecule.

In a number of soils prolonged treatment with sodium periodate results in the breakdown of most of the aggregate structure[22], indicating that polysaccharide is responsible for much of the binding of soil particles into aggregates. Periodate is considered to have a high degree of specificity for breaking carbon-carbon bonds carrying functional groups such as hydroxyl in suitable configuration. Most naturally occurring sugar structures are susceptible to periodate oxidation and the partially oxidized products are dissolved by a subsequent treatment with tetraborate.

Not all aggregated soils respond to the standard periodate treatment[78]. One reason for this is the presence of Ca which appears to inhibit the oxidation[48]. Latterly, it has been suggested that other bonding reagents are present in soil besides polysaccharide, part of which can be destroyed or released by treatment with sodium pyrophosphate subsequent to periodate treatment[25,107,113,114]. Recent studies on the effect of periodate on the stability of microaggregates (>45μ), however, suggest that for many soils, where it had been supposed there was more than one binding agent, the aggregation can now be accounted almost solely to periodate sensitive materials. This is because of the slowness of the periodate oxidation reaction. After the standard period of 6h of treatment of a cultivated podzol (Countesswells series) with 0.02M periodate, almost 50% of the polysaccharide remains and can be determined as reducing sugar after soil residue hydrolysis. Treatment for 48h results in a further 10% loss of polysaccharide, but even after 168h with 0.05M periodate 20% remains[22].

The relationship between the aggregation (particles <45μm) and
residual carbohydrate is shown in Fig. 1. This implies that 100%
disruption will occur only when all the reducing sugar has been destroyed.

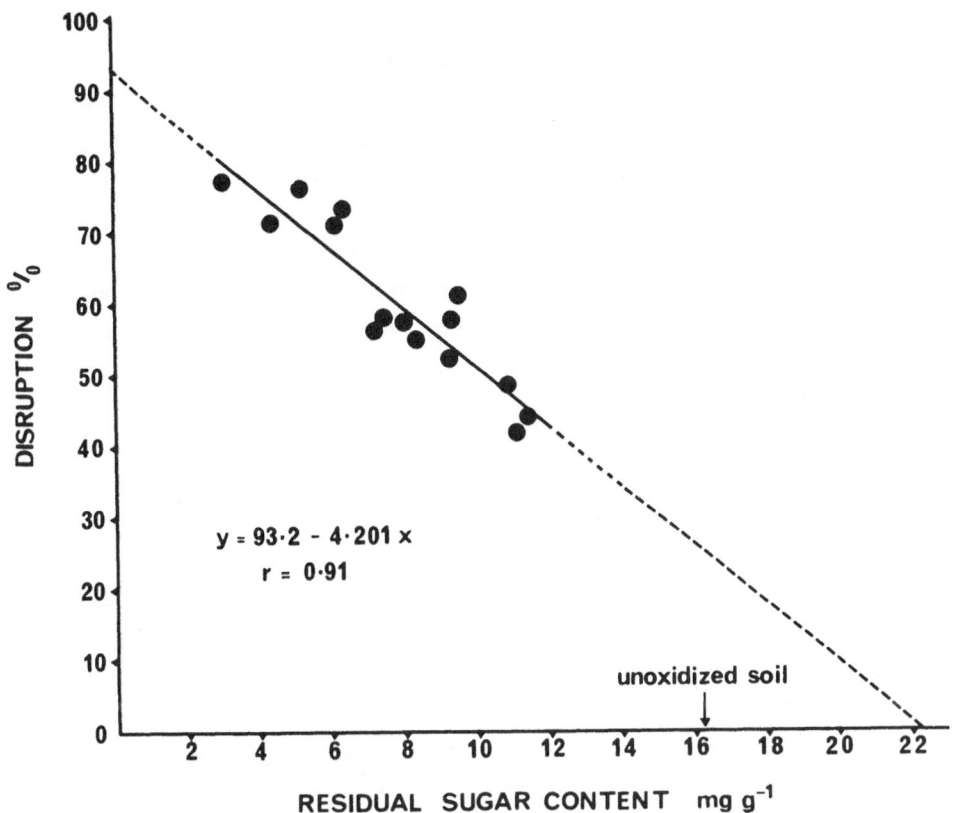

FIGURE 1. Correlation of residual sugar content with soil aggregate
stability.

Many other soils which have been tested in this way show a similar
relationship, but there are exceptions. For a range of soils from
North East Scotland, prolonged treatment with periodate followed by
tetraborate destroyed 80% of the carbohydrate and caused 75% increase in
the microaggregates <45μm. For some of these soils, extrapolation of the
line relating residual sugar content and disruption cuts the y axis at a

value significantly different from 100% i.e. disruption would be incomplete at the point of complete destruction of the carbohydrate. The linearity of the graph does not prove a causal relationship and could be interpreted to indicate that the polysaccharide is shielded by the aggregate and oxidized only as it is released when other binding agents undergo oxidation. Evidence which might appear to be in agreement with this is the differentiation undergone by the polysaccharide undergoing oxidation. The less-easily oxidized material remaining in the residue is much richer in glucose, arabinose and xylose than the rest. These sugars usually predominate in plant residues and might be expected to be present in relatively large particles. Their rate of oxidation may depend on their inaccessibility for that reason, but another mechanism may be envisaged.

Much carbohydrate, in the form of mucigel, is added to soil through the roots of living plants and is seen to interact with clay particles[50] which become aligned and adsorbed onto the mucigel face-on. The plant root mucigel, coated with clay particles, may thus become the nucleus for the formation of a microaggregate. As a result of this a large proportion of plant-like carbohydrate might be protected through being within the interior of the aggregate. Whether or not clays are active precisely in this way, it is known that they are effective in slowing down the rate of oxidation of carbohydrates by periodate[26,45] and this seems more likely to be the result of a clay-polysaccharide interaction than of any effect clay might have directly on the reagents. Encapsulated organisms in clay suspensions are considered to adsorb clay platelets in an edge-on position[76] because the polysaccharide has a negative charge. Clay domains linked both face-on and edge-on by organic matter are a feature of the model for a soil crumb proposed by Emerson[44].

Carbohydrates may also resist periodate oxidation because of the position of the bonds within the polymers. One example of this is 1 → 3 linked sugars, which, having no vicinal hydroxide groups, should remain unoxidized[12,41]. The same is true for vicinal hydroxyl groups in the trans position if there is not a certain degree of flexibility of the hydroxyl group about the bond. In addition, there is also the possibility of hemiacetal formation between the hydroxyl group of one sugar and its partially oxidized neighbour, rendering the chain stable to further periodate oxidation[88,89,90]. This is known to occur with 1 → 4 linked glucans and xylans.

An examination of the bonding of the polysaccharide in a soil by permethylation has shown a considerable proportion of 1 → 3 linked hexoses and pentoses[24]. The residue after periodate treatment of the soil, however, was rich in arabinose and 1 → 4 linked glucose and xylose, in keeping with the suggestion that it is predominantly plant material. Tisdall and Oades[115] distinguish aggregation which is transient, and dependent on polysaccharide formation involving readily available substrates such as glucose, from the more persistent aggregation engendered gradually by more recalcitrant substrates such as cellulose. The relatively insensitive reaction of the aggregates to periodate leads to a consideration of other binding agents, but it is conceivable that the protection against biological attack shown by the long term survival of some polysaccharides in soil is matched by a chemical protection against periodate. This protection may come partially from the humification of part of the substance. The polymer surfaces involved in bonding newly formed aggregates together would be inaccessible to the organisms in their initial attack on the organic matter, but the humification they would engender, either of the substrate or subsequently of themselves, at other surfaces, could make the residual material even more difficult to decompose and consequently the aggregates altogether more stable. This is not to deny a direct bonding function in aggregation for some of the more humified soil components, but in the past the effectiveness of short term periodate treatments has been overestimated, despite the known effects of soil components such as clays in slowing this reaction.

Although strong evidence has now been obtained that carbohydrate is the predominant bonding agent of microaggregates, there is still a need to explain the effective action of reagents such as sodium pyrophosphate in destroying aggregates. One explanation could be that some of the poly-saccharide is bonded to clay, silt and sand particles through iron and aluminium bridging atoms[42,43,60] which form complexes with pyrophosphate and this consequently releases the polysaccharide. Recent studies have related aggregate stability to iron and aluminium extractable with acetylacetone[98]. It is envisaged that **carboxyl** groups in the polysaccharide will react with aluminium $Al(OH)_2^+$ ions on the clay surface through bridging water molecules[17]. The extent of adsorption of charged polysaccharide on kaolinite or montmorillonite has been found to depend on the saturating ion[52,79,92].

Studies with neutral polysaccharides have shown that carboxyl groups
are not necessary for the formation of stable synthetic aggregates[27].
In these instances adsorption is probably achieved by hydrogen bonding
through the hydroxyl groups of component sugars.

4. INFLUENCE ON THE BEHAVIOUR OF WATER IN SOIL

Another way in which soil organic matter enhances fertility is by
enabling the soil to retain water.

Water which can be made use of by plants is called available water
and is considered to be the volume of water held in soil between 0.05 and
15 bar suction[112]. Soil factors which have been found to relate to
available water content are the soil bulk volume and the clay content[97].
In mineral soils the bulk volume is proportional to the organic matter
content[101].

Soil carbohydrate, accounting fairly consistently for about a tenth
of the organic matter of soils, and being relatively strong hydrophilic
must therefore make an appreciable contribution to water availability.
Furthermore, soil pore space is also a function of the aggregation of soil
particles[30] which, as described in section 3, is strongly dependent on
polysaccharide.

Water retention by soils is more dependent on the organic matter
than on the mineral constituents[67,104]. Chen and Schnitzer[20] examined
the water retention of the humic and fulvic acid of a podzol. The fulvic
acid was purified by dissolution in methanol which would have precipitated
some polysaccharide as well as insoluble salts. Humic and fulvic acids
from a number of tropical soils, prepared in the same way, had an
approximately similar range of carbohydrate contents (6.3 - 11.9%) as
determined colorimetrically by phenol/sulphuric[49] and such carbohydrate
components may have made a significant contribution to the water adsorption.
At low relative humidities, the adsorption was similar, but at high
relative humidities the fulvic acid adsorbed much more water. The
difference was accounted to the larger proportion of oxygen-containing
functional groups, particularly carboxyl, in the fulvic acid, about which
hydrogen bonded water molecules were envisaged to cluster. It is well
known that polysaccharides have a strong affinity for water[117].
Haworth et al.,[59] were the first to suggest that the highly viscous

polysaccharides produced by microorganisms might be important in conserving the moisture content of soil. Similarly the mucilage added to soil as plant root exudate is highly hydrated and serves to prevent desiccation of the root[85]. Soil polysaccharide extracted by alkali and purified using Polyclar and ethanol precipitation is very hydrophilic. The air-dried material may contain as much as 10% water.

Fig. 2 shows the relationship between water content at equilibrium and relative humidity for a polysaccharide in the salt form extracted with alkali from a cultivated podzol. The water retention is probably highly dependent on whether the material is in a salt or acid form[10].

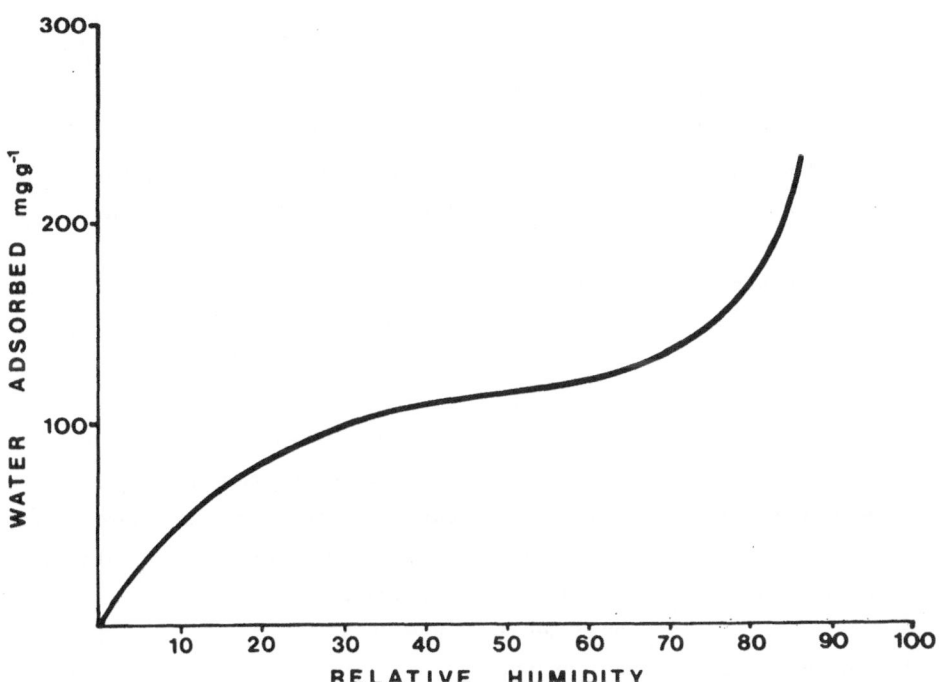

FIGURE 2.   Adsorption of water by soil polysaccharide.

The association of water with sugar molecules is dependent upon hydrogen bonding through the hydroxyl groups. This hydration depends on the conformation of the molecule; thus equatorial hydroxyl groups are hydrated to a greater extent than axial groups.

In polymers, intra-molecular interaction between monosaccharide residues governs the rotation about the glycosidic bond and modifies the hydration pattern. In gel formation, which is the result of crosslinking reactions, a small percentage of the water is hydrogen bonded, but most has physical properties indistinguishable from free water, that is, there is no large scale ordering of water molecules present within the gel. Gel formation appears to result from polymer-polymer interaction in water media which allows the polymers to take up the positions they do.

5. INFLUENCE ON THE ION EXCHANGE AND COMPLEXING PROPERTIES OF SOIL
5.1. Ion exchange

Organic matter makes a contribution to both the anion and cation exchange capacity of soils. The contribution to the latter has been shown by remeasuring the exchange capacity after destroying the organic matter, by hydrogen peroxide treatment for example[62], or by relating variation in exchange capacity to levels of clay and organic matter by regression analysis[31,35,61,106,118,121]. For many types of soil, particularly where there is any accumulation of organic matter, the organic matter makes a larger contribution to the CEC than does the clay. But even in soils dominated by clay it is still about 20% of the total. Absolute amounts of clay or organic components only give an approximation of the relative contribution of the component to the exchange capacity because of interaction. Calcinai and Sequi[18] found an increase in the exchange capacity accountable to clay when they removed some organic matter by periodate oxidation and pyrophosphate extraction, implying that some of the mineral matter sites had been occluded by the organic matter.

The contribution made by the soil carbohydrate to the cation exchange capacity will depend on the presence of uronic acids or closely related structures. Aggregation is thought to involve carboxyl groups of polysaccharides linked through metal ions such as aluminium to clay[60] and consistent with this is the fact that the isolation of soil polysaccharide usually requires the removal of associated exchangeable metal ions such as aluminium. The material purified by dialysis against acid has a titratable acidity of between 50 and 100 meq g^{-1}, a somewhat lower value than that found for humic substances. The humic substances of soil may make up 40-50% of the organic matter so that the contribution of the polysaccharide to the organic cation exchange sites is probably less than 8% of the whole. Free glucuronic acid has a pKa value of 3.20 and galacturonic acid one of 3.42. Substances are present in isolated soil polysaccharides which are more acidic than this. These may be saccharinic acids formed during the isolation[33,34], but they could occur naturally.

The anion exchange capacity of soils is much smaller than the cation exchange capacity. Values for several Australian soils ranged from 0.1 to 0.4 meq/100g in the surface horizon compared with 1 to 28 for the cation exchange capacity[47]. This is the electrovalent exchange capacity for ions such as Cl^- and NO_3^- and must be distinguished from adsorption processes which occur with some other anions such as $H_2PO_4^-$. Two types of anion exchange sites dependent on organic matter can be distinguished theoretically:- (1) Organic functional groups (2) Iron and aluminium atoms complexed by organic matter[91].

It is not known what contribution polysaccharide makes directly to the anion exchange. Potential anion exchange sites are present as amino groups (NH_3^+) in the amino sugars glucosamine and galactosamine, but in many soil organisms these substances are predominantly in the acetyl form in chitin[9]. However, a few free amino groups are present in chitin which account for the basicity and there is also a non-acetylated glucosamine polymer, chitosan, which occurs in the cell walls of some fungi (Zygomycotina). In bacteria the amino sugars are involved in the bridge linkage between the carbohydrate and peptide components, which is formed between the terminal amino group of the peptide chain and the carboxyl group in muramic acid (3-O-D-lactyl D-glucosamine). In higher plants the linkage between glucosamine and peptide is glycosidic, the glucosamine again being acetylated. Amino sugars comprise between 1 to 12% of the total soil N[94]. It might be considered that the contribution made by the unacetylated amino sugars as a source of positively charged sites for anion exchange is small, but other potential sites such as on the basic amino acids can be relatively unavailable. For instance, no evidence for sites on free NH_2 groups in humic acid could be found using fluorodinitro-benzene[15].

Anion exchange sites on iron or aluminium complexes of organic matter probably occur naturally in the polysaccharide fraction of soil. Soil polysaccharide isolated by adsorption on charcoal or Polyclar treatment of fulvic acid usually contains less than 1% aluminium[23], but problems arise with its re-solution after freeze-drying unless it is given a prior dialysis treatment with EDTA. This suggests that aluminium and iron are strongly bonded to the polysaccharide.

5.2 Complex formation

Complex formation between individual sugars and polyvalent cations can take place where three oxygen atoms in hydroxyl groups occur in an axial-equatorial-axial sequence[1]. These sites have been investigated for sugars; in polysaccharides they would normally only be available where a sugar, such as mannose, occurred as an end group. Complexing by polysaccharide, however, could involve oxygen atoms from more than one sugar ring. All hydroxy-acids form complexes with calcium, so that polymers containing uronic acids would be effective.

All the evidence for complex formation by soil carbohydrate in situ is circumstantial. In the laboratory, microbial polysaccharides react with metal ions to form complexes[64] which have subsequently been used to examine the effects of metal ions on the rate of decomposition in soil[63,75]. Secondly, it has been found that the purification of soil polysaccharide from inorganic substances such as aluminium, iron and silica is difficult, as already mentioned, indicating a strong bond between them. Soil polysaccharide isolated by Forsyth's technique using adsorption on charcoal and precipitation with acetone showed only a weak EPR signal with no discernible features[23], indicating a low content of transition metal complexes. The emergence of a sugar composition in which mannose is the predominant sugar, is a characteristic of long buried horizons in volcanic soils, and has been explained as the result of differences in the relative strengths of metal complexes[119,120].

A number of soil organisms produce 2-ketogluconic acid[38], a substance which occurs as a major microbial product in the rhizosphere[68,81] and is closely related to carbohydrate. It is an effective complexing agent for metal ions[11,39], particularly calcium. Its effectiveness in releasing soil P has been accounted to its acidity (pKa 2.66) rather than to any ability to chelate[80].

5.2.1. Boron complexing.

The availability of boron to plants increases with increasing concentrations of organic matter[53,82]. It has been suggested that boron as borate forms complexes with monosaccharides in soil[93]. It would also be likely to react with sugars in the form of polysaccharides, particularly

those with <u>cis</u> hydroxyl groups. Davis[32] studied the interaction of
soil polysaccharide from a cultivated podzol (Countesswells series) by
measuring the modification of titratable acidity following reaction with
boric acid. Four samples of polysaccharides were used. One, isolated
by the use of Polyclar, did show some change in the ionization of functional
groups with borate addition, but two others, prepared by charcoal
adsorption of successive alkali extracts of soil, and one which was a
hot water-extract of soil, did not. Infrared analysis of borate-
polysaccharide proved to be too insensitive to detect any complexation.
The correlation of boric acid adsorption with organic matter[53] in soils
may be related to exchange sites on Al and Fe held by the organic matter,
as boric acid adsorption is also correlated with extractable aluminium
in soils[56,58].

5.2.2. Complexing of organic substances,

Many common herbicides are weakly basic and will react with carbonyl
groups in carbohydrate substances[109]. However, it seems likely that the
main influence of carbohydrates on the behaviour of such substances in
soil is to increase the rate of degradation as discussed in section 2.2.
The interaction of soil carbohydrate with atrazine is only weak, whereas
humic substances show a strong affinity for this herbicide[40].

6. INFLUENCE ON PLANT ROOT GROWTH

Plant roots are coated with a layer of mucilaginous polysaccharide,
particularly at the root-tip. Several functions have been proposed for
this coating[85]. The mucilage may ease the penetration of the soil by
the growing root. It is said to protect the root from pathogens by
restricting the germination of fungal spores. The mucilage extends the
effective volume of the root and gives good contact with soil particles,
permitting a greater uptake of nutrients. This is facilitated by the
uronic acid content of the mucilage which gives it good ion-exchange
properties. It also prevents desiccation of the root.

The non-carbohydrate fulvic acid fraction of soil is biologically
active in stimulating the growth of excised tomato roots, but the polysaccharide fraction appears to be completely inactive[66], at least when
tested in a sugar-sufficient culture medium. This suggests that there is
no hormone-like effect of the carbohydrate (see Chapter 2).

There may be an indirect effect on plant growth from free sugars in
soil, since these encourage the activity of microorganisms which can

increase the turnover of organic substances, some of which may be
phytotoxic.

7. PROTECTION OF SOIL ENZYMES

Some extracellular enzymes produced by microorganisms are known to
be protected by association with co-secreted polysaccharide molecules[36].
This could form a basis for the observation that a range of enzymes,
including invertase, cellulase, phosphatase, proteases, β-glucosidase,
lysozyme and hyaluronidase exist in soil in complexes with carbohydrate
which serves to protect them against denaturation or proteolysis[5,102].
The complexes are not glycoproteins, but involve the adsorption of the
enzyme on polysaccharide. The evidence for the association comes from
the isolation from soil, using buffer extraction, of active enzyme
fractions which, on hydrolysis, yield the usual range of neutral sugars
common to soil. Dissociation of the enzymes (lysozyme and hyaluronidase)
from the carbohydrate occurs on treatment with sodium dodecyl sulphate,
which would not separate covalently linked components. This evidence
does not rule out the possibility that the association occurs during
coextraction.

The stability of a phenylesterase in soil was considered to be the
result of a carbohydrate-enzyme complex[103], although in this instance,
the existence of a carbohydrate-protein bond through N-acetylhexosamine-
tyrosine is postulated. Hyaluronidase treatment increases the activity
of the enzyme.

8. INFLUENCE ON THE GROWTH OF SOIL ORGANISMS

Many soil bacteria can produce copious amounts of extracellular
polysaccharide. A number of functions have been suggested for this
polysaccharide[36], which include adhesion, protection against desiccation,
ion exchange and selection, tolerance to metals and recognition and
immunological protection against predation. Some of these functions are
only of relevance to the particular organism from which the polysaccharide
is derived, but with others the effects extend beyond the interest of the
organism.

Extracellular polysaccharide enables organisms to adhere to particulate
surfaces. Part of the effect is physical, the polysaccharide increasing
the effective area of contact of the organism and reducing the contact
angle[99]. Various mechanisms may be operative in adhesion as mentioned

earlier in section 3 which deals with aggregation. Hydrogen bonding
would be universally applicable. More specific mechanisms might depend
on particular sugar residues such as uronic acids and these might be
revealed by comparison of compositions. The effectiveness of a polymer
is likely to involve conformational changes and these are unpredictable
without a full knowledge of the structure. Insoluble polysaccharides are
frequently involved in adhesion and the inhibition of adhesion by specific
cell-wall group polysaccharide antibodies is interpreted to be caused by
the obscuring of adsorption sites on the polysaccharide[99]. The adhesion
of aquatic bacteria to surfaces could not be accounted to any one
component of the extracellular polysaccharides[111]. Thus galacturonic
or glucuronic acid were only present in about half the materials which
also contained various mixtures of hexoses and sometimes rhamnose or
fucose. An equivalent study has not been made for soil organisms.

One function the extracellular polysaccharide does not have is that
of a storage polymer: very few organisms are able to digest their own
products.

9. CONCLUSION

This chapter has described several of the major functions of
polysaccharide in soil, such as the aggregating effect and water retention.
Many of these functions are also performed to some extent by other soil
components, but an important difference is that polysaccharide can be
synthesized by plants and this gives them the ability to modify and
improve their environment. In conclusion, the physical state in which
plant roots and many microorganisms flourish should be emphasized.
Frequently each type of organism has associated with it carbohydrate in
the form of a slime or gel which has the effect of increasing contact
with soil surfaces. Thus the gel, essentially a body of water, greatly
facilitates the movement and uptake of nutrients.

10. REFERENCES

1. ANGYAL S.J. 1980. Sugar-cation complexes - structure and
    applications. Chemical Society Reviews, 9, 415-427.
2. BACON J.S.D. 1967. The chemical environment of bacteria in soil. In
    The Ecology of Soil Bacteria, Ed. Gray T.R.G., Liverpool University
    Press. pp. 25-43.
3. BACON J.S.D., GORDON A.H., MORRIS E.J. and FARMER V.C. 1975. Acetyl
    groups in cell wall preparations from higher plants. Biochemical
    Journal, 149, 485-487.

4. BAIER R.E., SHAFRIN E.G. and ZISMAN W.G. 1968. Adhesion : mechanisms that assist or impede it. Science, 162, 1360-1368.
5. BATISTIC L., SARKAR J.M. and MAYAUDON J. 1980. Extraction, purification and properties of soil hydrolases. Soil Biology and Biochemistry, 12, 59-63.
6. BENERZET H.J. and MATSUMURA F. 1974. Factors influencing the metabolism of mexacarbate by microorganisms. Journal of Agriculture and Food Chemistry, 22, 427-430.
7. BERGER K.C. and PRATT P.F. 1963. In : Fertilizer Technology and Use. Eds. McVickar M.H., Bridger L.B. and Nelson L.B. pp 288-292.
8. BERGER K.C., and TRUOG E. 1945. Boron availability in relation to soil reaction and organic matter content. Soil Science Society of America Proceedings, 10, 113-116.
9. BERKELEY R.C.W. 1979. Chitin, Chitosan and their degradative enzymes. Chapter 9 in : Microbial Polysaccharides and Polysaccharases. Eds. Berkeley R.C.W., Gooday G.W. and Ellwood D.C. Academic Press. pp. 205-236.
10. BERKHEISER V.E. 1981. Comparison of water adsorption by monovalent exchange ion forms of soil humic material and synthetic exchanges. Soil Science, 131, 172-177.
11. BERROW M.L., DAVIDSON M.S. and BURRIDGE J.C. 1982. Trace elements extractable by 2-ketogluconic acid from soils and their relationship to plant contents. Plant and Soil, 66, 161-171.
12. BOBBITT J.M. 1956. Periodate oxidation of carbohydrates. Advances in Carbohydrate Chemistry, 11, 1-41.
13. BOND R.D. and HARRIS J.R. 1964. The influence of the microflora on physical properties of soils. 1. Effects associated with filamentous algae and fungi. Australian Journal of Soil Research, 2, 111-122.
14. BOUHOURS J.F. and CHESHIRE M.V. 1969. The occurrence of 2-O-methylxylose and 3-O-methylxylose in peat. Soil Biology and Biochemistry, 1, 185-190.
15. BREMNER J.M. 1957. Studies on soil humic acids II. Observations on the estimation of free amino groups. Reactions of humic acid and lignin preparations with nitrous acid. Journal of Agricultural Science, 48, 352-360.
16. BULL A.T. 1980. Biodegradation: Some attitudes and strategies of microorganisms and microbiologists. In : Contemporary Microbial Ecology. Eds. Ellwood D.C., Hedger J.N., Latham M.J., Lynch J.M. and Slater J.H. Academic Press. pp. 107-136.
17. BURCHILL S., HAYES M.H.B. and GREENLAND D.J. 1981. Adsorption, Chapter 6 in : the Chemistry of Soil Processes. Eds. Greenland D.J. and Hayes M.H.B. John Wiley & Sons Ltd. pp. 221-400.
18. CALCINAI M. and SEQUI P. 1977. Contribution of organic matter to cation-exchange capacity of soils. In : Soil Organic Matter Studies. IAEA Braunschweig. pp. 63-68.
19. CARBALLAS T., CARBALLAS M. and JACQUIN F. 1978. Biodegradation et humification de la matiere organique des sols humiferes atlantiques. Anales de Edafologia y Agrobiologia, 37, 205-212.
20. CHEN Y. and SCHNITZER M. 1976. Water adsorption on soil humic substances. Canadian Journal of Soil Science, 56, 521-527.
21. CHESHIRE M.V. 1979. Nature and origin of carbohydrates in soil. Academic press.

22. CHESHIRE M.V., SPARLING G.P. and MUNDIE C.M. 1983. Effect of periodate treatment of soil on carbohydrate constituents and soil aggregation. Journal of Soil Science, 34, 105-117.

23. CHESHIRE M.V., BERROW M.L., GOODMAN B.A. and MUNDIE C.M. 1977. Metal distribution and nature of some Cu, Mn and V complexes in humic and fulvic acid fractions of soil organic matter. Geochimica Cosmochimica Acta, 41, 1131-1138.

24. CHESHIRE M.V., MUNDIE C.M., BRACEWELL J.M., ROBERTSON G.W., RUSSELL J.D. and FRASER A.R. 1983. The extraction and characterization of soil polysaccharide by whole soil methylation. Journal of Soil Science, 34, 539-554.

25. CLAPP C.E. and EMERSON W.W. 1965. The effect of periodate oxidation on the strength of soil crumbs I Qualitative studies. Soil Science Society of America Proceedings, 29, 127-130.

26. CLAPP C.E. and EMERSON W.W. 1972. Reactions between Ca-montmorillonite and polysaccharides. Soil Science, 114, 210-216.

27. CLAPP C.E., DAVIS R.J. and WAUGAMAN S.H. 1962. The effect of rhizobial polysaccharides on aggregate stability. Soil Science Society of America Proceedings, 26, 446-469.

28. CLARKE A.L., GREENLAND D.J. and QUIRK J.P. 1967. Changes in some physical properties of the surface of an impoverished red-brown earth under pasture. Australian Journal of Soil Research, 5, 59-69.

29. COUGHLAN K.J., FOX W.E. and HUGHES J.D. 1973. A study of the mechanisms of aggregation in a krasnozem soil. Australian Journal of Soil Research, 11, 65-73.

30. CURRIE J.A. 1966. The volume and porosity of soil crumbs. Journal of Soil Science, 17, 24-35.

31. CURTIN D. and SMILLIE G.W. 1978. Estimation of components of soil cation exchange capacity from measurements of specific surface and organic matter. Soil Science Society of America Proceedings, 40, 461-462.

32. DAVIS H. 1978. Interactions between boric acid, borates and components of soil organic matter. PhD Thesis, University of Reading.

33. DAVIS H. and MOTT C.J.B. 1981. Titrations of fulvic acid fractions 1 : Interactions influencing the dissociation/reprotonation equilibria. Journal of Soil Science, 32, 379-391.

34. DAVIS H. and MOTT C.J.B. 1981. Titrations of fulvic and fractions 11 : Chemical changes at higher pH. Journal of Soil Science, 32, 393-397.

35. DRAKE E.H. and MOTTO H.L. 1982. An analysis of the effect of clay and organic matter content on the cation exchange capacity of New Jersey soils. Soil Science, 133, 281-288.

36. DUDMAN W.F. 1977. The role of surface polysaccharides in natural environments. Chapter 9 In : Surface Carbohydrates of the Prokaryotic cell. Ed. Sutherland I.W., Academic Press, pp. 357-414.

37. DUFF R.B. 1961. Occurrence of 2-O-methyl rhamnose and 4-O-methyl-galactose in soil and peat. Journal of the Science of Food and Agriculture, 12, 826-831.

38. DUFF R.B. and WEBLEY D.M. 1959. 2-ketogluconic acid as a natural chelator produced by soil bacteria. Chemistry and Industry, 1376-1377.

39. DUFF R.B., WEBLEY D.M. and SCOTT R.O. 1963. Solubilization of minerals and related materials by 2-ketogluconic acid-producing bacteria. Soil Science, 95, 105-114.

40. DUNIGAN E.P. and McINTOSH T.H. 1971. Atrazine-soil organic matter interactions. Weed Science, 19, 279-282.

284

41. DYER J. 1954. Use of periodate oxidation in biochemical analysis. In Methods of Biochemical Analysis 3, Ed. Glick D. pp 111-152.

42. EDWARDS A.P. and BREMNER J.M. 1965. Dispersion of mineral colloids in soils using cation exchange resins. Nature, 205, 208-209.

43. EDWARDS A.P. and BREMNER J.M. 1967. Microaggregates in soils. Journal of Soil Science, 18, 64-73.

44. EMERSON W.W. 1969. The structure of soil crumbs. Journal of Soil Science, 10, 235-244.

45. FINCH P., HAYES M.H.B. and STACEY M. 1967. Studies on soil polysaccharides and their interaction with clay preparations. Transactions of the International Society of Soil Science, Aberdeen, 19-32.

46. FORSTER S.M. 1979. Microbial aggregation of sand in an embryo dune system. Soil Biology and Biochemistry, 11, 537-543.

47. GILLMAN G.P. 1979. A proposed method for the measurement of exchange properties of highly weathered soils. Australian Journal of Soil Research, 17, 129-139.

48. GREENLAND D.J., LINDSTROM G.R. and QUIRK J.P. 1962. Organic materials which stabilize natural soil aggregates. Soil Science Society of America Proceedings, 26, 366-371.

49. GRIFFITH S.M. and SCHNITZER M. 1975. Analytical characteristics of humic and fulvic acids extracted from tropical volcanic soils. Soil Science Society of America Proceedings, 39, 861-867.

50. GUCKERT A., BREISCH H. and REISINGER O. 1975. Interface sol-racine 1 Etude au microscope electronique des relations mucigel-argile-microorganisms. Soil Biology and Biochemistry, 7, 241-250.

51. GUCKERT A., TOK H.H. and JACQUIN F. 1977. Biodegradation de polysaccharides bacterians adsorbes sur une montmorillonite. In Soil Organic Matter Studies. Proceedings of the IAEA meeting, Volume 1, Vienna. pp 403-411.

52. GUCKERT A., VALLA M. and JACQUIN F. 1975. Adsorption of humic acids and soil polysaccharides on montmorillonite. Pochvovedenie, 2, 41-47.

53. GUPTA U.C. 1968. Relationship of total and hotwater soluble boron, and fixation of added boron, to properties of podzol soils. Soil Science Society of America Proceedings, 32, 45-48.

54. HAIDER K. and DOMSCH K.H. 1969. Decomposition and transformation of lignified plant material by microscopic fungi in soil. Archives of Microbiology, 64, 338-348.

55. HAMZEHI E. and PFLUG W. 1981. Sorption and binding mechanism of polysaccharide cleaving soil enzymes by clay minerals. Zeitschrift für Pflanzenernahrung, Dungung und Bodenkunde, 144, 505-513.

56. HARADA T. and TAMAI M. 1968. Factors affecting behaviour of boron in soil 1. Some soil properties affecting boron adsorption of soil. Soil Science and Plant Nutrition, 14, 215-224.

57. HARRIS R.F., CHESTERS G. and ALLEN O.N. 1966. Dynamics of soil aggregation. Advances in Agronomy, 18, 107-169.

58. HATCHER J.T., BOWER C.A. and CLARK M. 1967. Adsorption of boron by soils as influenced by hydroxy aluminium and surface area. Soil Science, 104, 422-426.

59. HAWORTH W.N., PINKARD F.W. and STACEY M. 1946. Function of bacterial polysaccharides in the soil. Nature, 158, 836-837.

60. HAYES M.H.B. 1980. The role of natural and synthetic polymers in stabilizing soil aggregates. In : Microbial Adhesion to Surfaces. Eds. Berkeley R.C.W., Lynch J.M., Melling J., Rutter P.R. and Vincent B. Horwood, Chichester.

61. HELLING C.S., CHESTERS G. and CONEY R.B. 1964. Contribution of organic matter and clay to soil cation-exchange capacity as affected by the pH of the saturating solution. Soil Science Society of America Proceedings, 28, 517-520.

62. KAMPRETH E.J. and WELCH C.D. 1962. Retention and cation-exchange properties of organic matter in coastal plain soils. Soil Science Society of America Proceedings, 26, 263-265.

63. KIRK T.K., CONNORS W.J. and ZEIKUS J.G. 1976. Requirement for a growth substrate during lignin decomposition by two wood-rotting fungi. Applied and Environmental Microbiology, 32, 192-194.

64. LASIK Y. and GORDIYENKO S.A. 1977. (Complexing of soil bacteria polysaccharides with metals). Soviet Soil Science translated from Pochvovedenie, 1977, 92-98.

65. LASIK Y., GORDIYENKO S.A. and KALAKHOVA L. 1978. (Decomposition of bacterial polysaccharides in soil). Soviet Soil Science translated from Pochvovedenie, 1978, 151-153.

66. LINEHAN D.J. 1977. A comparison of the polycarboxylic acids extracted by water from an agricultural top soil with those extracted by alkali. Journal of Soil Science, 28, 369-378.

67. LINSER H.A. 1956. See Scheffer and Ulrich, 1960.

68. LOUW H.A. and WEBLEY D.M. 1959. A study of soil bacteria dissolving certain mineral phosphate fertilizers and related compounds. Journal of Applied Bacteriology, 22, 227-233.

69. LOWE L.E. 1965. Sulphur fractions of selected Alberta soil profiles of the chernozemic and podzolic orders. Canadian Journal of Soil Science, 45, 297-303.

70. LOWE L.E. 1968. Soluble polysaccharide fractions in selected Alberta soils. Canadian Journal of Soil Science, 48, 215-217.

71. MERTEL Y.A., DE KIMPE C.R. and LAVERDIERE M.R. 1976. Cation exchange capacity of clay rich soils in relation to organic matter, mineral composition and surface area. Soil Science Society of America Proceedings, 42, 764-767.

72. MARTIN J.P. 1971. Decomposition and binding action of polysaccharides in soil. Soil Biology and Biochemistry, 3, 33-41.

73. MARTIN J.P. 1982. Biodegradation, incorporation into biomass and stabilization in humus of polysaccharide carbons in soil. Abstract V-15 Xlth International Carbohydrate Symposium, Vancouver.

74. MARTIN J.P., ERVIN J.O. and RICHARDS S.J. 1972. Decomposition and binding action in soil of some mannose-containing microbial polysaccharides and their Fe Al Zn and Cu complexes. Soil Science, 113, 322-327.

75. MARTIN J.P., ERVIN J.O. and SHEPHERD R.A. 1966. Decomposition of the iron, aluminium, zinc and copper salts or complexes of some microbial and plant polysaccharides in soil. Soil Science Society of America Proceedings, 30, 196-200.

76. MARSHALL K.C. 1968. Interaction between colloidal montmorillonite and cells of rhizobium species with different ionogenic surfaces. Biochimica Biophysica Acta, 156, 179-186.

77. McCLURE G.W. 1970. Accelerated degradation of herbicides in soil by the application of microbial nutrient broths. Contributions from Boyce Thompson Institute, 24, 235-244.

78. MEHTA N.C., STREULI H., MULLER M. and DEUEL H. 1960. Role of polysaccharides in soil aggregation. Journal of the Science of Food and Agriculture, 11, 40-47.

286

79. MOAVAD H., GUSEV V.S., BABYEVA I.P. and ZUYAGINSTEV D.G. 1974. (Adsorption of extracellular polysaccharide of the yeast Lipomyces lipofer on Kaolinite). Soviet Soil Science translated from Pochvovedenie, 11, 79-84.

80. MOGHIMI A. and TATE M.E. 1978. Does 2-ketogluconic acid chelate calcium in the pH range 2.4 to 6.4? Soil Biology and Biochemistry, 10, 289-292.

81. MOGHIMI A., TATE M.E. and OADES J.M. 1978. Characterization of rhizosphere products especially 2-ketogluconic acid. Soil Biology and Biochemistry, 10, 283-287.

82. MORTVEDT J.J. and CUNNINGHAM H.G. 1971. In : Fertilizer Technology and Use, 2nd Edition Eds. Olson R.A., Army T.J., Hanway J.J. and Kilmer V.J. pp. 426-427.

83. MURAYAMA S. 1982. The monosaccharide composition of polysaccharides in ando soils. Journal of Soil Science, 31, 481-490.

84. NOVAKOVA J. 1977. Effect of bentonite on the substrate stabilization. In:Soil Biology and Conservation of the Biosphere. Proceedings of the Vllth meeting of the Soil Biology Section of the Society for Soil Science of the Hungarian Association of Agricultural Sciences held at Keszthely University of Agriculture, 1975. Ed. Szegi. pp. 271-275.

85. OADES J.M. 1978. Mucilages at the. root surface. Journal of Soil Science, 29, 1-16.

86. OGNER G. 1980. Analysis of the carbohydrates of fulvic and humic acids as their partially methylated alditol acetates. Geoderma, 23, 1-10.

87. PAGLIAI M., GUIDO G. and PETRUZZELLI G. 1978. Effect of molecular weight on dextran - soil interactions. Chapter 21 in : Modification of Soil Structure. Eds.Emerson W.W., Bond R.D. and Dexter A.R. John Wiley and Sons Ltd. pp. 175-180.

88. PAINTER T. and LARSEN B. 1970. Formation of hemiacetals between neighbouring hexuronic acid residues during the periodate oxidation of alginate. Acta Chemica Scandinavica, 24, 813-833.

89. PAINTER T. and LARSEN B. 1970. Transient hemiacetal structures formed during the periodate oxidation of xylan. Acta Chemica Scandinavica, 24, 2366-2378.

90. PAINTER T. and LARSEN B. 1973. A further illustration of nearest-neighbour auto-inhibitory effects in the oxidation of alginate by periodate ion. Acta Chemica Scandinavica, 27, 1957-1962.

91. PARFITT R.L. 1978. Anion adsorption by soils and soil materials. Advances in Agronomy, 30, 1-50.

92. PARFITT R.L. and GREENLAND D.J. 1970. Adsorption of polysaccharides by montmorillonite. Soil Science of America Proceedings, 34, 862-865.

93. PARKS W.L. and WHITE J.L. 1952. Boron retention by clay and humus systems saturated with calcium. Soil Science Society of America Proceedings, 16, 298-300.

94. PARSONS J.W. 1981. Chemistry and distribution of amino sugars in soils and soil organisms. Chapter 5 in : Soil Biochemistry, 5, Eds Paul E.A. and Ladd J.N. pp 197-227.

95. PFLUG W. 1982. Effect of clay minerals on the activity of polysaccharide cleaving soil enzymes. Zeitschrift für Pflanzenahrung und Bodenkunde, 145, 493-502.

96. REESE E.T. 1968. Microbial transformation of soil polysaccharides. In: Organic Matter and Soil Fertility. Pontificiae Academiae Scientiarum Scripta Varia, 32, 535-577.

97. REEVE M.J., SMITH P.D. and THOMASSON A.J. 1973. The effect of density on water retention properties of field soils. Journal of Soil Science, 24, 355-367.

98. REID J.B., GOSS M.J. and ROBERTSON P.D. 1982. Relationship between the decrease in soil stability effected by the growth of maize roots and changes in organically bound iron and aluminium. Journal of Soil Science, 33, 397-410.

99. ROGERS H.J. 1979. Adhesion of microorganisms to surfaces: some general considerations of the role of the envelope. In : Adhesion of Microorganisms to surfaces. Eds.Ellwood D.C., Melling J. and Rutter P. Special publication of the Society for General Microbiology, No. 2, pp.29-55.

100. ROVIRA A.D. and GREACEN E.L. 1957. The effect of aggregate disruption on the activity of microorganisms in the soil. Australian Journal of Agricultural Research, 8, 659-673.

101. SAINI G.R. 1966. Organic matter as a measure of bulk density of Soil. Nature (London), 210, 1295-1296.

102. SARKAR J.M., BATISTIC L. and MAYAUDON J. 1980. Les hydrolases du sol et leur association avec les hydrates de carbone. Soil Biology and Biochemistry, 12, 325-328.

103. SATANARAYANA T. and GETZIN L.W. 1973. Properties of a stable cell-free esterase from soil. Biochemistry, 12, 1566-1572.

104. SCHEFFER F. and ULRICH B. 1960. Humus and Humusdungung. Verlag.

105. SPARLING G.P., ORD B.G. and VAUGHAN D. 1981. Changes in microbial biomass and activity in soils amended with phenolic acids. Soil Biology and Biochemistry, 13, 99-104.

106. ST ARNAUD R.J. and SEPHTON G.A. 1972. Contribution of clay and organic matter to cation-exchange capacity of Chernozem soils. Canadian Journal of Soil Science, 52, 124-126.

107. STEFANSON R.C. 1971. Effect of periodate and pyrophosphate on the seasonal changes in aggregate stabilization. Australian Journal of Soil Research, 9, 33-41.

108. STEVENSON F.J. 1966. Lipids in soil. American Oil Chemists' Society, 43, 203-210.

109. STEVENSON F.J. 1971. Organic matter reactions involving herbicides in soil. Journal of Environmental Quality, 1, 333-343.

110. STOTZKY G. 1972. Activity, ecology and population dynamics of microorganisms in soil. Critical Reviews in Microbiology. Volume 2, CRC Press, The Chemical Rubber Co., Cleveland, Ohio, pp 59-127.

111. SUTHERLAND I.W. 1979. Polysaccharides in the adhesion of marine and fresh water bacteria. Chapter 18 in : Microbial Adhesion to Surfaces, Eds.Berkley R.C.W., Lynch J.M., Melling J., Ruther P.R. and Vincent B. Horwood, Chichester, pp.329-338.

112. THOMASSON A.J. 1978. Towards an objective classification of soil structure. Journal of Soil Science, 29, 38-46.

113. TISDALL J.M. and OADES J.M. 1979. Stabilization of soil aggregates by the root system of Ryegrass. Australian Journal of Soil Research, 17, 429-441.

114. TISDALL J.M. and OADES J.M. 1980. The effect of crop rotation on aggregation in a red-brown earth. Australian Journal of Soil Research, 18, 423-433.

115. TISDALL J.M. and OADES J.M. 1982. Organic matter and water-stable aggregates in soils. Journal of Soil Science, 33, 141-163.

116. VAUGHAN D. and ORD B. 1980. An effect of soil organic matter on invertase activity in soil. Soil Biology and Biochemistry, 12, 449-450.

117. WHISTLER R.L. 1973. Factor influencing gum costs and applications. Chapter 1 in : Industrial Gums Polysaccharides and Their Derivatives. Academic Press, pp. 5-18.
118. WRIGHT W.R. and FOSS J.E. 1972. Contributions of clay and organic matter to the cation exchange capacity of Maryland soils. Soil Science Society of America Proceedings, 36, 115-118.
119. YOSHIDA M. and KUMADA K. 1979. Studies on the properties of organic matter in buried humic horizon derived from volcanic ash III. Sugars in hydrolysates of buried humic horizon. Soil Science and Plant Nutrition, 25, 209-216.
120. YOSHIDA M. and KUMADA K. 1979. Studies on the properties of organic matter in buried humic horizon derived from volcanic ash IV. Characteristics of polysaccharides in hydrolysates of fulvic acid and in ethanol precipitates from fulvic acid in buried humic horizon. Soil Science and Plant Nutrition, 25, 217-224.
121. YUAN T.L., GAMMON N. and LEIGHTY R.G. 1967. Relative contribution of organic matter and clay fractions to cation exchange capacity of sandy soils from several soil groups. Soil Science, 104, 123-128.
122. ZUNINO H., BORIE F., AGUILERA S., MARTIN J.P. and HAIDER K. 1982. Decomposition of ^{14}C labelled glucose, plant and microbial products and phenols in volcanic ash-derived soils in Chile. Soil Biology and Biochemistry, 14, 37-43.

CHAPTER 8

SOIL NITROGEN: ITS EXTRACTION, DISTRIBUTION AND DYNAMICS

H.A. ANDERSON and D. VAUGHAN

The Macaulay Institute for Soil Research, Aberdeen, Scotland.

CONTENTS

1. INTRODUCTION

Nitrogen is a key component of soil organic matter. Apart from carbon, hydrogen and oxygen, it is the most abundant element in living tissue. It is also by far the most abundant element in the atmosphere. Yet, paradoxically, higher forms of life are incapable of drawing directly on this vast reservoir of nitrogen. In the soil, only certain micro-organisms possess the enzymes necessary to achieve the fixation of atmospheric N_2, converting it to forms that can be utilised by plants and animals. Soil nitrogen, therefore, is of paramount importance to the maintenance of life. Seen in this context, it is hardly surprising that there is a substantial literature on soil nitrogen reflecting its many and varied aspects. Because of the considerable scope of the subject it is, by necessity, desirable to define the topics to be discussed, albeit briefly, in this review. An excellent and up to date account of N in relation to biological processes and soil fertility has recently been published[150].

The distribution of N is such that soil N forms only about 0.02 per cent of the global N mass[145]. Nitrogen is present in many soil organic components such as humic and fulvic acids, amino acids and nucleic acids and indeed more than 90 per cent of all soil N is found in organic sub-stances. It is thus essential to consider, in some detail, the methods available for extracting and estimating nitrogenous components from the soil and soil microbial biomass and to assess whether these substances are altered during extraction.

The main input of labile N is in the form of plant litter which can be readily utilized by soil microbes. The manner in which soil micro-organisms utilize and transform this nitrogen, whether by anabolism or catabolism, is discussed briefly in this chapter. The series of transformations can be readily understood by reference to the "Nitrogen Cycle" and, as will be appreciated later, transformations can be purely chemical in addition to biological.   The abiotic processes contribute to a minor extent towards mineralization of organic N to ammonium ions and subsequent nitrification and production of nitrate, while chemo-denitrification leads to the loss of nitrate as N_2O.   Additionally, lightning is mainly responsible for the small abiotic N fixation of N_2, NH_3 and N_2O into mixed oxides, NO_x.

Finally, the role of N in soil fertility is also briefly discussed. The application of fertilizer N plays a key role in agriculture and it is the form in which N is applied, either organic or inorganic, which may determine the final yield of crop.   Although the importance of applying N in high or low energy forms is discussed in some detail in chapter 12, it is of crucial importance in this current chapter to consider the usefulness of balance sheets and the use of mathematical models in predicting N-fertilizer requirements to maximise crop production under different conditions.

2. THE EXTRACTION OF SOIL NITROGEN

Most N-containing components associated with living organisms are found in the soil and range in complexity from simple amino acids to the more complex nucleic acids and proteins.   Any extraction procedure should take account of this heterogeneity.   Generally these nitrogenous components are liberated from soils using hot mineral acids and bases[143].   Typically an acid hydrolysis extraction removes NH_3-N, amino acid-N, amino sugar-N and hydrolysable unknown-N (HUN) comprising some 20-30%, 30-45%, 5-10% and 10-20% of the soil N, respectively.   These fractions are described in detail below.   The residue, some 20-35% of the total soil N, is not extracted by hydrolysis and of that which is extracted only up to 50% is generally accounted for in known substances.

2.1. Total N, mineral forms and acid-hydrolyzable fractions

The classical Kjeldahl method of total N estimation is probably the most widely used and when allied with automated wet-chemical colorimetric analysis, forms a powerful means for the rapid estimation of numerous samples[41].   Autoanalytical methods are available for estimating ammonium[41], nitrate and nitrite[44,139], amino acid-N[15,16] and amino sugar-N[109] and these methods are arguably more useful in soil N studies than dedicated instruments. Such methods are also useful in determining fractions of soil N liberated during hydrolysis by mineral acid (eg. refluxing 6M HCl for 12-24h), which also can be estimated by the procedure of Bremner[23] based on the distillation of ammonia selectively released from amino sugars and amino acids etc treated under differing alkaline conditions.

Several forms of commercially available N analyzers are based on combustion analysis. This involves oxidative pyrolysis of samples, followed by chemical reduction of the resulting nitrogen oxides and the gas chromatographic estimation of the products and methods based on such principles are now used in several forms of commercial nitrogen analyzer and multi-element analyzer. Combustion in the presence of ozone can generate nitrogen dioxide in excited states, the phosphorescence of which is used to estimate N[121].

Soil solutions and salt extracts, eg. potassium or calcium chloride solutions used in "exchangeable" cation estimations, can be analysed using ion specific electrodes. Aqueous extracts and soil solutions can also be analysed for anions containing N by chromatography[133]. Here the anions are separated on a patented exchange resin and detected by conductance after passing through a hydrogen-form cation exchange resin to remove the background conductance of the eluent. Anions which show absorbance in the ultraviolet are more readily analysed using routine high performance liquid chromatography (HPLC) systems[132].

In addition to the above methods, the ammonium ion can be estimated by catalytic conversion to molecular N followed by microwave plasma emission spectrometry[58]. Nitrate can be estimated by conversion to nitrobenzene and subsequent gas chromatography[61]. Both procedures were originally intended for N isotope analysis and have the inherent advantage of speed and simplicity.

2.2. Amino acids and amino sugars

Amino acids are the building blocks of proteins and their presence in the soil has been known since the beginning of this century when Suzuki[143] reported in 1906 that aspartic and aminovaleric acids, alanine and proline were present in the acid hydrolysate of humic acids. Since that time all the amino acids known to occur in proteins have been detected in soils together with some which are not usual protein constituents.

Generally amino acids are extracted from the soil by refluxing in 6M HCl for up to 24h. The conditions for maximal extraction have been thoroughly investigated[18,128,143]. Once extracted, the amino acids are usually estimated by reaction with ninhydrin. Ammonium ions and amino sugars in the hydrolysate react with ninhydrin, but non-nitrogenous compounds such as levulinic acid interfere[15].

Amino acids can also be estimated by their conversion into volatile derivatives and separation by gas chromatography[16], but the sample preparation is tedious and subjected to interference from mineral soil components such as ferric iron[48].

The method of choice for the analysis of amino acids extracted from soils is ion exchange chromatography, based on the Moore and Stein methodology[153], and this is well illustrated in work relating to amino acid variation within profiles[141] and between climatic zones[138]. Ion exchange chromatography of soil amino acids has the advantage of allowing the concurrent estimation of ammonium and amino sugars, with separation of glucosamine and galactosamine[8].

During acid hydrolysis of peptides, amino acids are released quantitatively, but then some undergo extensive degradation. For example, tryptophan is lost completely with the partial release of ammonium, and the amides asparagine and glutamine release ammonium ions together with their parent α-amino acids. Alkaline hydrolysis has been used to estimate tryptophan, but components interfering with the subsequent analysis are also generated (Anderson and Bick, unpublished).

Most hydrolysates from soils contain cysteic acid, but after chromatography almost no cysteine or cystine is found on the chromatograms (Anderson and Bick, unpublished). Methionine, destroyed during acid hydrolysis[15], never appears. However, pre-treatment of soils with performic acid[49] allows the ready estimation of cysteine and methionine as cysteic acid and methionine sulphone respectively, both amino acids being important organo-sulphur compounds in soil[49,130].

In future, ion exchange amino acid analysers may be eclipsed by the more flexible solvent programmed HPLC systems. Rapid analyses of amino acid mixtures based on reverse phase HPLC of o-phthalaldehyde derivatives has been demonstrated with clinical samples[154]. It should be noted that the imino acids, proline and hydroxyproline can be converted into reactive products before chromatography[162].

The occurrence of amino sugars in soil was shown by Bremner[22] in the 1950's. Amino sugars occur as components of a wide range of substances such as mucopolysaccharides, mucopeptides, mucoproteins and some antibiotics. The amino sugar glucosamine occurs in chitin as a polymer of N-acetylglucosamine (see chapter 7). It has been claimed[55] that there is a direct relationship amino sugar-N and the Ca^{2+} content of a soil.

Sowden *et al*.[138] consider that there is a higher percentage of amino sugars in tropical soils when compared with arctic soils. Furthermore, Sowden[136] also showed that the ratio glucosamine : galactosamine varied between soils and was highest in a podsol under conifers and lowest in a prairie soil.

Amino sugars are generally obtained from soils by a procedure similar to that used for amino acids (hydrolysis with 3M or 6M HCl for 6 to 9h). Benzing-Purdie[12] reported that in general the higher yields were obtained with 6M HCl. Other extraction methods have also been used, for example concentrated HCl at room temperature for 48h[136]. Inevitably some degradation of the amino sugars occurs during extraction. Once extracted they can be estimated either by using the Elson-Morgan colorimetric method or by alkaline distillation[143,144].

The individual amino sugars in the soil hydrolysates may be separated from the other components using chromatography as described above for the amino acids[12,144]. As yet, no HPLC method has been developed to resolve both the amino acids and amino sugars. Among the amino sugars identified using such separation techniques are D-glucosamine, D-galactosamine, D-mannosamine, muramic acid and probably D-glucosamine and D-fucosamine[12,13,1] The importance of amino sugars in relation to soil polysaccharide is discussed in chapter 7. Recently Benzing-Purdie[13] has reported that there is a parallel between organic C, N and amino sugars with a definite enrichment of organic matter and amino sugars in the small particle size fractions in sandy loam and loamy sand soils.

2.3. Purines and pyrimidines

Nucleic acids occur in all living cells whether organised within a nucleus in eukaryotes or disorganized as in the prokaryotes. These acids comprise mononucleotide units of base-sugar-phosphate linked by the phosphoric acid ester to the sugar of the adjacent nucleotide. The base units are the purines adenine and guanine, and the pyrimidines cystosine and uracil (in RNA) or thymine (in DNA). In higher organisms the DNA also contains 5-methylcytosine. All these bases have been identified as components of soil organic matter[6,40] but the N in these bases comprises less than 1% of the total N in the soil[6,40].

These compounds are important components of microbial biomass in the soil (see chapter 9), and are, therefore, in a very dynamic state. Purines and pyrimidines are usually estimated after acid hydrolysis[6,40]. Nucleoside diphosphates[7], and apparently-intact bacterial DNA[152], have also been isolated from soil. Pretreatment of soils has been found necessary before one hydrolysis procedure (1M HCl at 100°C for 1h)[156]. The bases released were chromatographed on ion exchange resins[40], but, as in the case of amino acids, HPLC methods are now proving faster. Extraction of the starting material with HF before hydrolysis with HCl has been recommended for the analysis of nucleic acids in lake sediments[156].

2.4. Minor N-containing substances

Although theoretically amino acids, amino sugars and nucleic acid derivatives usually account for over 95% of identifiable soil organic N, trace amounts of many other N-containing compounds have been reported. Chromatographic evidence showed the presence of mono- and poly-amines in anaerobic soils[54] and phospholipids in arable soils[66].

An unsatisfactory approach to the estimation of porphyrin compounds, based on one-point visible absorption spectrometry of acetone-water extracts of soil[75], has been rectified by partial fractionation and characterization of individual porphyrins using multi-point absorption spectrometry[36]. Many other N-containing compounds have been identified in soils, but in amounts which are considered to be relatively unimportant at present.

2.5. Unidentified nitrogenous components

The soluble N in the soil hydrolysate which cannot be accounted for either as NH_3-N or in known compounds is referred to as the HUN fraction[143]. This is frequently measured as the difference between the total hydrolysable N and the sum of the fractions containing ammonium, α-amino N and amino sugar N. However, this will give an overestimate because non α-amino N in the basic amino acids arginine, lysine and histidine would not be included, nor would nucleic acid derivatives. Nevertheless, even after taking the non α-amino N into account, HUN remains substantial. The non-hydrolysable fraction of the soil N remaining in the residue after 6M HCl hydrolysis has not been identified.

Hydrolysis methods for the degradation of polymers are usually imperfect and some destruction or modification of the resultant monomers can complicate attempts to examine the unidentified fraction of soil N. Some may ultimately even prove to be artifacts of the extraction.

Nitrogenous components of humic substances have recently been categorised as those with distinct chemical identities (Group 1) and "N compounds that have become integral constituents of humic materials" (Group 2)[127]. This definition appeals because of its simplicity but a precise estimation of the two groups is difficult because of the possible formation of degradation products during extraction[122].

The products of the degradation of amino acids such as tryptophan, methionine, serine and threonine and other easily identified components, should be amenable to isotope-tracing using [15]N. Such techniques could also distinguish between "old" organic-N and the relatively-new easily-decomposed materials.

2.6. Nitrogen isotope measurements

There are six known isotopes of N, but of these, only the two stable species ^{14}N and ^{15}N are of interest here. Nitrogen, under natural conditions, usually has an approximate isotopic composition of 99.634 atom per cent ^{14}N and 0.336 atom per cent ^{15}N and any compound containing N, whose isotopic composition differs significantly from this, may be used as a tracer. Substances enriched with ^{15}N are the more commonly used but the accuracy which can be accomplished with modern mass spectrometry also allows the use of ^{15}N-depleted materials, prepared in quantity by high-efficiency cryogenic distillation.

Before analysis, because the N components in a sample are rarely present as N_2 either (a) all nitrogen is converted to ammonium ion and then, by oxidation by hypobromite[123,161] into N_2 or (b) a Dumas combustion on a briquette of CuO-CaO mixture (1:1), in a sealed evacuated tube, converts the sample-N into N_2 gas. The latter method has been found extremely useful in studying amino acids separated by thin layer chromatography[81].

Although the Rittenberg hypobromite oxidation remains the method of choice in mass spectrometry, labelled nitrate has been converted into nitrobenzene, which has then been analysed by gas chromatography-mass spectrometry using selective ion monitoring[61]. Similarly, simplified procedures have been used in the emission spectrometric method[31,164], but the most elegant method for the conversion of aqueous ammonium chloride to N_2 is based on a reverse Haber reaction[58]. The calculation of [14]N : [15]N ratios is fully discussed elsewhere[61,143,164].

The efficiency of any isotopic tracer technique depends primarily upon the establishment of a relatively constant "natural background" level of the isotope and upon non-discrimination between isotopes in the system under investigation. Slight variation occurs in the [14]N : [15]N isotope

composition in unlabelled soil nitrogen components as a result of
discrimination between isotopes resulting in fractionation processes
occurring in soil.   There exists in many soils a slight ^{15}N enrichment
due to selective loss of the lighter isotope during bacterial denitrification[143] and it has also been shown that isotope fractionation during
plant uptake is possible[100].   But such effects are so small that they
can be ignored in most tracer experiments.   It has been suggested that
the natural abundance of ^{15}N in herbage can be used to measure symbiotic
N fixation[120], but this is a good example of a situation in which a
significant error could be introduced by natural fractionation, an error
minimised by the use of ^{15}N enriched tracer.

3. DISTRIBUTION OF SOIL NITROGEN

Because of the diversity of nitrogenous compounds in soil organic matter,
and their close associations with the biomass, it would be surprising if
the distribution of N, both qualitatively and quantitatively, did not vary
between different soils or indeed the same soil with increasing depth down
the profile.   Surprisingly, however, little is known of the factors which
influence the distribution of nitrogenous compounds in soils and substantial
variations in the distribution pattern have been reported.

3.1. Distribution in soil

The total N content of soils generally decreases with depth.   The
contributions of the various forms of N show the same decrease, with the
possible exception of nitrate.   In certain semi-arid regions, where the
total annual rainfall is not greater than potential evaporation, and where
there is no drainage beyond the deep subsoils, nitrate may be leached into
these subsoils and used by deep-rooting crops;   concentrations of 1-7 mg
kg^{-1} nitrate-N in the top-most metre can rise to 5-40 mg kg^{-1} in the metre
below[39].

All other forms of N show an absolute decrease with depth, but in 6M HCl
soil hydrolysates, amino acid N and ammonium show an apparently-inverse
relationship, with an increasing proportion of the total N being released
as ammonium.   But problems arise in assessing data on the distribution
pattern of N in soils.   These include the somewhat inaccurate methods of
analyses for amino-acid N and the liberation of NH_3 produced by the partial
destruction of amino sugar-N during acid extraction.   Early investigators
concluded that, although some of the released ammonium originated in organic

compounds such as amino acid amides (glutamine and asparagine), the large amount of ammonium released could only be explained by the concurrent release of "fixed" ammonium held in clay-ammonia complexes[141].

Detailed analysis of amino acids in soils has proved useful when comparing different soil types or different environments. Thus the results of soil hydrolyses from different climatic zones show that the proportion of the total N attributable to amino acid increases on passing from arctic to tropical zones, whereas the hydrolysed ammonium behaves inversely yet again[138]. Intensive cultivation usually results in a slightly lower amino acid contribution to a depressed total N, but to a degree which is relatively insignificant. In contrast, clay-fixed ammonium N is generally increased by cultivation[82,96].

Many attempts have been made to demonstrate the presence of free amino acids in soils, but, mainly due to the limits of detection, this was first realized only with acid peat soils[42]. The analysis of aqueous expressed solutions from seven peats gave free α-amino N values within the range 0.4–5.0 mg.l^{-1}, the concentration varying inversely with soil pH. These figures correspond to a maximum of 6 ng α-amino N-g^{-1} dry peat, or 0.4% of the total N. Improvements in analytical methods have now allowed the detection of free amino acids in nearly every type of soil, based upon extraction into 70–80 (v/v) ethanol[113,115,118,131], increasing the yield over that from water extracts. Extraction with 0.05M barium hydroxide and 0.5M ammonium acetate has also allowed the detection of more amino acids in increased yields of 50–500 ng.g^{-1} of soil[111]. With improvements in technique, minimizing losses by rapid sample extraction at low temperatures and avoiding cell-lysing solvents, 3–16 mg α-amino N.g^{-1} soil was detected in water extracts of rhizosphere soils. These values are between 10- and 30-times greater than those for the corresponding non-rhizosphere soils[76].

None of the mineral soil values relate to the intrinsic soil solution because all were obtained by extraction methods. But recently soil solutions recovered by centrifugation at 2,000 g have been examined directly (Anderson and Linehan, unpublished), and have contained much lower levels of free amino acids (Table 1) than previously reported for other soils[138].

While "free" amino acids comprise only a small proportion of the total N, combined amino acids form the major component. Earlier methods underestimated the nitrogen associated with amino acids[63], but modern techniques,

though more effective, have not yet been used to identify the various
individual forms of amino acids and telomers present in soil organic matter.
Instead, aggregate values are still being presented for both soils and
organic matter.

Table 1.   Free amino acids (µg 100 g^{-1} soil) extracted from soils by
different media.

Extraction medium:-	Soil solution[a]		80% ethanol aqueous[b]		Water[c]		20% ethanol aqueous[c]		Water/carbon tetrachloride[c]	
Soil(*):-	N	R	N	R	N	R	N	R	N	R
ASP	0.2	0.4	2.8	1.3	8	178	9	342	31	573
GLU	0.2	0.6	3.1	2.0	10	249	29	514	48	780
CYS	ND	ND	1.6	ND	ND	ND	ND	ND	ND	ND
ASN	0.3	1.2	ND	ND	3	47	11	48	24	25
GLN	1.4	13.4	ND	ND	2	67	10	72	15	45
SER	ND	ND	ND	ND	40	333	53	499	178	542
HIS	ND	ND	1.7	tr	3	19	4	19	8	39
GLY	tr	tr	ND	ND	19	185	29	211	77	295
THR	ND	ND	ND	ND	11	133	27	183	70	203
ARG	ND	ND	1.4	ND	3	tr	7	22	8	30
ALA	0.1	0.1	ND	ND	13	182	43	272	81	402
TYR	ND	ND	ND	ND	4	84	12	87	18	100
VAL	tr	0.3	0.7	0.2	6	77	19	117	38	353
PRO	ND	ND	ND	ND	5	59	9	89	23	150
PHE	tr	tr	tr	0.3	4	53	21	90	32	126
ILE	tr	tr	ND	ND	4	35	12	69	26	108
LEU	tr	tr	1.2	0.2	5	42	36	130	61	199
LYS	ND	ND	2.4	0.2	3	tr	3	tr	6	tr

Notes:

(a) Sandy loam Ap horizon - Anderson H. and Linehan D.J. (unpublished).

(b) Silt loam Ap horizon - Putnam and Schmidt[118].

(c) Sandy loam Ap horizon - Ivarson and Sowden[76].

(*) N = non-rhizosphere,   R = rhizosphere soil
     ND = not detected,   tr = trace.

A relationship between C and organic N contents of soils has been recognized for a long time, and a relationship between organic N and organic phosphorus and sulphur has been observed more recently[10,14,85].

The total N content of mineral soils is also closely related to the clay content, and it has been shown that when different soils from similar climatic regions have their N contents adjusted for differences in soil texture, the calculated values are very similar[67,79]. In a silt loam, the fraction less than 2 µm in size, forming 17.5% of the surface soil, contained 41.4% of the total soil N. The corresponding subsoil fraction comprised 13.9% of the mass and contained 50.4% of the soil N[35]. However, reservations must be expressed concerning the possible synthesis of clay-organic N complexes by the fractionation and extraction processes. Clays can adsorb, from aqueous suspension, their own weight or more of a range of organic compounds[62], the amount adsorbed varying with the solution concentration[59,60,99] and pH[62,108], as well as with changes in the cation saturating the clay and the molecular size of the adsorbed molecule[68]. Nevertheless, the stability of organic N in clay complexes has been amply demonstrated[59,60,135].

The products from extractions of soils with alkali suggest that certain organic N fractions are linked to silicates and can only be extracted by hydrolysing the bonds at high pH or by destruction of the silicate framework with HF[110]. The use of anhydrous formic acid as an extractant gave a fraction low in N, probably not linked to silicate material. The residue, following HF treatment, gave a second formic acid-soluble fraction much richer in N, which is possibly adsorbed to silicate surfaces.

It is widely known that clays, particularly the swelling clays such as montmorillonite, protect adsorbed nitrogenous organic molecules. For example, proteins adsorbed in clays are less readily attacked by protease than those in solution[99]. Some of the soil organic N may even be protected by inclusion in micropores inaccessible to micro-organisms[143].

3.2. Nitrogen in humic substances

A number of studies have been made on the occurrence of N in the various humic substances and the distribution between them. In many studies soil organic matter is separated into humic (HA) and fulvic (FA) acids by the traditional technique of alkaline extraction followed by acidification. Much work has been done on the qualitative and quantitative aspects of N in these humic substances and fractions derived from them.

During the last century there was much controversy as to whether N did
in fact occur in humic substances (see Introductory chapter). During the
1930's it was shown[71] that about half of the total N of a humic acid (HA)
was present in "some polypeptide form". The same workers also deduced[72]
that the distribution of "non-humic N" in the soil was approximately 30-40%,
12%, 5% and 40-50% in the peptide N, ammonia N, free amino N and other forms
of N respectively. In soils receiving additions of different organic
manures over 30 years, higher proportions of the total N were present in the
fulvic acid (FA) fraction of the green-manured soils (33-42%), than the FYM-
treated soils (28-29%). The HA nitrogen of the FYM-treated soil correspondingly
contributed a proportionately higher share of the soil N than
that of the green manure treatment. Very low values, about 2% of total N,
remained in the soil residue. It has been reported that amino acids
constitute higher proportions of the total N in HA's than in FA's whereas
the contribution from NH_4 is greater in FA[117], although this difference has
not been noticed by other workers[129]. It has been shown[8] that, in
cultivated topsoils, about 80% of the total FA amino acid N is lost on
dialysis against dilute HCl and water.

A method for the fractionation of soil extracts has recently been
advocated by the International Humic Substances Society[2], in which the
total acid-soluble material is divided into FA and a "hydrophilic acid
fraction" by passage through XAD-8 ion-exchange resin. The FA is adsorbed
at low pH values and eluted in dilute alkali, whereas the hydrophilic acid
soluble fraction remains unadsorbed and passes through the resin.

The N content of HA's increases in the early stage of humification[86],
and this has been interpreted as a release of "dead biomass" rich in N at
a point of exhaustion of readily available substrates. The "degree of
humification", a scale based on the colour of the HA[88], increases during
the subsequent stage, with a concomitant release of N. Nevertheless, the
amino acids released by hydrolysis of HA's are very similar in composition
regardless of origin or degree of humification[87].

The fractionation of soil organic extracts by most chemical, chromatographic or electrophoretic techniques met with little success in earlier
days[24], but latterly, the fractionations achieved by gel chromatography
have begun to appear promising. Most workers have preferred to avoid
adsorption effects by saturating adsorption sites with electrolyte in the
eluting solvent. Unless this is done, the use of gel permeation

chromatography as a molecular size fractionation is virtually precluded[32].
Under these conditions, there is usually considerably more N, P and S in
the higher MW fractions than in the lower[57], although the S content is
dependent upon the age of the soil.   Using distilled water as the eluting
solvent it was found that salt boundary effects improved chemical
fractionation by increasing adsorption of some of the organic material[9,147].
Following this observation, adding a salt to the sample prior to gel
chromatography eliminates the differences in salt content between various
samples[8].

The use of tandem columns of methacrylate (Bio-Gel) and dextran (Sephadex)
gels[8] gives characteristic salt boundary chromatograms (Fig. 1).   In this
case, the first three fractions have separated according to molecular size.
Comparison of the distributions of nucleic acid bases and amino acids in
6M HCl hydrolysates of the humic fractions (Table 2) indicate that the

Table 2.   Distribution of amino acid - and nucleic acid-N in 6M HCl
            hydrolysates of the fractions of alkali-soluble humic sub-
            stances, isolated from a brown forest soil, and separated
            by "salt-boundary" gel chromatography.

Horizon	Fraction	% weight distribution	% N distribution in:	
			Amino acids	Nucleic acids
LF	A	33	40	40
	B	9	17	8
	C	22	17	6
	D	36	26	46
AH	A	26	37	37
	B	8	11	10
	C	30	41	32
	D	36	12	24
AE	A	27	36	37
	B	4	9	5
	C	24	33	32
	D	43	22	24
BC	A	12	29	13
	B	5	10	10
	C	35	39	51
	D	47	22	24
CH	A	13	36	4
	B	5	13	7
	C	43	41	37
	D	39	9	52

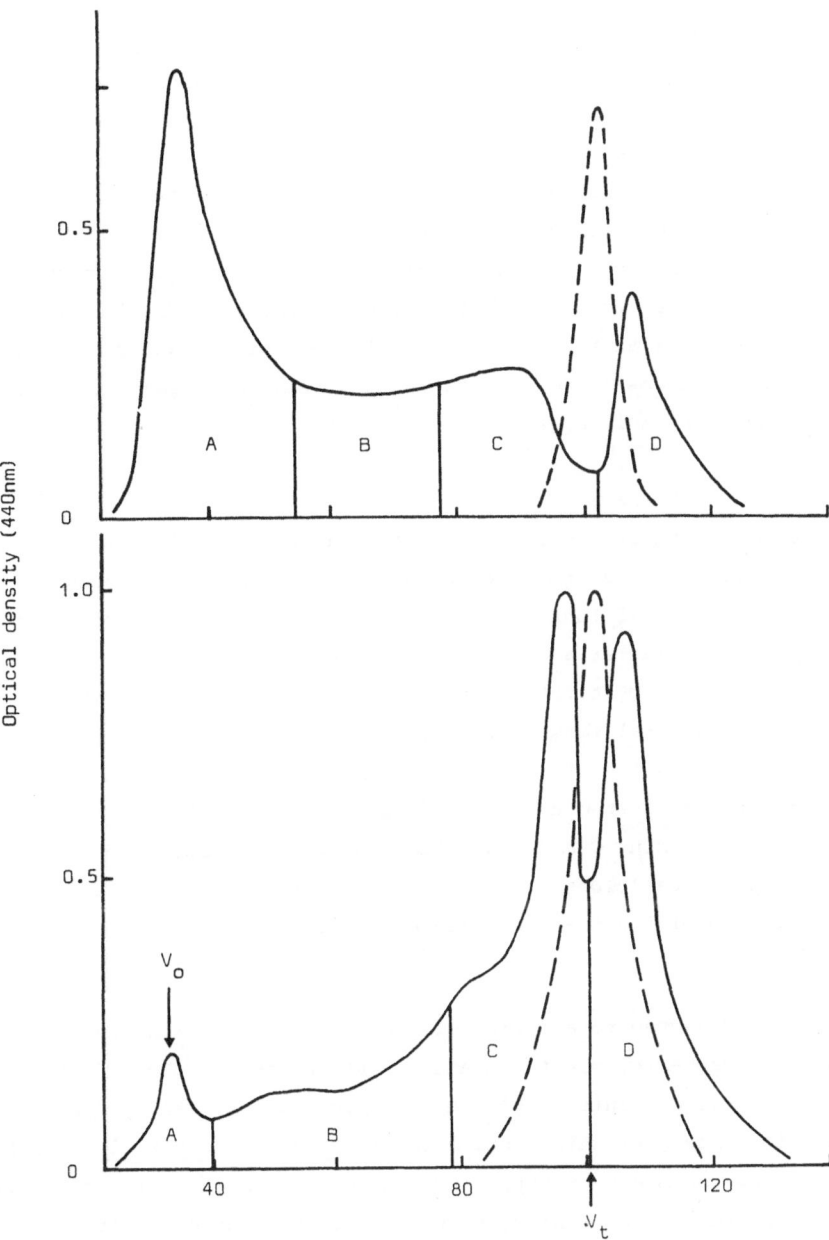

FIGURE 1. Typical "salt boundary" gel chromatograms of sodium humates. V_o and V_t mark the exclusion and total internal volumes of the gel column. The broken line indicates the high conductivity of the co-eluted sodium acetate. A-D indicate the usual fractionation of the chromatograms.

relationship, if any, between the nucleic acid and protein content is not
a simple one in such extracts.   It has been stated that the ease of
hydrolysis of the nucleic acid bases during a brief hydrolysis with 1M HCl
indicates that they are probably not integral parts of the HA and FA
structures[127].

The use of these newer chromatographic fractionation techniques shows
some promise and a greater degree of chemical fractionation should now be
forthcoming.   Using an improved chromatographic procedure, Stevenson[144]
has recently reported that the distribution of amino sugars in a HA
preparation was similar to that of the original soil.   Nevertheless, it
remains to be seen whether milder extraction techniques can provide high
yields of unaltered soil organic matter.

4. SOIL NITROGEN DYNAMICS

Soil organic matter, containing substantial amounts of N, plays a key
role in agriculture (see Introductory chapter).   Grassland soils
frequently contain 5-15 tonnes N ha^{-1} in the root zone, most of which is
present in organic matter and arises from death and decay of roots and
herbage[74].   Because cultivation of increasingly larger areas of land has
resulted in substantial losses of organic matter and N[126], it is essential
to consider the dynamics of the soil system under various conditions.
The dynamics of the system depend on such factors as the inputs of plant
constituents containing C and N and also on energy sources such as
cellulose and hemicellulose[112].   In this section the various aspects of
soil dynamics involving N are considered briefly.

4.1. The N cycle

Nitrogen is subjected to a series of transformations which are dependent
upon soil, atmospheric, plant and animal interactions.   These main
divisions, or compartments, may themselves contain several cyclic processes,
but the main concern of this section is the Soil Nitrogen Cycle, which can
be regarded as the means of supplying mineral N from the soil to growing
plants.   The concept of a nitrogen cycle was formulated in the early part
of the present century, and it essentially takes the form of an interconnected series of biochemical and chemical transformations (Fig. 2) under
biological control.   Campbell and Lees[33] have emphasized that these
processes are in no way interdependent and there is no orderly progression
from one compound to the next as occurs in the operation of the citric

acid cycle.   Nevertheless, the concept of the N cycle is useful in so far
as it enables us to understand and predict the transformations of N between
various forms.

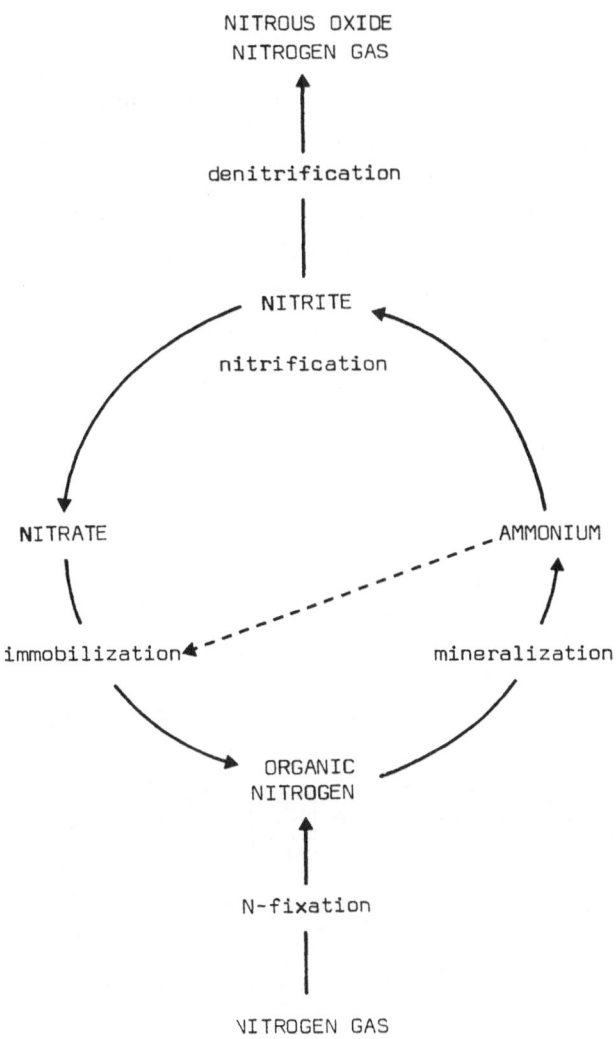

FIGURE 2. A process nitrogen cycle.

In the absence of added fertilizer N, the nitrogen reaching the soil comes, essentially, from plant, animal and microbial residues or from atmospheric N fixed by certain soil organisms. Nitrogen is lost from the soil either by leaching (mainly as nitrate) or in the form of gases produced in a process called denitrification. Within the soil there is an internal N cycle involving (a) the production of mineral N, ie ammonium and nitrate ions, from decomposing soil organic nitrogen - a process described as mineralization, and (b) the removal or immobilization of mineral N by organic matter synthesis.

The direction of N transformation depends on a number of factors, but one of the most important is the C:N ratio of the soil organic matter (see Introductory chapter). Soil micro-organisms use the organic matter as a substrate to satisfy their own energy and nutrient requirements. In this process, if N is not limiting, ammonium will accumulate and, under suitable conditions, be oxidized to nitrate (nitrification). If N is limiting, as in the presence of excess carbon, mineral N is quickly re-used by micro-organisms, returned to the organic form and temporarily immobilized. Under anaerobic conditions, NO_3-N may be reduced to NO_2^-. The NO_2^- may be further reduced to N_2O or N_2 and lost to the atmosphere or reduced to NH_4^+. Thus fertilizer N losses, such as leaching of nitrate and denitrification during anaerobiosis, can readily be envisaged; but less obvious is the situation of soil crumbs with aerobic outer sectors in which nitrification can take place, and an inner anaerobic core from which N is lost through denitrification[48,84,149].

Denitrification is an important process in that N is thereby lost from the soil/plant system[38] and returns to the atmosphere as outlined in the scheme below:

$$\boxed{NH_4^+} \underset{\longleftarrow - - -}{\overset{\text{Nitrosomonas}}{\longrightarrow}} \boxed{NO_2^-} \underset{\longleftarrow - - -}{\overset{\text{Nitrobacter}}{\longrightarrow}} \boxed{NO_3^-}$$

(Denitrification)

$$\boxed{N_2 \text{ plus } N_2O}$$

Nitrite is very important in N transformations because it is an intermediate product in aerobic nitrification by the bacteria Nitrosomonas and Nitrobacter as well as in anaerobic denitrification processes[155]. Although the above scheme emphasises the important role of nitrite, this anion is rarely found in nature[155]. Only occasionally does it accumulate in soils, depending on soil characteristics and agricultural practices.

These are discussed more fully in section 4.3. It should however be borne in mind that under anaerobic conditions, which often dominate in flooded or poorly drained soils, denitrification and fermentation processes are common[149].

Micro-organisms are mainly responsible for the processes of N-fixation, nitrification and denitrification and contribute largely towards immobilization. But not all processes involving N fixation are biological and within the soil N cycle, abiotic processes contribute towards mineralisation and volatilization of ammonia, although chemo-denitrification leads to the loss of nitrite as N_2O.

Nitrogen must be regarded as a dynamic soil component and its vital role can be assessed not only by fractionation of soil into biologically and chemically related components, but also by studying the many transformation processes in which it is involved. Such processes can now be followed directly by incorporating N tracers into the soil (see section 2.6.).

4.2. The fate of fertilizer N

Although the first recorded use of ^{15}N tracers in soil involved the mineralization of plant residues[106], much of the work thereafter has concerned the evaluation of the fate of fertilizer N and the soil-plant relationship, in addition to drawing up balance sheets of N inputs and outputs. This is illustrated by the recent reports for some Scottish soils by Smith et al.[134]. These workers used ^{15}N-labelled $Ca(NO_3)_2$ to investigate the relative uptake by barley of fertilizer-N and soil-N. More fertilizer-N and soil-derived-N were taken up by winter barley than by spring barley reflecting the longer growth period and higher dry matter yields of the former crop. The previous conventional methods of fertilizer study, ie. the measurement of differences and net effects, could not explain certain features of N behaviour in soil that were revealed by the tracer techniques. But as late as 1963, it was claimed[78] that ^{15}N techniques were of "limited use for the evaluation of fertilizer N from the practical point of view".

However, in 1962 the joint FAO/IAEA Division of Atomic Energy in Food and Agriculture began a series of cooperative field experiments throughout several countries studying ^{15}N tracers on rice, maize and wheat[46]. In these experiments the applied fertilizers contained one atom per cent ^{15}N or less and it was later reported[119] that in 15 maize experiments (in 8 countries), an atom excess of 0.3 per cent ^{15}N would have been adequate

308

for the analysis of the data within acceptable limits of precision.   Four
or more replications were used, however, and given treatment effects of
substantial magnitude, the variability inherent in small plots[11] was not a
significant drawback.

In attempts to understand the cycling of N in soil, the preparation of
a balance sheet comprises, in itself, a difficult objective.   The use of
[15]N compounds has provided invaluable evidence for many parts of the N
cycle and has also shown that at every stage in the cycle, the dynamics
dictate that alteration of one compartment directly affects all other
compartments.   The fate of fertilizer N is determined by both plant uptake
and the interaction of the applied N with soil.   In cases where the tracer
uptake is used to define the use of fertilizer by plants, serious under-
estimates will result when exchange between added N and native soil N is
appreciable. In investigations into the fate of applied N over several
growing seasons and the potential long term value of nitrogenous organic
residues, tracer methods give invaluable information not obtainable
otherwise.

4.3. Chemical N transformation

Microbial activity is important for the conversion of ammonium,
resulting from mineralization of organic-N or from fertilizer input, into
nitrite and then nitrate.   These ions are reactive chemical species which
can combine with soil organic matter to produce "fixed" N which may be
unavailable to plants.

The chemical reactions involving ammonium fixation may yield
unidentified forms of organic N but the available data on this aspect are
meagre.   It has been known for some twenty years that fixation is favoured
by high pH values[28], is accompanied by the uptake of oxygen[105], and
approaches a "saturation" level over a prolonged period[27].   Even before
these facts were reported, coupling reactions involving simple quinones,
oxygen and ammonia at high pH values had been invoked as mechanisms leading
to ammonium fixation[47,94].   These reactions would lead to substituted
phenoxazine ring systems.   When it was realized that even these intractable
heterocycles could not account for the C:H:N ratios of the soil products[30],
more simple quinones were shown to yield amino-phenoxazines which
theoretically could polymerize to produce appropriate products.   To date,
however, there has been no experimental proof of these or similar structures
in humic substances.   It has long been postulated that lignin is a probable

source of quinonoid centres for chemical fixation of ammonium[94], a postulate
sustained by the fixation of N during the autoxidation of lignin in ammonium
hydroxide solution. However, it has also been shown that in ammonia
fixation by sugar-cane bagasse, the reaction involved mainly the hemicellulose fraction[34].

The possibility that carbohydrate-ammonia reactions may participate in
non-biological processes of humus formation under natural soil conditions
received much attention during the period 1950-1960. Unlike the Maillard
or "browning" reaction, in which melanoidin pigments are formed in mixtures
of sugars and amino acids in acidic media[90,91] the ammonia reaction was
examined in aqueous solutions at high pH[73]. Nevertheless, the Maillard
reaction involving amino acids and reducing sugars is still regarded as of
importance in the chemical immobilization of soil N[143]. Although the
starting-materials for the reaction are doubtless present in abundance in
soil solutions, normal soil temperatures in temperate latitudes are too
low for the reaction to proceed but slowly. However, it has been pointed
out that catalysis may occur on mineral surfaces[127], to facilitate the
condensation.

The agricultural importance of the chemical immobilization of ammonia
can be judged from estimates of annual sorption of atmospheric ammonia of
25-50 kilos of N per hectare[92], which does not appear to be reflected in
crop response[4]. It seems likely that the sorbed ammonia is quickly
immobilized in slowly-available forms, which would have little effect on
current crop response[101].

In addition to ammonia immobilization, equally significant chemical
reactions can take place between nitrite and soil organic matter[53]. For
example, anomalous results have been obtained in the Van Slyke determination
of amino N contents of lignins and soil humic acids in which nitrite is
added as a reagent[19,20]. In addition to the expected evolution of N_2,
small amounts of HCN were also produced with simultaneous N-fixation in the
organic substrates[21]. About one-third of the fixed N was released as
ammonia by acidic or alkaline hydrolysis, and hydroxylamine was also
detected after acid hydrolysis. In most soils, the microbial oxidation
of nitrite ion to nitrate is faster than the production of nitrite from
ammonium. Even so, high application rates of ammonia or ammonium based
fertilizers can lead to zones with locally-high pH values and accumulation
of nitrite[70]. The situation can lead to chemodenitrification, ie the

non-biological loss of N in the form of N_2 and nitrogen oxides. Just as losses of 33-79% of added nitrite-N occurred when soils were air-dried immediately after the addition of nitrite solutions[53], so the drying-out of soil following addition of fertilizer favours the conversion of nitrite into gaseous products, whose slow evolution would lead to significant losses of fertilizer N[143]. About 3 decades ago, Allison[4] suggested that denitrification could account for 15% loss of N from applied N-fertilizers. Recent studies have suggested that 0-20% of the N may be lost from fertilizer N in arable land to less than 7% loss on grasslands[38]. There is also strong evidence that nitrate concentration is an important factor in deciding the ratio of evolved $N_2O:N_2$, because it has been observed[17] that nitrate inhibits the reduction of nitrous oxide in soils.

5. AGRICULTURE AND NITROGEN

The importance of N in relation to soil fertility is discussed in chapter 12. Agricultural productivity may be influenced by various factors such as climate, plant cultivars and pesticides, but total dependence ultimately rests on the amount of photosynthesis and the supply of essential inorganic materials. Combined or "fixed" N is often limiting for crop productivity and when this is so N must be supplied externally. It is sobering to realize that less than 0.001% of the global N is cycling at any one time between its usable fixed form in the soil pool and its inert form in the atmospheric pool[104]. It is obvious, by reference to the global N cycle, that N fixation controls the atmosphere-to-terrestrial flow while nitrification and denitrification convert ammonia to nitrate and then to N_2 gas which is lost to the atmosphere; and leaching and erosion transfer fixed-N from the land to the sea. Interestingly the biological world apparently stays ahead of an N deficiency only because the fixation rate is marginally greater than the denitrification rate[103]. Any adverse changes in this delicate balance could have serious consequences for agriculture.

5.1. Practical implications of N fixation

Nitrogen fixation occurs naturally by chemical and/or biological processes, or by commercially-important chemical syntheses. Estimates place the biological processes as by far the major contributor of some 122×10^6 tonnes per annum globally, while lightning and combustion add a further 30×10^6 tonnes, with industrial fixation realizing some 50 x

10^6 tonnes. Without fully understanding the derived benefit, the Greeks and Chinese used biological N-fixation thousands of years ago in the form of legumes as green manures. In the S-E Asia rice-growing areas, traditionally no fertilizer is used, but a water fern Azolla is grown as a green manure in the fallow season and can be used to provide fixed-N to rice when grown in dual culture. Nitrogen fixation is totally dependent on the symbiotic association of the Azolla with its blue-green algal partner Anabaena azolla. Indeed the use of blue-green algae to fix N is common in many tropical countries.

Although the family Leguminoseae includes tropical and temperate plants, most of these can be infected by bacteria of the Rhizobium spp, which colonize the plant roots within nodules. It has been estimated that root nodules may supplement soil N by more than 200 kg ha^{-1} yr^{-1} under favourable conditions and this is probably greater than any other biological N-fixing system[33]. Agriculturally important symbioses occur with peas, beans, clovers and alfalfa, and these associations show some specificity with certain bacteria being not only specific to a particular plant species but frequently to a particular cultivar. Rhizobia are not all effective as N-fixers, so effective strains are used for inoculation of the plant seed. It has been observed that infection by one strain effectively protects the plant from invasion by other strains. Inhibition of infection may also be caused by the soil already containing significant levels of fixed N[65]. The degree of effectiveness of nodulation after inoculation with Rhizobium can often depend on the soil. Indeed, Newbould and Rangeley[102] have recently reported that nodule production and N fixation in a deep peat was enhanced but this did not occur in a brown earth soil. Nevertheless, the symbiotic associations between legumes and Rhizobia will probably continue to be the primary source of N fixation in agriculture in the immediate future.

Actinorhizae are root nodules, found in woody trees and shrubs, harbouring Actinomycetes. In the N-W United States, the growth of Douglas fir is apparently stimulated by intercropping with alder, which possess nodules containing Actinorhizae. These nodules thus differ markedly from those containing Rhizobium found on legumes[151] but are also effective N fixers. Non-symbiotic associations between various grasses and certain bacteria are also of great agricultural interest. Azotobacter paspali live and fix N in a mucilaginous sheath surrounding the root of the tropical grass

Paspalum; the grass Digitaria'is invaded by Azospililillum brasilense but in this case no nodulation occurs[43] (see chapter 6 for associative N-fixation involving non-symbiotic N-fixing bacteria). The contribution of these associative symbioses to agriculture may be restricted, similar to that of free-living, N-fixing organisms.

5.2. Agricultural importance of N dynamics

In most areas there is as yet no satisfactory laboratory soil test for predicting crop requirements for fertilizer-N. The total required by the crop is known but the contribution from natural soil sources varies from season to season depending on such factors as temperature, rainfall, residual fertilizer N and the nature of previously incorporated organic residues. The addition of crop or animal residues to soils, whether as tops or stubble, green manures or composted manures, introduces a variety of substances which differ in their availability to micro-organisms and which, with time, gradually undergo changes which decrease their availability. The nitrogenous organic matter in soil can conveniently be subdivided into an active fraction[77] comprising easily decomposable N such as proteins, amino sugars and nucleic acids, and a passive organic phase[80]. The latter has been subdivided by Juma and Paul[80] into two components, the first with a half life of about 35 years and a second, very stable component of about 600 years based on the age of the associated carbon. In the tracer experiment used, the two components comprised 36% and 50% respectively of the total N. The active component is likely to be the major source of plant-available N[77].

Measurements of mineral N release during long-term incubation experiments, and subsequent application of kinetic equations, have been used to estimate the size and decay rate constant of the mineralisable N pool[140]. Mineralization and immobilization are both continuous processes, governed by the growth and activity of the microbial biomass which, in turn, are dependent on the availability of suitable substrates containing C, N, P etc. As explained in the Introductory chapter, the net effect of the mineralization and immobilization processes is strongly influenced by the relative amounts of C and N (the C:N ratio) in the soil (Fig. 3). Recently incorporated organic residues are, therefore, very important particularly with reference to their ease of breakdown. For example, where the residues are N deficient, the microbes use any available mineral N and, initially, it is rapidly immobilized. As decomposition proceeds, this will be followed by a comparatively slow re-mineralization with some of the immobilized N joining the more passive soil fractions[25].

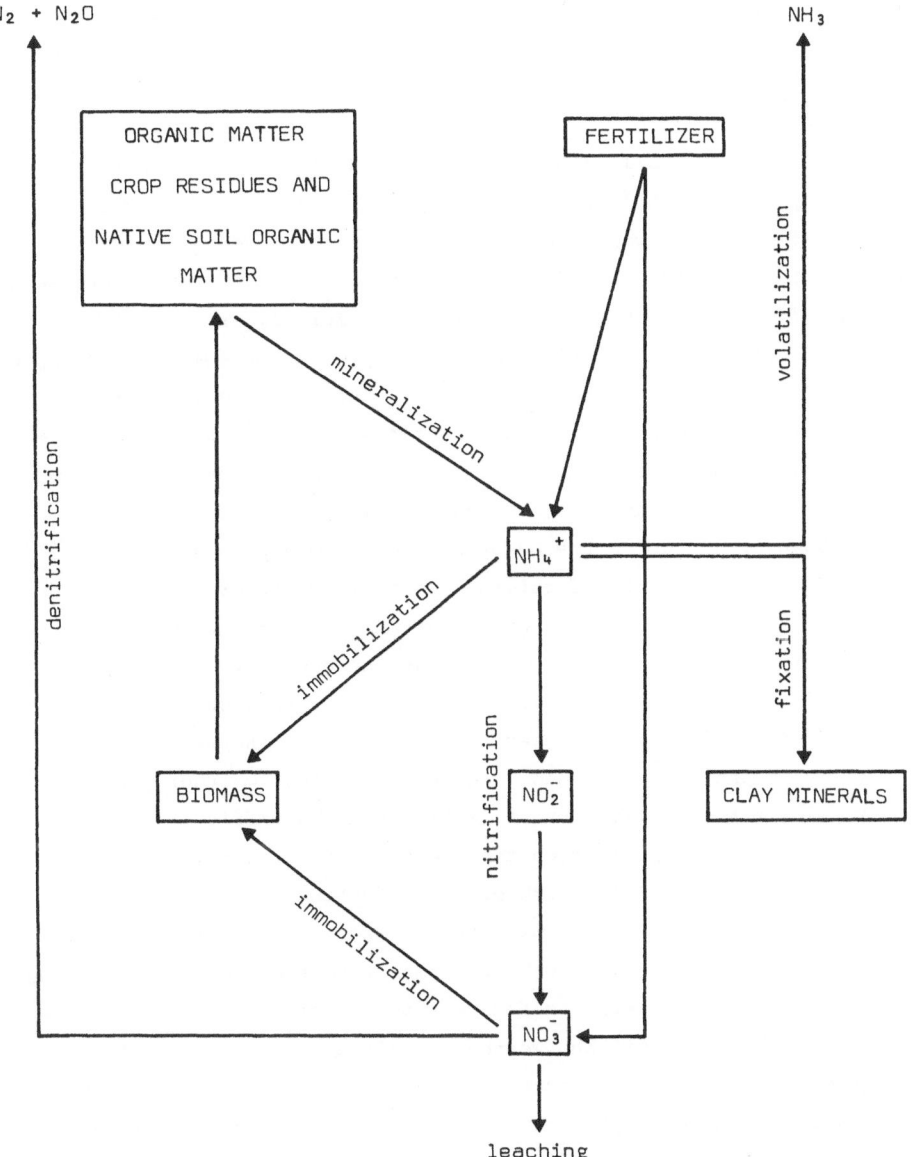

FIGURE 3. General outline of the van Veen and Frissel (1979) model.
(By permission of Pudoc).

In general unfertilized arable soils are only able to provide a fraction of the N required for high-yielding crops, often in the region of 20% or or less[64]. In recent years the most promising solution to predicting the rate of N fertilizer application has been the use of mathematical models based on soil type, rainfall and residual N levels. The use of such models is just as advantageous to developed countries as to the third world.

6. MODELLING

Since the 1950's, agricultural production in the UK has increased consistently. About half of this increase is accounted for by the production of new crop varieties through the work of geneticists (see chapter 12). Nevertheless, among the other reasons for this increase has been a five-fold increase in the input of fertilizer N, which has to be compared with almost constant levels of applied potassium and phosphate fertilizer levels over the same period[56]. This picture is repeated throughout the developed agricultural nations and represents a vast capital investment in industrial production facilities (approximately US $150 x 10^6 per unit), which are dependent on fossil fuels[104].

Although crop yields in the under-developed countries are often restricted by the lack of N-fertilizers (see Introductory chapter and chapter 12), a lack of understanding concerning optimal levels can often lead to wasteful over-application of these increasingly expensive substances. In the latter case, yield and quality are generally lowered, with the added environmental hazard of unacceptable increases of nitrate in groundwater and water courses. Considering that weather also plays a major role in the utilization of fertilizer, the difficulties become apparent in choosing application times and rates which favour the production of maximal yields with minimal additions.

Major efforts are now being made in the area of modelling the utilization of fertilizer N, both in terms of comprehending the ongoing processes, and in attempting to forecast fertilizer requirements in various combinations of crop, soil and weather. Although many mathematical descriptions of C and N transformations in soils have been developed in the last decade only a few explicitly model the activity and mass of soil organisms[159]. Considerable difficulties still exist in formulating the basic parameters governing the kinetics of microbial turnover[159]. Basically, models which have been postulated fall into three main categories in which:

(i) Physical processes such as transport and erosion predominate,

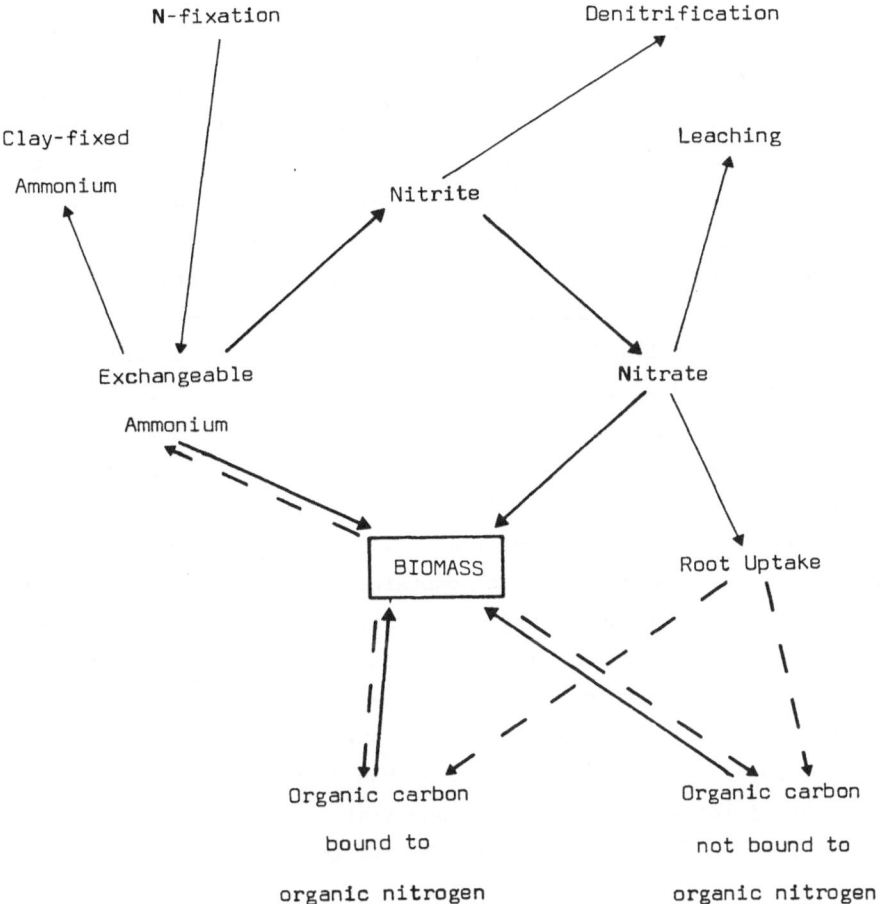

FIGURE 4. Flow diagram for the interaction of N and C in soil.

    Bold arrows   - decomposer biomass pathways;
    Broken arrows   - biomass (roots and decomposers)
                               mortality pathways;
    Unaccented arrows - N utilization outwith decomposer biomass.

(ii) Biological processes such as mineralization or nitrification predominate,

(iii) The uptake of N by plants and the fate of the plant are considered[159].

One major benefit of the modelling approach has been to identify areas in which quantitative data is lacking, or unusable. The input data for models should ideally be obtained from in situ measurements, or from undisturbed soil cores, with a sufficient number of measurements being made to provide an estimate of the precision of each coefficient used[1]. Because of the number of distinguishable factors involved, it is unlikely that an all-embracing model will be made in the near future, and indeed those existing models are generally simplified structures. Due to experimental interest being concentrated upon narrow segments, models tend to be expanded into the full N cycle with only superficial treatment of the related aspects[52]. It has also become obvious that the development of mechanistic models has lagged behind that of statistically based models and that the two types must be combined for the best practical use to be made of these systems.

One such model is that formulated in 1979 by van Veen and Frissel[158] and shown in general outline in figure 4. This scheme shows transformation processes which are a composite of separate sub-models, namely, mineralization and immobilization, nitrification, denitrification, ammonia volatilization, NH_4^+ fixation on clay minerals, and leaching. Components considered as variables within the scheme include total N, C, O and micro-organisms. Each separate sub-model contains descriptions of the coupling interrelationships of C and N, where necessary. These processes are central to both this model and that proposed by McGill et al.[98], which separates fungi from bacteria and actinomycetes, assuming differing responses in the two groups to soil temperature and moisture, and in their ability to digest different substrates. Van Veen and Frissel do not separate the microbial classes, but concentrate instead on groups of micro-organisms, classed according to their ability to decompose the soil fractions considered.

In more recent work based on further laboratory and field data, van Veen et al.[159] updated the previous model of van Veen and Frissel. Their updated model fitted the experimental data well, and indicated the contrasting metabolism of both C and N in a clay versus a sandy soil which could be explained by differences in the capacities of the two soils to preserve organic matter and micro-organisms. Furthermore, they considered that biomass and its immediate organic matter products of decay form a mainly closed system from which only small proportions of the products leak out as stabilized materials.

One of the major areas of uncertainty in the attempted simulation of N-cycling concerns the soil microbial biomass (see chapter 6). While there is much data available on the growth rates of isolated organisms and their metabolism and transformation of N-substrates in the presence of other compounds, the control mechanisms in soils are likely to be quite different. Reliable data are not available for the form and quantity of N- and C-compounds being released into a soil from the biomass. Indeed there are considerable variations between the results of different methods of biomass estimation, with chemical techniques, such as the fumigation technique and ATP measurements (see chapter 6) giving higher values for viable biomass than direct microscopy or plate-counting[29,114]. In a recent model simulating the C, N and biomass interrelationships in the decomposition of wheat straw in sand[83], it was concluded that the initial concentrations of available N and C determined the course of subsequent decomposition.

Apart from the types and amounts of N- and C-compounds released into the soil medium by the biomass, lack of knowledge surrounds the protection mechanisms which operate in soils. Although much detail is known about the N-dynamics, very little is known about the reason why the greater part of soil N cycles so slowly. Until this area is thoroughly investigated, we shall remain unable to predict the degree of immobilization of added fertilizer-N and the effect of added fertilizer-N on the existing soil organic matter.

7. CONCLUSIONS

Given the vital importance of N in agriculture, it is hardly surprising to find that an extensive and often discordant literature has evolved. On closer examination, one can still find important areas where information is badly needed, despite the fact that their importance has been recognised for a very long time. For example, our knowledge of the chemical changes in N components in the soil solution under different conditions is negligible. The introduction of modern chromatographic and spectroscopic techniques has provided a means of examining these, and many areas should be better-understood in the near future.

A major future research need is for the detailed identification and characterization of the non-identified soil N. It remains a source of concern that roughly half of the organic N of many agricultural soils occurs in unknown forms. The present crude extraction and fractionation methods will have to be replaced by some milder techniques, or extraction abandoned in favour of the use of ^{15}N tracing by high resolution NMR methods. There is much to favour the extraction of biologically meaningful fractions rather than those simply derived for chemical investigations per se. Some meticulous work is presently being carried out on extraction and fraction-ation methods[127,128,129], but it remains to be seen whether the "unknown" nitrogen, or at least some proportion of it, can be an artefact of the isolation techniques employed.

The methods of bonding of organic N, especially protein, onto or into soil humic substances remain obscure. Once again, general observations need to be followed up in detail, for example the removal of proteinaceous material of recent microbial origin by extraction with phenol[97]. The possible bonding of amino residues via quinones has not been examined much beyond the observation that an increase in amino acid N, at the expense of unidentified N, occurred when a humic hydrolysate was further hydrolysed with acid containing hydrogen peroxide[3]. Indeed, it remains to be seen whether our present means of characterizing and quantifying known forms of nitrogen are giving optimum yields of unmodified products. Only when we have succeeded in characterizing the major part of organic N will we be able to understand fully the many and varied processes which occur in transformation reactions in the soil, reactions which are of critical importance in determining the ultimate stability and fertility of the soil.

8. REFERENCES

1. ADDISCOTT T.M., DAVIDSON J.M., HARMSEN K., LEFFELAAR P.A., PARTON W.L., RAO P.S.C., RAYNER J.H., SMITH K.A. and WAGENET R.T. 1981. Migration processes in soils. In Simulation of nitrogen behaviour of soil-plant systems. Eds. Frissel M.J. and van Veen J.A. Pudoc, Wageningen, The Netherlands.
2. AIKEN G.R. 1983. Isolation techniques for aquatic humic substances. First International Meeting of the International Humic Substances Society. Denver, U.S.A.
3. ALDAG R.W. 1977. Relations between pseudo-amide nitrogen and humic acid nitrogen released under different hydrolysis conditions. In Soil organic matter studies. IAEA, Vienna.
4. ALLISON F.E. 1955. The enigma of soil nitrogen balance sheets. Advances in Agronomy, 7, 213-250.

5. ALLISON F.E. 1973. Soil organic matter and its role in crop production. Developments in Soil Science. Volume 3. Elsevier Press, New York.

6. ANDERSON G. 1967. Nucleic acids, derivatives, and organic phosphates. In Soil Biochemistry. Eds. McLaren A.D. and Petersen G.H. Marcel Dekker, New York.

7. ANDERSON G. 1970. The isolation of nucleoside diphosphates from alkaline extracts of soil. Journal of Soil Science, 21, 96-104.

8. ANDERSON H.A. and HEPBURN A. 1977. Fractionation of humic acid by gel permeation chromatography. Journal of Soil Science, 28, 634-644.

9. ANDERSON H.A., BICK W., GAULD J.H. and STEWART M. 1983. Organic nitrogen in "natural" profiles. First International Meeting of the International Humic Substances Society. Denver, U.S.A.

10. BARROW N.J. 1961. Phosphorus in soil organic matter. Soils and Fertilizers, 24, 169-173.

11. BARTHOLOMEW W.V. 1964. Guides in extending the use of tracer nitrogen in soils and fertilizer research. Proceedings of the Nitrogen Research Symposium. TVA Report number T64-45F, 81-96.

12. BENZING-PURDIE L. 1981. Glucosamine and galactosamine distribution in a soil as determined by gas liquid chromatography of soil hydrolysates. Effect of strength and cations. Journal of the Soil Science Society of America Proceedings, 45, 66-70.

13. BENZING-PURDIE L. 1984. Amino sugar distribution in four soils as determined by high resolution gas chromatography. Journal of the Soil Science Society of America Proceedings, 48, 219-222.

14. BLACK C.A. and GORING C.A.I. 1953. Organic phosphorus in soils. In Soil and fertilizer phosphorus. Eds. Pierre Witt and Norman A.G. Academic Press, New York.

15. BLACKBURN S. 1968. Amino acid determination. Edward Arnold, London.

16. BLACKBURN S. 1983. CRC Handbook for Chromatography : Amino acids and amines, Volume 1. CRC Press, Inc., Boca Raton.

17. BLACKMER A.M. and BREMNER J.M. 1978. Inhibitory effect of nitrite on NO to N by soil microorganisms. Soil Biology and Biochemistry, 10, 187-191.

18. BREMNER J.M. 1949. Studies on soil organic matter. Part 1. The chemical nature of soil organic nitrogen. Journal of Agricultural Science, 39, 183-193.

19. BREMNER J.M. 1952. The nature of soil-nitrogen complexes. Journal of the Science of Food and Agriculture, 3, 497-500.

20. BREMNER J.M. 1955. Recent work on soil organic matter at Rothamsted. Zeitschrift für Pflanzernährung, Düngung und Bodenkunde, 69, 32-38.

21. BREMNER J.M. 1957. Studies on soil humic acids; observations on the estimation of free amino groups; reactions of humic acid and lignin preparations with nitrous acid. Journal of Agricultural Science, 48, 352-360.

22. BREMNER J.M. 1958. Amino sugars in soil. Journal of the Science of Food and Agriculture, 9, 528-532.

23. BREMNER J.M. 1965. Organic forms of nitrogen. In Methods of soil analysis. Eds. Black C.A. et al. American Society of Agronomists, Madison, U.S.A.

24. BREMNER J.M. 1965. Organic nitrogen in soils. In Soil nitrogen. Eds. Bartholomew W.V. and Clark F.E. American Society of Agronomists, Madison, U.S.A.

25. BREMNER J.M. and SHAW K. 1957. The mineralization of some nitrogenous materials in soil. Journal of the Science of Food and Agriculture, 8, 341-347.

26. BROADBENT F.E. 1973. Sources and sinks of nitrate in soils. In Proceedings of the first annual trace contaminants conference. Pp 108-119. Oak Ridge National Laboratory, National Science Foundation, Washington, D.C.

27. BROADBENT F.E. and STEVENSON F.J. 1966. Organic matter interactions. In Agricultural anhydrous ammonia. Eds. McVickar M.H. et al. American Society of Agronomists, Madison, U.S.A.

28. BROADBENT F.E., BURGE W.D. and MAKASHIMA T. 1960. Factors influencing the reaction between ammonia and soil organic matter. Transactions of the Seventh International Congress of Soil Science, 2, 509-516.

29. BROOKES P.C., TATE K.R. and JENKINSON D.S. 1983. The adenylate energy change of the soil microbial biomass. Soil Biology and Biochemistry, 15, 9-16.

30. BURGE W.D. and BROADBENT F.E. 1961. Fixation of ammonia by organic soils. Soil Science Society of America Proceedings, 25, 199-204.

31. BURRIDGE J.C. and HEWITT I.J. 1980. Ammonia-filled discharge tubes for optical emission spectrometric determination of ^{15}N:^{14}N ratios. Analytica Chemica Acta, 118, 11-28.

32. CAMERON R.S., SWIFT R.S., THORNTON B.H. and POSNER A.M. 1972. Calibration of gel permeation chromatography materials for use with humic acid. Journal of Soil Science, 23, 342-349.

33. CAMPBELL N.E.R. and LEES H. 1967. The nitrogen cycle. In Soil Biochemistry, Volume 1. Eds. McLaren A.D. and Peterson G.H. Edward Arnold, London.

34. CHANG C.D., KONONENKO O.K. and HERSTEIN K.M. 1961. The ammoniation of sugar-cane bagasse. Journal of the Science of Food and Agriculture, 12, 687-693.

35. CHICHESTER F.W. 1969. Nitrogen in soil organo-mineral sedimentation fractions. Soil Science, 107, 356-363.

36. CHOPRA N.M. 1976. Investigations into the fate of plant pigments in some Canadian soils. Soil Science, 121, 103-113.

37. CLARK F.E. 1981. The nitrogen cycle, viewed with poetic licence. In Terrestrial nitrogen cycles. Eds. Clark F.E. and Rosswall T. Ecological Bulletins (Stockholm), 33, 1-24. Swedish National Research Councils.

38. COLBOURN P. and DOWDELL R.J. 1984. Denitrification in field soils. Plant and Soil, 76, 213-226.

39. COOKE G.W. 1981. The fate of fertilizers. In The chemistry of soil processes. Eds. Greenland D.J. and Hayes M.H.B. John Wiley and Sons Ltd., London.

40. CORTEZ J. and SCHNITZER M. 1979. Nucleic acid bases in soils and their association with organic and inorganic soil components. Canadian Journal of Soil Science, 59, 277-286.

41. CROOKE W. and SIMPSON W. 1971. Determination of ammonium in Kjeldahl digests of crops by an automated procedure. Journal of the Science of Food and Agriculture, 22, 9-10.

42. DADD C.C., FOWDEN L. and PEARSALL W.H. 1953. An investigation of the free amino acids in organic soil types using paper partition chromatography. Journal of Soil Science, 4, 69-71.

43. DOBEREINER J. and DAY J.M. 1976. Associative symbiosis in tropical grasses. In The proceedings of the first international symposium on nitrogen fixation. Eds. Newton W.E. and Nyman C.J. Washington State University Press, Pullman.

44. DOWNES M.T. 1978. An improved hydrazine reduction method for the automated determination of low nitrate levels in freshwater. Water Research, 12, 673-675.

45. EDMEADES D.C. and GOH K.M. 1979. The use of the ^{15}N dilution technique for field measurement of symbiotic nitrogen fixation. Communications in Soil Science and Plant Analysis, 10, 513-520.

46. FAO/IAEA (Joint Division of Atomic Energy and Agriculture).
1970. Rice fertilization. Technical Report 108.
1970. Fertilizer management practices for maize. Technical Report 121.
1974. Isotope studies on wheat fertilization. Technical Report 157.

47. FLAIG W. 1950. Zur Kenntis der Huminsauren. 1. Zur chemischen Konstitution der Huminsauren. Zeitschrift für Pflanzenernährung, Düngung und Bodenkunde, 51, 193-212.

48. FLUHER H., STOLZY L.H. and ARDAKANI M.S. 1976. A statistical approach to denitrification. Soil Science, 122, 115-123.

49. FRENEY J.R., STEVENSON F.J. and BEAVERS A.H. 1972. Sulphur-containing amino acids in soil hydrolysates. Soil Science, 114, 468-476.

50. FRIED M. 1978. Nitrogen in the environment. In Nitrogen behaviour in field soils. Volume 1. Eds. Nielsen D.R. and MacDonald J.E. Academic Press, New York.

51. FRISSEL M.J. (Editor). 1978. Cycling of mineral nutrients in agricultural ecosystems. Elsevier Press, Amsterdam.

52. FRISSEL M.J. and van VEEN J.A. 1982. A review of models for investigating the behaviour of nitrogen in soil. In The nitrogen cycle. The Royal Society, London.

53. FUHR F. and BREMNER J.M. 1964. Beeinflussende Fakoren in der Fixierung des Nitrit-Stickstoffs durch die organische Masse des Bodens. Atompraxis, 10, 109-113.

54. FUJII K., KOBAYASHI M. and TAKAHASHI E. 1974. On the alteration of microflora during the decomposition of plant residues. Journal of the Science of Soil and Manure (Japan), 43, 155-159.

55. GALLALI T., GUCKERT A. and JACQUIN F. 1975. Etude de la distribution des sucres aminés dans la matiere organique des sols. Bulletin de l'Ecole Nationale Superieure d'Agronomie et des Industries Alimentaires, XVII, 53-59.

56. GASSER J.K.R. 1982. Agricultural productivity and the nitrogen cycle. In The nitrogen cycle, The Royal Society, London.

57. GOH K.M. and WILLIAMS M.R. 1982. Distribution of carbon, nitrogen, phosphorus, sulphur and acidity in two molecular weight fractions of organic matter in soil chronosequences. Journal of Soil Science, 33, 73-87.

58. GOULDEN J.D.S. and SALTER D.N. 1979. Automatic emission spectrometer for the determination of nitrogen-15. Analyst, 104, 756-765.

59. GREAVES M.P. and WILSON M.J. 1970. The degradation of nucleic acids and montmorillonite-nucleic acid complexes in soils by micro-organisms. Soil Biology and Biochemistry, 2, 257-268.

60. GREAVES M.P. and WILSON M.J. 1973. Effects of soil micro-organisms on montmorillonite-adenine complexes. Soil Biology and Biochemistry, 5, 275-276.

61. GREEN L.C., WAGNER D.A., GLOGOWSKI J., SKEPPER P.L., WISHNOK J.S. and TANNENBAUM S.R. 1982. Analysis of nitrate, nitrite and (N-15) nitrate in biological fluids. Analytical Biochemistry, 126, 131-138.

62. GREENLAND D.J. 1965. Interactions between clays and organic components in soils. Soils and Fertilizers, 28, Part I, 414-425 : also Part II, 521-532.

63. GREENFIELD L.G. 1972. The nature of the organic nitrogen of soils. Plant and Soil, 36, 191-198.

64. GREENWOOD D.J. 1982. Nitrogen supply and crop yield : the global scene. Plant and Soil, 67, 45-59.

65. HAM G.E. 1980. Inoculation of legumes with Rhizobium in competition with naturalized strains. In Nitrogen fixation. Volume 2. Eds. Newton W.E. and Orme-Johnson W.H. University Park Press, Baltimore.

66. HANCE R.J. and ANDERSON G. 1963. Identification of hydrolysis products of soil phospholipids. Soil Science, 96, 157-161.

67. HARRADINE F. and JENNY H. 1958. Influence of parent material and climate on texture and nitrogen and carbon contents of virgin California soils. Soil Science, 85, 235-243.

68. HARTER R.D. and STOTSKY G. 1971. Formation of clay-protein complexes. Soil Science Society of America Proceedings, 35, 383-389.

69. HAUCK R.D. 1978. In Nitrogen in the environment. Eds. Nielsen D.R. and McDonald J.G. Academic Press, London.

70. HAUCK R.D. and STEPHENSON H.F. 1965. Nitrification of nitrogen fertilizers. Effect of nitrogen source, size and pH of the granule, and concentration. Journal of Agriculture, Food and Chemistry, 13, 486-492.

71. HOBSON R.P. and PAGE H.J. 1932. Studies on the carbon and nitrogen cycles in the soil. VII. The nature of the organic nitrogen compounds of the soil : "humic" nitrogen. Journal of Agricultural Science, 22, 497-515.

72. HOBSON R.P. and PAGE H.J. 1932. The nature of organic nitrogen compounds of the soil : "non-humic" nitrogen. Journal of Agricultural Science, 22, 515-526.

73. HOUGH L., JONES J.K.N. and RICHARDS E.L. 1952. The reaction of amino-compounds with sugars. Part I. The action of ammonia on D-glucose. Journal of the Chemical Society, 3854-3857.

74. HENZELL E.F. and ROSS P.J. 1973. The nitrogen cycle of pasture eco-systems. In Chemistry and biochemistry of herbage. Volume 2. Eds. Butler G.W. and Bailey R.W. 227-246.

75. HOYT P.B. 1971. Fate of chlorophyll in soil. Soil Science, 111, 49-53.

76. IVARSON K.C. and SOWDEN F.J. 1969. Free amino acid composition of the plant root environment under field conditions. Canadian Journal of Soil Science, 49, 121-127.

77. JANSSON S.L. 1958. Tracer studies on nitrogen transformations in soil with special attention to mineralization-immobilization relation-ships. Annals of the Royal Agricultural College, Sweden, 24, 101-361.

78. JANSSON S.L. 1966. Nitrogen transformations in soil organic matter. In The use of isotopes in soil organic matter studies. Report of the FAO/IAEA Meeting, Vienna. Pergamon Press, Oxford.

79. JENNY H. and RAYCHAUDHURI S.P. 1960. Effect of climate and cultivation on nitrogen and organic matter reserves in Indian soils. Indian Council for Agricultural Research, New Delhi.

80. JUMA N.G. and PAUL E.A. 1981. Use of tracers and computer simulation techniques to access mineralization and immobilization of soil nitrogen. In Simulation of nitrogen behaviour of soil-plant systems. Eds. Frissel M.J. and van Veen J.A. Pudoc, Wageningen, The Netherlands.

81. KANO H., YONEYAMA T. and KUMAZAWA K. 1981. Emission spectrometric ^{15}N analysis of the amino acids and amides in plant tissues separated by thin layer chromatography. Analytical Biochemistry, 67, 327-331.

82. KEENEY D.R. and BREMNER J.M. 1964. Effect of cultivation on the nitrogen distribution in soils. Soil Science Society of America Proceedings, 28, 653-656.

83. KNAPP E.B., ELLIOT L.F. and CAMPBELL G.S. 1983. Carbon, nitrogen and microbial biomass inter-relationships during the decomposition of wheat straw : a mechanistic simulation model. Soil Biology and Biochemistry, 15, 455-461.

84. KNOWLES R. 1978. Common intermediates of nitrification and denitrification and the metabolism of nitrous oxide. In Microbiology 1978. pp 367-371. American Society for Microbiology.

85. KOWALENKO C.G. 1978. Organic nitrogen, phosphorus and sulphur. In Soil organic matter. Developments in soil science. Volume 8. Eds. Schnitzer M. and Khan S.U. Elsevier Press, New York.

86. KUMADA K. 1955. Physico-chemical studies on the formation of humic acids. Part 6. Elementary composition of humic acids. Journal of the Science of Soil and Manure, Japan, 26, 19-22.

87. KUMADA K. 1978. Chemical studies on soil humic acids. III. Nitrogen distribution in humic acids. Soil Science and Plant Nutrition, 24, 561-570.

88. KUMADA K., SATO O., OHSUMI Y. and OHTA S. 1967. Humus composition of mountain soils in central Japan with special reference to the distribution of P type humic acid. Soil Science and Plant Nutrition, 13, 151-158.

89. LINDBECK M.R. and YOUNG J.L. 1965. Polarography intermediates in the fixation of ammonia by p-quinone-aqueous ammonia systems. Analytica chimica Acta, 32, 73-80.

90. MAILLARD L.C. 1912. Action des acides amines sur les sucres; formation des melanoidines par voie methodique. Comptes rendus hebdomadaires des séances de l'Académie des Sciences, 154, 64-68.

91. MAILLARD L.C. 1917. Identite des matieres humiques de synthese avec les matieres humiques naturelles. Annales de Chimie, 9e serie, 7, 113-152.

92. MALO B.A. and PURVIS E.R. 1964. Soil absorption of atmospheric ammonia. Soil Science, 97, 242-247.

93. MASON V.C., BECH-ANDERSON S. and RUDEMO M. 1980. Hydrolysate preparation for amino acid determinations in feed constituents. Zeitscrift für Tierphysiologie, Tierernährung und Futtermittelkunde, 41, 226-235.

94. MATTSON S. and KOUTLER-ANDERSSON E. 1942. The acid-base condition in vegetation, litter and humus : V. Products of partial oxidation and ammonia fixation. Annals of the Royal Agricultural College, Sweden, 10, 284-332.

95. MATTSON S. and KOUTLER-ANDERSSON E. 1943. The acid-base condition in vegetation, litter and humus : VI. Ammonia fixation and humus nitrogen. Annals of the Royal Agricultural College, Sweden, 11, 107-134.

96. MEINTS V.W. and PETERSON G.A. 1977. The influence of cultivation on the distribution of nitrogen in soils of the Ustoll sub-order. Soil Science, 124, 334-342.

97. McGILL W.B. and PAUL E.A. 1976. Fractionation of soil and ^{15}N nitrogen to separate the organic and clay interactions of immobilized N. Canadian Journal of Soil Science, 56, 203-212.

98. McGILL W.B., HUNT H.W., WOODMANSEE R.G. and REUSS J.O. 1981. Dynamics of C and N in grassland soils. In Terrestrial nitrogen cycles. Eds. Clark F.E. and Rosswall T. Ecological Bulletins, 33, 49-115. Swedish National Research Councils.

99. McLAREN A.D. and PETERSON G.H. 1965. Physical chemistry and biological chemistry of clay mineral-organic nitrogen complexes. In Soil nitrogen. Eds. Bartholomew W.V. and Clark F.E. American Society of Agronomists, Madison, U.S.A.

100. MOORE A.W. and CRASWELL E.T. 1976. Non-uniformity of ^{15}N labelling in plant material. Communications in Soil Science and Plant Analysis, 7, 335-344.

101. MORTLAND M.M. and WOLCOTT A.R. 1965. Sorption of nitrogen compounds by soil materials. In Soil nitrogen. Eds. Bartholomew W.V. and Clark F.E. American Society of Agronomists, Madison, U.S.A.

102. NEWBOULD P. and RANGELEY A. 1984. The effect of lime, phosphorus and mycorrhizal fungi on growth, nodulation and nitrogen fixation by white clover (Trifolium repens) grown in U.K. hill soils. Plant and Soil, 76, 105-114.

103. NEWTON W.E. 1981. Nitrogen fixation. In Kirk-Othmer encyclopedia of chemical technology. Third edition, Volume 15. Wiley-Interscience, New York.

104. NEWTON W.E. and BURGESS B.K. 1983. Nitrogen fixation : its scope and importance. In Nitrogen fixation - the chemical-biochemical-genetic interface. Eds. Muller A. and Newton W.E. Plenum Press, London.

105. NOMMIK H. and NILSSON K.O. 1963. Fixation of ammonia by the organic fraction of the soil. Acta agriculturae Scandinavica, 13, 371-390.

106. NORMAN A.G. and WERKMAN C.H. 1943. The use of the nitrogen isotope N-15 in determining nitrogen recovery from plant material decomposing in soil. Journal of the American Society of Agronomy, 35, 1023-1025.

107. ORION RESEARCH INC. Cambridge, Massachusetts, U.S.A.

108. PARFITT R.L. and GREENLAND D.J. 1970. Adsorption of polysaccharides by montmorillonite. Soil Science Society of America Proceedings, 34, 862-866.

109. PARSONS J.W. 1981. Chemistry and distribution of amino sugars in soils and soil organisms. In Soil Biochemistry. Volume 5. Eds. Paul E.A. and Ladd J.N. Marcel Dekker, New York.

110. PARSONS J.W. 1981. Clay-organic nitrogen complexes in soils. In Colloque humus-azote. Eds. Dutil P. and Jacquin F. ENSAIA-INPL, 54000 Nancy, France.

111. PAUL E.A. 1959. Extraction and quantitative estimation of free amino acids in soils. Dissertation Abstracts, 19, 2419.

112. PAUL E.A. 1984. Dynamics of organic matter is soils. Plant and Soil, 76, 275-286.

113. PAUL E.A. and SCHMIDT E.L. 1960. Extraction of free amino acids from soil. Soil Science Society of America Proceedings, 24, 195-198.

114. PAUL E.A. and van VEEN J.A. 1978. The use of tracers to determine the dynamic nature of organic matter. Transactions of the eleventh International Congress of Soil Science, 3, 61-102.

115. PAYNE T.M.B., ROUATT J.W. and KATZNELSON H. 1956. Detection of free amino acids in soil. Soil Science, 82, 521-524.

116. PORTER L.K. 1975. Nitrogen transfer in ecosystems. In Soil biochemistry. Volume 4. Eds. Paul E.A. and McLaren A.D. Marcel Dekker, New York.

117. PRESTON C.M., MATHUR S.P. and RAUTHAN B.S. 1981. The distribution of copper, amino compounds and humus fractions in organic soils of different copper content. Soil Science, 131, 344-352.

118. PUTNAM H.D. and SCHMIDT E.L. 1959. Studies on the free amino acid fraction of soils. Soil Science, 87, 22-27.

119. RENNIE D.A. and FRIED M. 1971. An interpretive analysis of the significance in soil fertility and fertilizer evaluation of ^{15}N-labelled fertilizer experiments conducted under field conditions. Proceedings of the International Symposium on Soil Fertility, New Delhi, 1, 639-656.

120. RENNIE D.A.; PAUL E.A. and JOHNS L.E. 1976. Natural ^{15}N abundance of soil and plant samples. Canadian Journal of Soil Science, 56, 43-50.

121. RENNIE D.A., WARD M.W.N., OWENS C.W.I., JACKSON M.J., GARROD P., ROUND J.M. and LEWIN M.R. 1981. Chemiluminescent nitrogen analysis. In Amino acid analysis. Ed. Rattenbury J.M. Ellis Horwood, Chichester.

122. RIFFALDI R. and SCHNITZER M. 1972. Electron spin resonance spectrometry of humic substances. Soil Science Society of America Proceedings, 36, 301-305.

123. RITTENBURG D., KESTON A.S., ROSEBURY F. and SCHOENHEIMER R. 1939. Studies on protein metabolism. II. The determination of nitrogen isotopes in organic compounds. Journal of Biological Chemistry, 127, 291-299.

124. ROCHUS W. 1977. The organic nitrogen content of peat soils combined with different fractions of humic substances. In Soil organic matter studies. Report of the IAEA Meeting, Vienna.

125. ROSSWALL T. 1982. Microbial regulation of the biogeochemical nitrogen cycle. Plant and Soil, 67, 15-34.

126. ROSSWALL T. and PAUSTIAN K. 1984. Cycling of nitrogen in modern agricultural systems. Plant and Soil, 76, 3-21.

127. SCHNITZER M. 1983. The nature of nitrogen in humic substances. First International Meeting of the International Humic Substances Society. Denver, U.S.A.

128. SCHNITZER M. and HINDLE D.A. 1980. Effect of peracetic acid oxidation on N-containing components of humic materials. Canadian Journal of Soil Science, 60, 541-548.

129. SCHNITZER M., MARSHALL P.R. and HINDLE D.A. 1983. The isolation of soil humic acid fulvic acid components rich in "unknown nitrogen". Canadian Journal of Soil Science, 63, 425-433.

130. SCOTT N.M., BICK W. and ANDERSON H.A. 1981. The measurement of sulphur-containing amino acids in some Scottish soils. Journal of the Science of Food and Agriculture, 32, 21-24.

131. SIMONART P. and PEETERS F. 1954. Free amino acids in humus. Transactions of the fifth International Congress of Soil Science, 3, 132-135.

132. SKELLY N.E. 1982. Separation of inorganic and organic anions on reversed-phase liquid chromatography columns. Analytical Chemistry, 54, 712-715.

133. SMALL H., STEVENS T.S. and BAUMAN W.C. 1975. Novel ion exchange chromatographic method using conductimetric detection. Analytical Chemistry, 47, 1801-1809.

134. SMITH K.A., ELMES A.E., HOWARD R.S. and FRANKLIN M.F. 1984. The uptake of soil and fertilizer nitrogen by barley growing under Scottish climatic conditions. Plant and Soil, 76, 49-57.

135. SORENSEN H. 1972. Stabilization of newly formed amino acid metabolites in soil by clay minerals. Soil Science, 114, 5-11.

136. SOWDEN F.J. 1959. Investigations on the amounts of hexosamines found in various soils and methods for their determination. Soil Science, 88, 138-143.

137. SOWDEN F.J. 1969. Effect of hydrolysis time and iron and aluminium removal on the determination of amino compounds in soil. Soil Science, 107, 364-372.

138. SOWDEN F.J., CHEN Y. and SCHNITZER M. 1977. The nitrogen distribution in soils formed under widely differing climatic conditions. Geochimica et Cosmochimica Acta, 42, 1524-1526.

139. STAINTON M.P. 1974. Simple, efficient reduction column for use in the automated determination of nitrate in water. Analytical Chemistry, 46, 1616.

140. STANFORD G. and SMITH S.J. 1972. Nitrogen mineralization potentials of soils. Soil Science Society of America Proceedings, 36, 465-472.
141. STEVENSON F.J. 1957. Distribution of the forms of nitrogen in some soil profiles. Soil Science Society of America Proceedings, 21, 283-287.
142. STEVENSON F.J. 1957. Investigation of amino polysaccharides in soil. II. Distribution of hexosamines in some soil profiles. Soil Science, 84, 99-106.
143. STEVENSON F.J. 1982. Humus chemistry - genesis, composition, reactions. John Wiley and Sons, New York.
144. STEVENSON F.J. 1983. Isolation and identification of amino sugars in soil. Journal of the Soil Science Society of America, 47, 61-65.
145. STEWART W.F.P. and ROSSWALL T. 1982. The nitrogen cycle. The Royal Society, London.
146. SYNGE R.L.M. 1975. Interactions of polyphenols with proteins in plants and plant products. Qualitatus Plantarum, 24, 337-350.
147. TATE K.R. and ANDERSON H. 1978. Phenolic hydrolysis products from gel chromatographic fractions of soil humic acids. Journal of Soil Science, 29, 76-83.
148. THOM B. 1976. A study of the composition and properties of some organic nitrogen compounds occurring in soils. Ph.D. Thesis, University of Aberdeen, Scotland.
149. TIEDJE J.M., SEXSTONE A.J., PARKIN T.B., REVSBECH N.P. and SHELTON D.R. 1984. Anaerobic processes in soil. Plant and Soil, 76, 197-212.
150. TINSLEY J. and DARBYSHIRE J.F. 1984. Editors of the Proceedings of the conference on biological processes and soil fertility. Plant and Soil, Volume 76.
151. TORREY J.G., TJEPKEMA J.D., TURNER G.L., BERGERSON F.J. and GIBSON A. H. 1981. Dinitrogen fixation by cultures of Frankia SP Cp 11 demonstrated by dinitrogen (^{15}N) incorporation. Plant Physiology, 68, 983-984.
152. TORSUIK V.L. 1980. Isolation of bacterial DNA from soil. Soil Biology and Biochemistry, 12, 15-21.
153. TRISTRAM G.R. and RATTENBURY J.M. 1981. The development of amino acid analysis. In Amino acid analysis. Ed. Rattenbury J.M. Ellis Horwood, Chichester.
154. TURNELL D.C. and COOPER J.D.H. 1982. Rapid assay for amino acids in serum or urine by pre-column derivatization and reversed-phase liquid chromatography. Clinical Chemistry, 28, 527-531.
155. VAN CLEEMPUT O. and BAERT L. 1984. Nitrite : a key compound in N loss processes under acid conditions? Plant and Soil, 76, 233-241.
156. VAN DER VELDEN W. and SCHWARTZ A. 1974. Purine and pyrimidines in sediments from Lake Erie. Science, 185, 691-693.
157. VAN VEEN J.A. and FRISSEL M.J. 1979. Mathematical modelling of nitrogen transformations in soil. In Modelling nitrogen from farm wastes. Ed. Gasser J.F.K. Applied Science Ltd., London.
158. VAN VEEN J.A. and FRISSEL M.J. 1981. Simulation model of the behaviour of N in soil. In Simulation of nitrogen behaviour of soil-plant Systems. Eds. Frissel M.J. and van Veen J.A. Pudoc, Wageningen, The Netherlands.
159. VAN VEEN J.A., LADD J.N. and FRISSEL M.J. 1984. Modelling C and N turnover through the microbial biomass in soil. Plant and Soil, 76, 257-274.
160. VAN VEEN J.A., McGILL W.B., HUNT H.W., FRISSEL M.J. and COLE C.V. 1981. Simulation models of the terrestrial nitrogen cycle. In Terrestrial nitrogen cycles. Eds. Clark F.E. and Rosswall T. Ecological Bulletins, 33, 25-48. Swedish National Research Councils.

161. VOLK R.J. and JACKSON W.A. 1979. Preparing nitrogen gas for nitrogen-15 analysis. Analytical Chemistry, $\underline{51}$, 463-464.
162. WIEGELE M., DE BERNARDO S. and LEIMGRUBER W. 1973. Fluorometric assay of secondary amino acids. Biochemical and Biophysical Research Communications, $\underline{50}$, 352-356.
163. WOODMANSEE R.G., DODD J.L., BOWMAN R.A., CLARK F.E. and DICKINSON C.E. 1978. Nitrogen budget for a shortgrass prairie ecosystem. Oecologia (Berlin), $\underline{34}$, 363-376.
164. YAMAMURO S. 1981. The accurate determination of nitrogen-15 with an emission spectrometer. Soil Science and Plant Nutrition, $\underline{27}$, 405-419.

CHAPTER 9

SOIL PHOSPHORUS

K.R. TATE

N.Z. Soil Bureau, D.S.I.R., Lower Hutt, New Zealand

CONTENTS

1. INTRODUCTION

Phosphorus (P) has a unique metabolic role in all living systems as the energy transducer in the enzymic generation of ATP (adenosine-5'-triphosphate) and ADP (adenosine-5'-diphosphate) from inorganic orthophosphate (P_i). Phosphate derivatives are also involved in almost all significant metabolic pathways, as well as being present as a structural component of many bio-chemicals including nucleic acids, coenzymes, phosphoproteins and phospholipids[190]. As P_i, P also participates in the chemical processes of precipitation-solubilisation and adsorption-desorption in soil that largely control its availability to plants and soil organisms. Compared to these processes, the rates and pathways of P through soil organic matter (used in its broadest sense to include both living and dead soil con-stituents) are relatively little understood. Recent reports[45,55,57,85] however, indicate that organic phosphorus (P_o) has an important role to play in the nutrition of plants grown in temperate soils, as well as in tropical soils[6].

This review is an attempt to bring together the current knowledge on the soil P cycle, emphasising particularly the rates and pathways of P through soil organic matter, and the availability of P_o to plants.

2. GLOBAL CYCLE

The global movement of phosphorus occurs through three interconnected cycles; the main compartments, and their P contents and annual fluxes based on data of Pierrou[141], are given in Fig. 1.

331

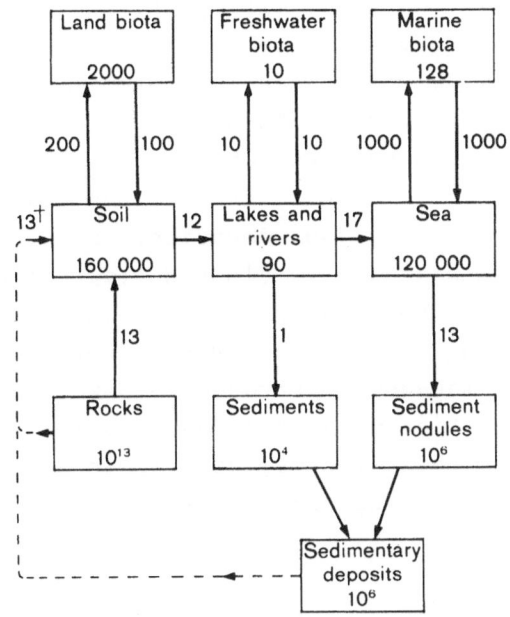

FIGURE 1. Global P compartments and annual fluxes between them (Mt P). † Mined for fertiliser use. From Emsley[71]. Reproduced by permission of Springer-Verlag.

The primary cycle is inorganic, in which phosphate rock formation, mainly from diagenesis of P-enriched sediments in marine and fresh waters, takes place over very long periods of time (ca. 10^8-10^9 y[46]).

The much faster land-based organic cycle is sustained by application of this rock phosphate to soils as fertiliser, as well as contributions from natural weathering.

According to Pierrou[141], about 13 Mt P reach the soils of the world annually by each of these pathways, while a similar amount (ca. 12 Mt) is lost through soil erosion and leaching. Plants have been estimated to remove somewhere between 178 and 240 Mt P, while biological returns amount to about 100 Mt annually. An even faster water-based P cycle, with turnover times measured in months rather than years as for the land-based cycle, is essentially closed, with the very small quantity of P entering the oceans in rivers being approximately balanced by that removed in sediments[46].

Despite the difficulty of estimating the sizes of the various P pools on a global scale, and of the fluxes through these pools, available evidence indicates that P reserves are far from nearing exhaustion. Instead, the main problem facing agriculturalists, particularly where soils are naturally P deficient, is the escalating costs of recovering fertiliser P from phosphate rock and guano deposits, and of applying this P to soil to maintain agricultural production at commercially acceptable levels. The solution depends on a number of factors, including the development of new fertiliser technology, alternative land management strategies, and the introduction of new plant varieties with lower demands for P. Successful implementation of these factors will above all depend on a much better understanding than at present of the soil P cycle.

3. P AND PEDOGENESIS

As can be seen in the global cycle (Fig. 1), the amounts of P occurring in soils greatly exceeds that found in plants. On a global basis, 59×10^{10} t of carbon are located in the plant biomass[43], mainly in trees, and if the P content is taken as 0.2%, then only about 1200 Mt P reside in plants compared to 160,000 Mt in soil. However, the P_i concentration in plants which is in the range 5-10 nM if plant tissues are regarded as equivalent to aqueous solutions[37], is several orders of magnitude larger than that in soil solution (ca. 1.5 μM[24]). Consequently, the important question concerning P in soil is not - how much does the soil contain - but instead - how available is this P to plants, and what factors limit its availability?

A consideration of the P cycle in soil on a pedogenetic time scale of several thousand years is a useful starting point in attempting to answer these questions.

Soil formation begins as a result of natural weathering processes operating on rock minerals to release P and other nutrients at a rate governed by the weathering intensity[182]. Organic matter accumulates near the soil surface through biological activity, to an extent that is largely controlled by the amount of P present in the underlying parent rocks[181].

Table 1. The amounts of organic C, total P and P_O and 'available' (0.5 M H_2SO_4)P in a development sequence of New Zealand zonal soils in tussock grassland.

Soil name (Classification*)	Depth cm	Carbon mg g^{-1}	Total P mg g^{-1}	'Avail' %	P/P_T %	P_O/P_T %
Conroy (Xeralfic Haplargid)	0-7	35	1.06	62		31
Cluden (Typic Ustochrept)	0-18	21	0.94	48		49
Tima (Udic Haplustalf)	0-13	23	0.89	55		43
Obelisk (Haplic Cryohumod)	0-13	93	1.04	18		84
Carrick (Typic Cryorthod)	1-4	49	0.88	19		74
Tawhiti (Dystric Cryochrept)	0-11	48	0.84	20		71
Lammerlaw (Typic Placaquept)	0-15	73	0.52	13		83
McKerrow (Typic Dystrochrept)	0-5	75	0.51	10		92
Maungatua (Typic Placaquod)	0-5	187	0.43	9		93

* Soil Taxonomy; see [166].
Adapted from Molloy and Blakemore [128].

As a soil develops, P appears in different forms, and both the distribution of P among these forms and the total amount of P present in the soil provides a useful index of the extent of soil development [156,182]. In a climosequence of New Zealand soils in tussock grassland, for example, total P (P_T) mainly declined with increasing soil development (Table 1) with lowest levels in the three podzolised members [128]. The proportion of 'available' P (0.5 M H_2SO_4-soluble P) also declined across the sequence, while the proportion in organic forms increased. Levels of P_O that can occur in soils vary widely from near zero to over 0.2% [15]. In some New Zealand soil chronosequences, P_O levels have been shown [182] to reach a maximum eventually which is maintained up to soil ages of about 20,000 years. Levels then decline as a result of leaching losses, such as occur in the humid tropics.

Ultimately only small amounts of P_O and strongly Fe-bound P_i (occluded P) remain[189].

In contrast to pedogenetic processes where a time scale of thousands of years is involved, the transfer of P between different pools in the surface horizons of soil occurs on a much shorter time scale of days, to years.

4. FORMS OF SOIL P AND THEIR CHARACTERISATION
4.1. Chemical characterisation of soil P

Soil P occurs in a wide variety of different inorganic and organic forms. About 200 phosphate-containing minerals are known[74], and in soils the orthophosphate anion also occurs in combination with aluminium and iron oxides and hydroxides, clay minerals, solid carbonates and soil organic matter. Small amounts of condensed P_i of mainly microbial origin, are also found in soil as pyrophosphate and higher polyphosphates[131].

In the surface horizons of most soils, P also occurs in organic forms in amounts that can vary widely from about 20 to 90% of P_T[120]. Estimates of total soil P_O can only be made indirectly by ignition or extraction methods, and at best these indirect methods can be regarded as approximate since they often suffer from analytical problems (for reviews, see e.g.[14,87,153]).

Soil P_O mainly occurs as mono- and diesters (i.e. $(RO).PO_3H_2$ and $(RO)(R'O).PO_2H)$ and their chemical diversity is due mainly to the organic moieties R and R'. Several excellent reviews have appeared on the composition of soil P_O[14,15,66,87,111]. The predominant form of P_O in soils are the inositol hexa- and pentakisphosphates, which together constitute up to 60% of the total P_O. In some cases, the inositol phosphates may be over-estimated, however, because other stable P_O compounds can interfere in the commonly used anion exchange method[123] unless alkaline hypobromite oxidation is used to destroy them[100]. Instead of isolating inositol phosphates from alkali extracts, ester P including inositol phosphates has been precipitated from soil with Ti to give recoveries comparable with more commonly used procedures[124].

Relatively small amounts of phospholipids (usually < 5% of

total P_o) and nucleotides (up to 9%[107]) are found in soil. The
former are mainly phosphatidyl choline and phosphatidyl
ethanolamine, and are isolated by solvent extraction methods[21,113].
Nucleotide-P has mainly been inferred by the appearance in soil
hydrolysates of pentose sugar, purines or pyrimidines and
phosphates[111]. Nucleoside diphosphates have, however, been isolated from soil[13], and recently, up to 1.5 mg bacterial DNA[175]
of high purity was recovered from 37 g (dry weight) of soil[174].
Apart from traces of sugar phosphates, phosphoprotein, and
glycerophosphate[58], and small amounts of phosphonate[131], the
chemical nature of a large proportion of the remaining P_o in
soil is still unidentified, probably occurring as insoluble
complexes with clay minerals and organic matter. Several separation techniques such as anion exchange[65], gel filtration
e.g.[165,178], and high pressure liquid chromatography[79] have been
used on soil extracts with mixed success in attempts to characterise soil P_o which occurs in high molecular weight forms that
are not always intimately connected with the other constituents
of soil organic matter[111].

An alternative approach to the problem of characterising
soil P_o has been to use ^{31}P- nuclear magnetic resonance (nmr) to
obtain both qualitative[131] and quantitative[166] estimates of the
various forms of P in soil alkaline extracts.

Sodium hydroxide is the most effective extractant of soil
P_o[161] but its use inevitably causes changes to soil organic
matter, including the hydrolysis of alkali-labile P_o like RNA[12],
phospholipids and phosphonolipids[131]. In one soil, myo-inositol
hexakisphosphate was confirmed as the main orthophosphate
monoester[131].

Extensive hydrolysis of labile P_o can be avoided by using
less drastic reagents like acetylacetone and a cation exchange
resin[96]. This technique not only recovered much of the P_o from
a Japanese Andosol, but also enabled the P_i held on organic
matter by Fe and Al to be uncoupled, thereby avoiding its
measurements as P_o. The greatest diversity of P_o compounds
occurred in the lowest molecular size fraction of the fulvic
acid from this soil while inositol phosphate was exclusively

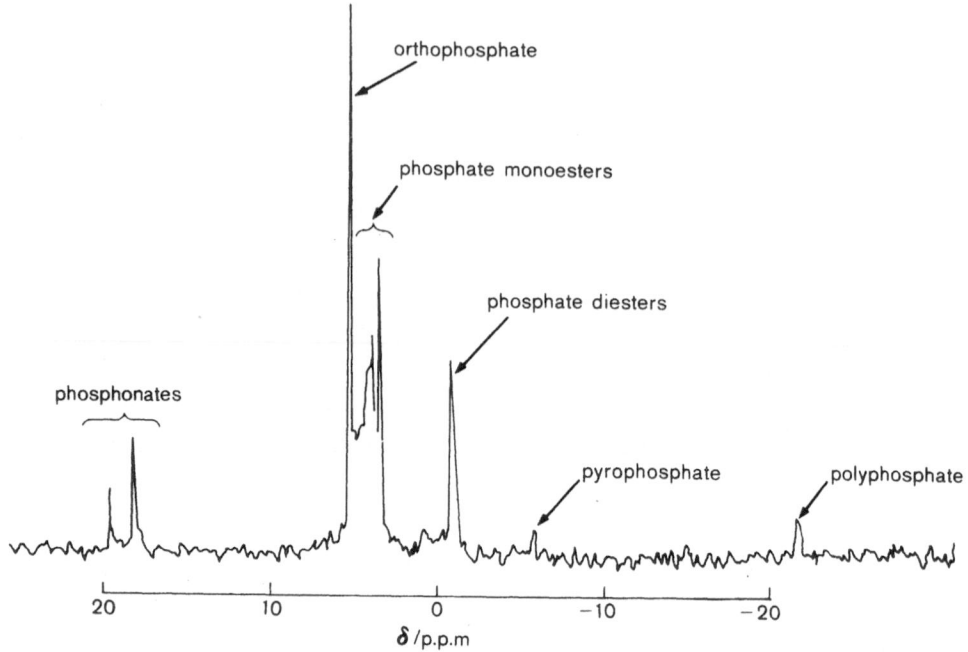

FIGURE 2. ^{31}P-nmr spectrum of an álkali extract of McKerrow soil (Table 1) showing the different forms of P. From Newman and Tate [131].

associated with the highest molecular size fraction, and the humic acid[97] .

Despite the continuing progress being made towards identification of the different chemical forms of P_o in soil extracts, future prospects for the quantitative separation of unaltered P_o from soil are not very encouraging. Complicating factors include the chemical complexity of soil P_o, the susceptibility of some P_o compounds to hydrolysis during extraction, strong sorption of P_o by clays, and insoluble salt formation with metal cations.

Similar difficulties have been encountered in attempts to characterise P_i in soil; again, solubility in various extractants has frequently been used[31]. As for P_o, however, the extractants used have limited ability to extract discrete P_i compounds without alteration[135]. The amounts of Fe-, Al- and Ca-bound P_i

measured by these methods have nevertheless been observed to change in a predictable way with soil development, indicating general trends in the composition of the soil P_i [31].

Direct application to whole soil of a cross polarisation-^{31}P-nmr technique was recently used[191] to indicate the chemical form of the soil P_i. It may become possible with the continuing development of this powerful analytical technique to extend its application beyond characterising the C chemistry of whole soils[193] to include also the direct characterisation of soil P.

4.2. Microbial P

The microbial biomass, defined as the living part of the soil organic matter excluding plant roots and soil animals larger than 5×10^3 μm^3 [104], represents a small but labile pool of soil P, mainly in the form of RNA and DNA, polyphosphates and phospholipids, with minor amounts of other P-containing cell biochemicals like adenosine phosphates. Measurement of the P content of the microbial biomass is an essential prerequisite for assessing its role in the P cycle, and its effect on plant nutrition. Until very recently, estimates of microbial P have had to be made from microbial biomass measurements, and literature values for the P contents of laboratory cultured micro-organisms[16,85] which can vary widely depending on growth conditions[33,177].

Two new methods for measuring soil microbial P have been described[47,92], based on the widely used $CHCl_3$- fumigation procedure[105] for measuring the soil microbial biomass. In these methods, the P released from microbial cells following lysis with $CHCl_3$ is extracted from the soil with $NaHCO_3$. Some of this P will be immediately fixed by soil colloids, and in one method[47], a spike of P_i is used to correct for this. The other method[92] is calibrated for recovery of microbial P for each soil type. Non-microbial P was not affected to any extent by fumigation.

The two methods, while similar in principle, differ in some respects. Microbial P is calculated from the difference in extractable P_i [47] or P_T [92] between fumigated and unfumigated soil, using a correction factor (kp) for the fraction of the microbial P extracted after fumigation.

In one method[92], removal of labile P_i by anion exchange resin from air dried, ground soil provided for a precise estimate of microbial P on small soil samples, and the change in resin-extractable P during extraction indicated microbial P uptake. Larger samples (10 g) of field moist soils were needed to compensate for greater soil heterogeneity in the other method[47], but microbial P contents were believed to relate more closely to field soils.

The kp factors for both methods were obtained using either fresh[92] or lyophilised laboratory grown[47] microorganisms, giving values of 0.37 and 0.47 respectively. The validity of both methods rests on how closely these kp's are to that of a native soil population, and this is unknown at the present time. A kp of 0.4 has therefore been provisionally recommended for use in determining soil microbial P[47,92].

Both these methods, which were mainly developed using near neutral to alkaline soils, provide an opportunity of more clearly defining the role of the microbial biomass in the P cycle. They now need to be tested on a wider range of soil types, including acid soils, to confirm their value for general use. The results recently obtained for some New Zealand topsoils under high producing pasture, however, appear promising. The microbial P contents of 21 topsoils (0-75 mm; pH range 4.9-6.8), measured by the method of Brookes et al.[47], ranged from 20 to 88 μg P g^{-1} dry soil (mean 51 μg P g^{-1} soil)[140]. These values are equivalent to 11-57 kg P ha^{-1}, representing 0.5-11.7% of the P_T, which is similar to the range reported for some British soils (0.5-11.9%)[47].

4.3. Characterisation by chemical extractant

The use of chemical extractants to characterise soil P fractions is not new. For example, the concept of a labile pool of soil P was defined operationally by relating results from chemical extraction and anion exchange methods to those obtained by isotope exchange, rather than on the basis of an

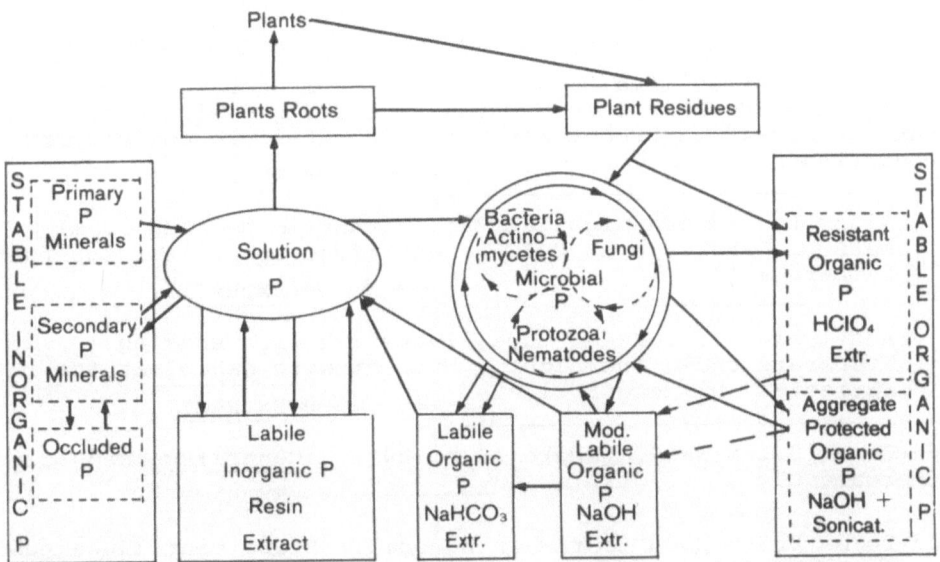

FIGURE 3. A schematic representation of the P cycle showing the measurable components. From Chauhan et al.[57].

understanding of soil P chemistry[150].

Research on P cycling in grassland soils[63,85] stimulated interest in methods for separating soil P_o fractions according to their activity in the P cycle[44] and led to the development of a fractionation scheme (Fig. 3) based on the use of different soil extractants[45]. Four P pools were identified in this scheme as follows:

1. A labile pool extracted with 0.5 M $NaHCO_3$ (pH 8.5); added RNA, nucleotides and glycerophosphate were recovered from soil with this extractant.

2. A moderately labile pool, consisting of acid-soluble P_o and alkali-soluble P_i.

3. A moderately resistant fraction (fulvic acid-P).

4. A highly resistant fraction (humic acid-P).

This scheme (Fig. 3) was used recently[56,57] with indices of labile P_i (resin extractable)[10,155], more stable P_i fractions[139,151] and microbial P to follow the rate of P movement between different pools in response to C additions in soil incubation

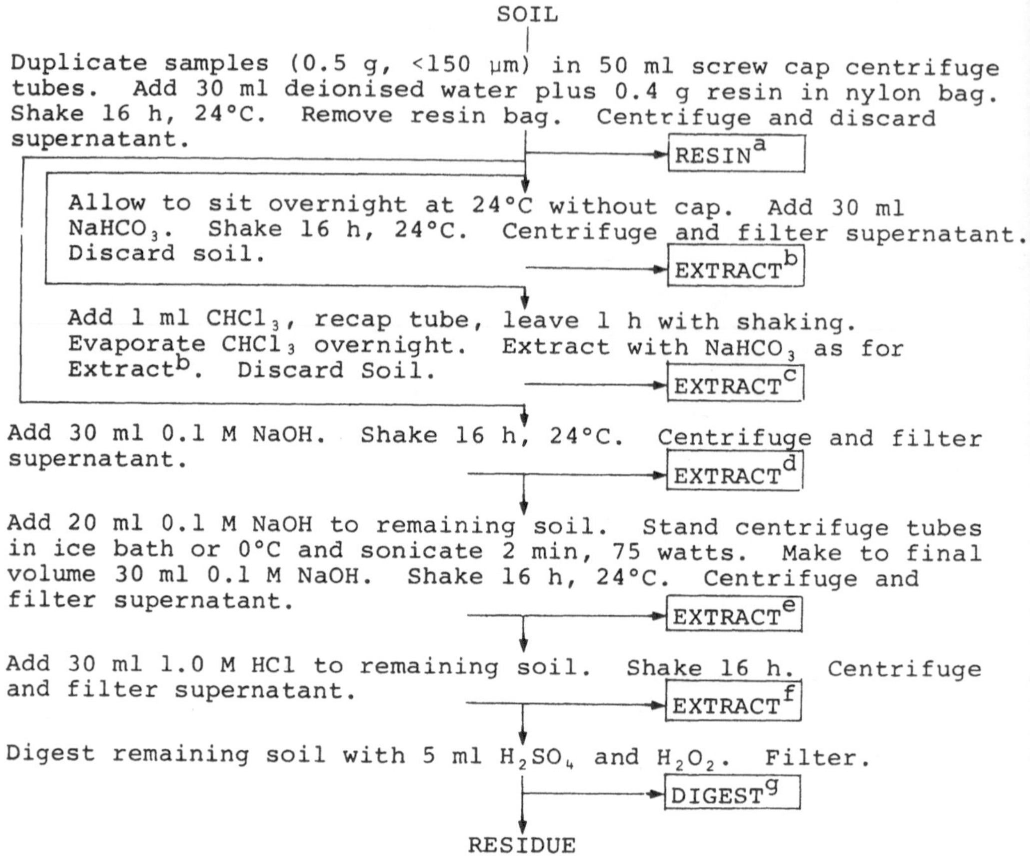

SOIL

Duplicate samples (0.5 g, <150 µm) in 50 ml screw cap centrifuge tubes. Add 30 ml deionised water plus 0.4 g resin in nylon bag. Shake 16 h, 24°C. Remove resin bag. Centrifuge and discard supernatant.

→ RESIN[a]

Allow to sit overnight at 24°C without cap. Add 30 ml NaHCO₃. Shake 16 h, 24°C. Centrifuge and filter supernatant. Discard soil.

→ EXTRACT[b]

Add 1 ml CHCl₃, recap tube, leave 1 h with shaking. Evaporate CHCl₃ overnight. Extract with NaHCO₃ as for Extract[b]. Discard Soil.

→ EXTRACT[c]

Add 30 ml 0.1 M NaOH. Shake 16 h, 24°C. Centrifuge and filter supernatant.

→ EXTRACT[d]

Add 20 ml 0.1 M NaOH to remaining soil. Stand centrifuge tubes in ice bath or 0°C and sonicate 2 min, 75 watts. Make to final volume 30 ml 0.1 M NaOH. Shake 16 h, 24°C. Centrifuge and filter supernatant.

→ EXTRACT[e]

Add 30 ml 1.0 M HCl to remaining soil. Shake 16 h. Centrifuge and filter supernatant.

→ EXTRACT[f]

Digest remaining soil with 5 ml H₂SO₄ and H₂O₂. Filter.

→ DIGEST[g]

RESIDUE

FIGURE 4. Flow chart of the fractionation of soil P into various P_i and P_o fractions. Extractable P fractions measured are: a. Most biologically available (resin-P_i); b. Labile P_i and P_o (NaHCO₃-P); c. P in microbial cells (CHCl₃/NaHCO₃-P); d. Moderately labile P_i and P_o (NaOH-P); e. Aggregate-protected P_i and P_o (sonicate/NaOH-P); f. Apatite-type minerals and occluded P_i (HCl-P); g. Stable P_i and P_o (residual P). Microbial P is calculated from c-b[92]. Adapted from Hedley et al.[93].

experiments. Although the fractionation procedure was reproducible and indicated P movement between different fractions, further refinements were considered necessary[56].

A similar but more comprehensive fractionation scheme was proposed recently[93] (Fig. 4). This allows the determination of recognised P_i fractions (labile, secondary, occluded, and

primary minerals), the separation of labile and stable forms
of P_o, and the measurement of microbial P[92]. The new fractionation scheme was used recently on a calcareous Chernozem soil to
study seasonal P transformations during simulated fallow under
different management practices[93]. It has yet to be tested on
a range of soil types. In a slightly modified form, however,
it has already proved useful for identifying P fractions in a
rhizosphere soil that had suffered depletion due to plant
uptake[94]. Research on the P cycle has received a new impetus for
progress from the development of these new techniques for
fractionating soil P using different extractants. Their use in
conjunction with techniques like ^{31}P-nmr to indicate the
different chemical forms and quantities of P extracted with
alkali[131], should in future lead to a much better understanding
of those P transformations that cause the long-term changes in
soil P.

5. SOIL P CYCLE

   Most of the P in uncultivated soils is concentrated at or
near the soil surface. Inorganic orthophosphate (P_i), unlike
nitrate-N, is relatively immobile in soil, with the apparent
diffusivity factors being in the ranges of 10^{-7}-10^{-11} cm^2 sec^{-1}
and 10^{-5}-10^{-6} cm^2 sec^{-1} respectively[26]. While movement of P_i
in the soil profile by diffusion is very slow, some trans-
location and deposition can occur further down the profile
via plant roots[148]. Loss of P as P_i from roots on desiccation
can be substantial[119], and root exudation of P can represent a
major pathway for transfer into the soil profile under trees[89].
Compared to P_i, P_o can move greater distances into the
soil[66,144] but eventually, after mineralisation, this P will
become immobilised by chemical precipitation and physico-
chemical adsorption on to soil surfaces, and by microorganisms.

   Additions of organic matter to soil as plant, animal and
microbial residues provide the energy needed to sustain the
cycling of nutrients in the soil, as well as returning P to the
soil for recycling. The mechanisms by which the P enters soil
are consequently of considerable importance, and, as already

mentioned, include the incorporation of roots as well as aerial parts of the plant. Leaching of P from dead or dormant vegetation is an important pathway, although microbial activity in plant tissues substantially reduces the amounts of water soluble P released[121]. The availability to plants of P in plant litter is largely governed by temperature[75,171] and moisture[76], which can adversely affect the rate of P release unless grazing animals are introduced. The role of grazing animals in nutrient cycling[30] and the return to soil of P through plant and animal residues[121] have recently been reviewed.

Invertebrates have a key role in detrital processing and incorporation of residues into soil. The activity of surface-casting earthworms can increase the short term availability of P in plant residues two or three fold[117] mainly by releasing the P_i in plant material by physical disruption. Their casts may contain as much as 9 and 13 kg P·ha^{-1} as P_i and P_o respectively in a permanent New Zealand pasture[154]. The higher bacterial concentrations and greater availability of N and P in the casts than the surrounding soil[70] partly explain the faster decomposition rates associated with earthworm activity. The agronomic significance of this earthworm activity is likely to be especially important for supplying P to soils of low P status, but the effectiveness of P fertiliser can also be increased by earthworm activity[116]. The ecological niche occupied by earthworms in many temperate soils of medium to high fertility has recently been compared with that of termites in warm temperate and tropical regions[164].

The extremely complex interrelationships that exist in the rhizosphere between detrital processing, mineralisation of organically-bound nutrients and nutrient uptake, makes field investigation of these processes difficult. Consequently, an alternative approach has been to use sterilised soil and added consumer and decomposer organisms (microcosms) in various combinations in the laboratory to investigate the role of the microfauna in the soil P cycle.

In the example shown (Fig. 5), rhizospheres were simulated using combinations of bacterial, amoebal and nematode

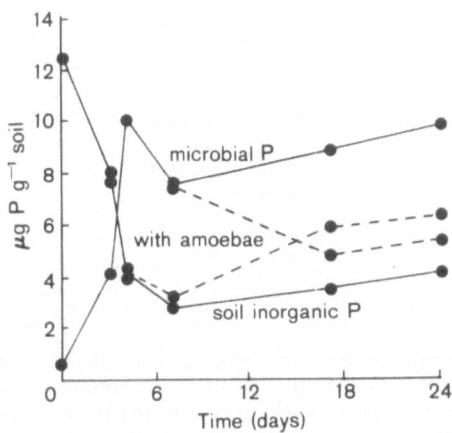

FIGURE 5. Effects of amoebae on levels of NaHCO$_3$-P and microbial P over a 24 day incubation in soil microcosms. From Cole et al.[61].

populations[61]. The bacteria rapidly immobilised labile P$_i$ from the soil as carbon substrates were metabolised, but most of the bacterial P was subsequently mineralised to P$_i$ by the amoebae. In the field, 60% decreases in bacterial populations have been attributed to amoebal grazing[53]. Similarly, nematodes also increase the amount of nutrients in circulation by short-ening decomposition cycles[17,126] and enhancing P$_O$ mineralisation[18]. Invertebrate consumption of detritus and microorganisms is therefore a key process in the cyling of P in soil. The regulatory role of invertebrates in P cycling has recently been reviewed[99]. The microbial cell contents of dead micro-organisms, or those released by amoebal and nematode grazing comprise a range of P compounds including RNA (30-50%), DNA (5-10%), acid-soluble P$_i$ and P$_O$ (15-20%) (e.g. P$_i$, poly P$_i$, sugar and adenosine phosphates, as well as phosphorylated co-enzymes) and phospholipids (<10%). These P$_O$ compounds will all interact to varying degrees with soil colloids, and their mineralisation rates will depend on phosphatase activity as well as on the type of P$_O$ released[125,166].

Consequently, microorganisms in soil comprise an important source and sink for nutrient elements like P, as well as being the main agents of transformation of these elements[16,137,138].

Table 2. Potential P_i uptake rate of the soil flora.

	Estimated fresh weight biomass in 0-10 cm[a] (g m^{-2})	Calculated P_i uptake rate in 10 μM P_i[b] (μmol P^- min^{-1})	Ratio of calculated uptake to that of plant roots
Bacteria	103	359	9.0
Fungi	260	71	1.8
Plant roots	5000	40	1.0

a. Bacterial and fungal biomass based on Clark and Paul[60] assuming a dw/fw ratio of 0.2. Root biomass based on value of 50 cm^{-2} for root length per unit volume[26] and a conversion factor of 1 mg fw cm^{-1} root length.
b. Calculated[32] assuming all fungi behave like A. nidulans, bacteria like E. coli, and plant roots like millet.
From Beever and Burns[32].

Until very recently, no reliable methods were available for determining their P contents[47,92], and the contribution that microbial P makes to plant nutrition still remains to be determined. Although difficult to measure, estimates of the comparative efficiencies of P_i uptake from soil by microorganisms and plant roots have indicated that microorganisms, and especially bacteria, have very efficient P uptake mechanisms (Table 2). The uptake, storage and utilisation of P by fungi have recently been reviewed[33]. Decomposer uptake of P exceeded plant uptake 3 to 5 fold in a simulation model for P cycling in a semi-arid grassland[63]. Competition for the small amounts of P in the soil solution is most intense in the rhizosphere, where substrates from root exudates, sloughed off root cells, tissues and mucigels sustain a larger and more active microbial population than in the bulk soil[83].

The part of the P cycle of most immediate importance to plants and microorganisms can be summarised by the following reaction:

$$\text{soil solution P} \rightleftharpoons \text{labile } P_i \rightleftharpoons \text{non labile P}$$

Uptake of soil solution P (ca. 1.5 μM[24]) results in a several thousand-fold concentration of P_i by plants (ca. 0.5-25 mM P_i[37]) for example and is usually limited by the slow

rates of P diffusion through soil[23] creating depletion zones
close to roots[36,115]. The P removed from solutions by plants and
microorganisms is rapidly replenished from the solid phase
labile P_i, defined as the P fraction that can enter the soil
solution by isotopic exchange in an appropriate time scale[114].
This labile P_i is held on clay minerals, the oxides and
hydroxides of Al and Fe, and on organic matter, by relatively
weak bonds compared to the non-labile P_i and P_o. The non-
labile P_o forms part of recalcitrant humic polymers that may
require depolymerisation before enzyme catalysed cleavage of
phosphate esters can occur. Some P may not be readily desorbed
because of strong bond formation with reactive Fe and Al
gels[180] and through Al, and to a lesser extent Fe, in metal-
humus complexes[95,179]. In neutral and calcareous soils, calcium
phosphates are formed and if calcite ($CaCO_3$) is present,
insoluble hydroxyapatite and calcium phosphates are adsorbed
onto the calcite surfaces.

The physico-chemical adsorption-desorption mechanisms that
involve exchange of the P between the soil solid and solution
phase are complex[184], and participation in these mechanisms of
plants and microorganisms, through the production of organic
acids for example[127] could help to maintain the P_i concentration
of the soil solution in the rhizosphere (see Grinsted et al.[84]).
Several excellent reviews of soil P chemistry provide further
information on P_i in soil, and its reactions[28,114,135,184].

Although P_o may be directly taken up by plants (e.g.[125]) and
microorganisms (e.g.[32]), there is general agreement that the
organic forms of P must first be mineralised through the action
of extracellular phosphatases. Their activity is usually
highest when soil solution P concentration is low[158], because
phosphatases are adaptive enzymes produced in response to a
need for P by soil microorganisms and plant roots[35]. Higher
phosphatase activities are also usually found in rhizosphere
compared to non-rhizosphere soil because of higher microbial
populations, higher phosphatase activities or rhizosphere
organisms, and the presence of plant root phosphatases[82].

As well as the production of extracellular phosphatases,
plants have developed other strategies, such as root branching,
root hairs, and symbiotic relationships with mycorrhizal fungi
to enable a large volume of soil to be exploited for P and
other nutrients. Mycorrhizal fungi in particular perform an
important function in improving the P nutrition of plants of
economic importance[3,133], by taking up P from a larger soil
volume than is possible for uninfected plants. Mycorrhizal
fungi can also mineralise P_o[81].

In summary, the equilibrium concentration of P in the soil
solution, and the capacity of the soil to maintain this
concentration against uptake by plants and microorganisms, are
governed by complex physico-chemical and biochemical mechanisms
that release P_i from inorganic-bound forms as well as from
organic matter. McGill and Cole[122] recently argued that these
biochemical mechanisms differed from those responsible for the
mineralisation of C, N and C-bound S from organic matter where
the need of soil organisms for energy is the driving force.
The latter process was termed biological mineralisation. The
release of P and S in esters on the other hand is strongly
controlled by the supply of, and need for the element which
is released from soil organic matter by extracellular enzymes
(biochemical mineralisation). These concepts, portrayed in
Fig. 6, can accommodate the variability of $C:N:S:P_o$ ratios and
the greater variability for P_o than for S in soil organic
matter, and provide a rationale for understanding the
stability of P_o in different soils.

6. MINERALISATION AND IMMOBILISATION

The mineralisation and immobilisation of P in soil constitutes the major biological processes controlling its
availability to plants. These are principally microbial
processes and have been identified, by simulation modelling
of the P cycle in undisturbed[63] and managed[41] grassland soils,
as parts of the P cycle where major gaps in knowledge occur.

The rate of mineralisation rather than the amount of P_o in
soil is the main factor that determines its availability to

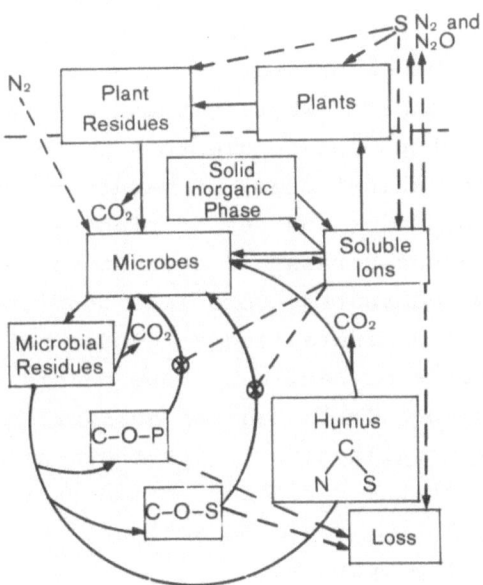

FIGURE 6. Interrelation of C, N, S and P cycling within soil-plant systems. A schematic illustration from McGill and Cole[122].

plants. Measurements of mineralisation rates have usually been made by monitoring the change with time of P_i or P_o in soil after incubation; generally only net effects are measured unless radio-isotopes are used, e.g.[80]. Various factors can affect the precision of these measurements, including soil heterogeneity, the effect of strong extractants on labile P compounds, and the difficulty of accurately estimating recoveries of P_i and P_o from some soils[90]. A new method was recently developed[90] for measuring the mineralisation rates of labile P_o in soil. This involves the incubation of [^{32}P] RNA in soil at 13°C for 24 h, and enables immobilisation and mineralisation to be measured separately. Comparative rates of mineralisation in woodland soils were measured satisfactorily by this technique[90], whereas a more conventional procedure using $NaHCO_3$ extraction of P[57,85] was not sufficiently precise.

The factors controlling the balance between mineralisation and immobilisation are those that control the population dynamics and activities of soil microorganisms, and include temperature, moisture, aeration and the soil reaction. The composition of the P_O in soil also affects the overall mineralisation rate, as demonstrated recently in a New Zealand soil development sequence[166], where climate was the main soil forming variable. Orthophosphate diesters including choline phospholipids and phosphonolipids, of mainly microbial origin, were considered to be major sources of plant-available P in these undisturbed tussock grassland ecosystems; orthophosphate monoesters, which predominated, were less sensitive to differences in climate and mineralised at slower rates.

Mineralisation rates of labile P_O were mainly related to soil pH, and extractable Ca content of 50 English woodland soils[91], where P_O mineralisation limits the availability of P to plants. Over 90% of the variation in mineralisation rates, measured using $[^{32}P]RNA$[90], could be accounted for by the soil chemical and physical properties.

Several of the more important factors that influence the balance between, and the rates of, mineralisation and immobilisation processes in soil are included in the discussion that follows.

7. SOME IMPORTANT FACTORS INFLUENCING P TRANSFORMATIONS
7.1. Temperature

The balance between the mineralisation and immobilisation of P in soil is strongly influenced by temperature, which causes the marked seasonal fluctuations observed in labile P levels of temperate[152] and cool-temperate grassland soils[85], and of total P_O levels of cool-temperate cultivated soils[69]. Under cultivation however, P_O is apparently more stable than C, N or S in temperate climates but not in tropical climates[66].

High P_O mineralisation rates are observed above 30°C[66], coinciding with optimum temperatures for growth of many bacteria[160]. However, some soil phosphatases (e.g. phytases) apparently have optimum temperatures around 45°C so that cell-free phosphatases may make a contribution to the mineralisation process at higher soil temperatures[19].

The availability to crops of the P in crop residues is governed largely by temperature[171]. At high temperatures and adequate available moisture levels, these residues can be just as effective as fertiliser at supplying P to crops[1]. Consequently, in tropical soils where P_o levels are generally lower than in temperate soils, P_o has been used as an index of P availability[6]. The nature of the organic residues can, however, influence P_o mineralisation rates in tropical[20] as well as in cool-temperate soils. For example, mineralisation of the P in sheep faeces was less adversely affected by low temperatures than the P in plant residues in Scottish hill soils[75], suggesting that increased grazing by sheep would increase nutrient cycling.

At temperatures below 30°C, immobilisation of the major plant nutrients by soil microorganisms is increasingly favoured, with cold-tolerant species like yeasts playing a more important role[59].

In Arctic tundra soil, temperatures remain low for much of the year, and annual P fluxes through the organic matter are only 15-40% of those in temperate grassland soils. Here, temperature indirectly limits decomposition by affecting the rate of microbial biomass increase, following the periodic crashes in population that occur seasonally during freeze-thaw cycles[52].

7.2. Moisture

In general, decomposition of organic residues occurs at maximum rates over a wide range of water contents[103]. The P in sheep faeces and in plant material also mineralises at similar rates over a wide range of moisture contents[76], although inconsistent effects of moisture on the release of P from plant material added to soil have also been observed[40]. However, the mechanisms controlling seasonal fluctuations in soil P_o levels can have a very narrow tolerance to soil moisture[183]. The small amount of data available on the effect of soil moisture on P transformations[66] indicates that drying soil increases the amount of extractable P_i without having a consistent effect on P_o[39].

Desiccation of soil results in the release of significant

amounts of both P_i and P_o from dead microbial cells[47] and plant
tissues[130], as well as P_o leaked from still viable microbial
cells[48] and autolysed plant material[107,119]. Fungi and actinomycetes are both better able to function under stress conditions
than bacteria, and are the predominately active group in dry
soil (i.e. between -8 and -20 bar[192]); bacterial activity
becomes important above -8 bar.

When soil is rewetted following desiccation, P in the
assortment of cell debris and root tissues, and discrete P_i
and P_o, should rapidly re-enter the P cycle to be utilised by
microorganisms and plants, and adsorbed by soil colloids.
Peaks in bacterial population occur soon after rainfall[50,54],
indicating that the available substrates that accumulated
during the preceding dry spell are rapidly being utilised, and
P and other nutrients immobilised in fresh microbial tissue.
In a Canadian dryland fallow loam soil, soil moisture increases
following rainfall were shown[51] to cause bacteria to proliferate
and soil P_o to mineralise in the 0-2.5 cm depth of the first
year fallow. Large and rapid increases in microbial activity
that accompany rewetting of dry soil are also well known[5,136],
although the mechanisms involved are not well understood.

Wetting and drying cycles greatly stimulate the mineralisation of organic matter. In a laboratory incubation experiment,
repetition of these cycles eventually led to complete
mineralisation of P_o, but only partial mineralisation of
organic C and N[38].

7.3. Oxygen

Waterlogging of soil, unless only temporary, rapidly causes
a general decline in microbial activity[195], because the amounts
as well as the diffusion rates of oxygen are much lower in
water than in air. Bacteria become the predominant active group
of microorganisms as anaerobic pathways replace aerobic
pathways.

The mineralisation of P_o is generally encouraged, however,
when conditions become anaerobic[7,25,66,146], so that poorly drained
soils often contain less P_o than comparable well drained

soils. In poorly drained acid soils, the pH may eventually stabilise at about pH 6.5 if sufficient Fe is present[142,149], providing conditions more favourable to microbial activity than in the undrained soils. However, P_o mineralisation rates are usually slower in anaerobic soils[68], possibly because the increased solubility of Fe- and Al-bound P_i inhibits phosphatase production and activity. There are, however, some reports of higher phosphatase activities in saturated than in dry soils[158].

The increased availability of P_i in waterlogged soils is well known[108]. Lowland rice, through a special physiological mechanism, is able to regulate the availability of the P_i near its roots by controlling the degree of aeration in the rhizosphere[9].

P deficiency in dryland crops can be induced by draining previously flooded soils[187], and this has been attributed[186] to an increase in P sorption capacity of the soil after drainage rather than to immobilisation of P_i.

Even in well aerated soils, anaerobic microsites may develop for a time in the vicinity of decomposing organic matter, for example, where oxygen demands from intense microbial activity may exceed the ability of the soil solution to supply oxygen by diffusion. To take advantage of the increased nutrient availability from anaerobic microsite formation in agricultural soils[157], aerobic microbial activity should be encouraged to stimulate mineralisation processes, but not excessive nitrification rates which limit Fe reduction by the production of high nitrate concentrations. This could be achieved through the use of minimum tillage, better use of organic amendments, and the chemical inhibition of nitrification[157].

The effects of aeration on the transformation of P in soil are clearly very complex, and many of the interacting factors, including the effects of aeration on microbial biomass and activity, need close attention before the reasons for the greater availability of P in anaerobic soils and sites can be fully understood.

7.4. Clays

Humus stabilised in organo-mineral complexes appears from modelling studies[106,137] to be of considerable importance for the medium-term cycling of nutrients. As for S[11], some potentially labile P_o, including nucleic acid-P for example[168], is stabilised with more recalcitrant forms (e.g. inositol phosphates) in these organo-mineral complexes, to be released when the soil is cultivated and aggregates are broken open[188].

The P_o content of a number of Canadian and Ghanaian soils were found[20] to be highly correlated with their clay contents, and in general, clay soils usually contain higher levels of P_o than comparable lighter textured soils[109]. Andosols, for example, with their high amorphous aluminosilicate (e.g. allophane) contents, accumulate large amounts of very stable organic matter due to the strength of the organo-mineral bonding[167,196]. The recalcitrance of inositol polyphosphates is usually attributed to their propensity for salt formation with metal cations, especially Fe and Al, in acid soils, because in the absence of clays the P in these compounds is readily assimilated by plants indicating rapid mineralisation[15].

The metabolic activities of soil microorganisms are also significantly altered by surface adsorption on clays[49], although very little is known yet about the interactions that take place between the indigenous population and soil surfaces. Soil phosphatases are also intimately associated with the soil organo-mineral complex[129] and their adsorption on clay surfaces usually but not invariably[147] reduces enzyme activities. The mechanisms of adsorption of proteins, including enzymes, on clays and the effects on their properties have recently been reviewed[168].

The influence of the amounts and types of clays and clay colloids in soil on the biological transformations of P have received little attention to date, mainly because of the complexity of the soil system, and a lack of suitable techniques.

In an aggregation model proposed[173] for Australian red-brown earths, organic binding agents were classified into

(a) transient, mainly polysaccharides (b) temporary, roots and fungal hyphae and (c) persistent, resistant aromatic components associated with polyvalent metal cations, and strongly sorbed polymers. Apart from the inositol polyphosphates that are known to reside in the persistent fraction, a proportion of the remaining P in most soils is also bound to carbohydrate, such as deoxyribose in DNA and its nucleotides, and acidic poly-saccharides[58]. By visualising the distribution of soil P_o in physical terms as described by Tisdall and Oades'[173] aggregate model, experiments could be designed to provide a better appreciation of the way land cultivation affects mineralisation of P_o.

7.5. Carbon

The C, N and S contents of the organic matter in most soils are quite well correlated[111] whereas the P content can vary widely[66], largely because the P_i group as well as the organic moiety is responsible for the stability of P_o in soil[122]. Consequently, the various attempts at using critical $C:P_o$ ratios of soil organic matter[72] and plant material[73] to indicate the balance point between net mineralisation and immobilisation have, with few exceptions[110] been unsuccessful. Critical $C:P_o$ ratios can range[122] from <100 for soils well supplied with available P to over 200 for P deficient soils. Although ratios over 300 usually indicate net immobilisation, these can vary from site to site[29,171].

A very wide range of $C:P_o$ ratios, from <1 to >1000 is theoretically possible for soil organic matter, depending on its chemical composition, and the amount of P_i associated through cation bridges with humic substances[185]. The ratio for a 1:1 mixture of inositol hexa- and pentakisphosphate for example, both of which are not readily mineralised in soil, is 0.42, whereas for hydroxy Al- organic matter complexes that make a significant contribution to P_i adsorption in acid soils[42], values range from about 100 to over 1000[185].

White[185] argued recently that the balance point between mineralisation and immobilisation of P during organic matter decomposition, whether as fresh residual or recalcitrant humic substances, depends on the C:P ratio of the substrate actually being utilised by microorganisms, as well as the microbial growth yield (CF) and the C:P ratio of the microbial cells (Cm:Pm). A critical C:P ratio of the substrate of between 50 and 70 was estimated from the equation:

$$P:C \text{ substrate} = (1-CF)\frac{Pr}{Cr} + \frac{CF}{Cm:Pm}$$

using literature values of 0.5 for CF[102], 50 for Cm:Pm[8] and 0.009-0.02 for Pr:Cr[67,169]. Pr:Cr is the ratio of the P mineralised to the C respired as CO_2 in a given time. The range of critical C:P ratios predicted from the equation needs further testing, but it agrees well with the ratio of 55 reported[27] for the mineralisation of sodium β-glycerophosphate. The equation can also indicate the extent to which the P_i in soil solution is buffered by desorption of P_i from soil surfaces during decomposition of a simple substrate of known C:P ratio.

The influence of added carbon as dried grass, and as cellulose (with and without added P) on the transfer rates of P between P_i, P_o and biomass P was investigated recently[56] in a laboratory incubation experiment, using a Chernozemic black soil. Periodic additions of C every 30 d caused immobilisation of P by microorganisms, and concomitant decreases in labile P_i and P_o (Fig. 7). At the end of 3 months, 39 and 22% respectively of the P added in grass and with cellulose were found in organic forms. Although extrapolations from laboratory experiments to the field should be viewed with caution, the marked uptake of P by microorganisms indicated in Fig. 7, in response to added C, may help to explain the substantial seasonal changes observed in some field soils[69,85]. In a subsequent greenhouse incubation experiment with two Chernozem soils, lack of labile P_i eventually slowed down the decomposition of added C, but this effect was partially compensated by increased mineralisation of P_o[57], especially in the soil having low P_i status. Microbial P and solution P values were significantly correlated in this experiment, in which microbial C:P ratios ranged from 12 for the soil having a large reserve

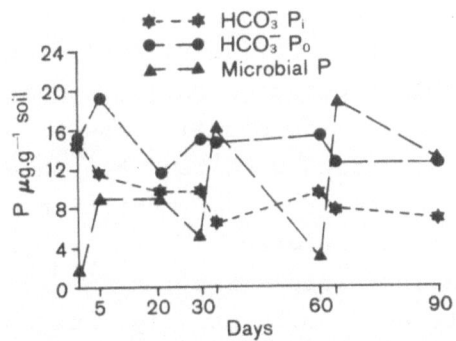

FIGURE 7.Changes in HCO_3^--extractable P_i and P_O and in microbial P in a Chernozemic black soil (A_p horizon) to which
cellulose and N were added every 30 d. From Chauhan et al.[56] .

of available P_i to 45 for the soil with limited reserves of
P_i.

These laboratory investigations have recently been extended
using a sequential extraction technique (Fig. 4) to investigate
the mechanisms involved in changes in P_i and P_O resulting from
long-term cropping in a wheat-wheat-fallow rotation[93], compared to an adjacent permanent pasture. Most of the P lost as
a result of cropping (74% of P_T lost) was P_O and residual P.
Of the P_T lost, 22% was from extractable P_O and 52% from stable
forms, indicating considerable microbially mediated mineralisation of these P_O forms.

These investigations have demonstrated the central role of
microorganisms in P mineralisation-immobilisation processes,
and the effect that additions of C with and without added P
can have on them. Further field investigations are now needed
to test the hypotheses formulated in these laboratory studies,
so that the influence on P cycling and P supply to crops of
seasonal changes, and different systems of management, can be
better understood.

7.6. pH
As for organic matter in general, soil organic P is less
stable in neutral or alkaline than in acid soils[72] because the
diversity and activities of soil organisms decline with

increasing soil acidity. Variations in the levels of soil P_o in calcareous soils have been[176] significantly correlated with the P contents of plants growing on them, whereas no such correlation was observed for acid soils. The P in animal manure was more effective in calcareous soils[2] than in acid soils because of a combination of greater penetration of P_o into the profile[88], high microbial activity, and the rapid cycling of P between organic and inorganic forms[1].

Addition of lime to acid soils usually, but not invariably[4], enhances the release of P_i from organic matter by encouraging microbial activity and increasing rates of C, N and P mineralisation[86]. In soils developed under pasture, liming commonly increases earthworm numbers, so that more plant available P would be expected at the soil surface in casts[154] as a result of greater burrowing activity.

Despite these benefits of lime, the P_i released as a result of heightened microbial activity may not be immediately available to plants. During the course of a long-term field investigation of the annual superphosphate requirement of an irrigated New Zealand pasture soil for example (see Fig. 8), lime was added (equivalent of 4 t ha^{-1}), to raise the pH in the top 75 mm from 5.8 to 6.5[145]. Net immobilisation of P as P_o then ceased and a build-up in P_i occurred without a corresponding increase in dry matter production. The lime-induced mineralised P_o accumulated as insoluble Ca-bound P, which was expected to release P_i gradually by dissolution as the pH again declined. Forms of P in NaOH extracts of this topsoil before and after liming were found to be similar by ^{31}P-nmr (K.R. Tate and R.H. Newman, unpublished results), and mainly comprised orthophosphate monoesters. These results suggest that mineralised P_o came mainly from recently incorporated organic matter, but probably also from the more recalcitrant native P_o by increasing its solubility[145].

Soil pH may not be a reliable index of the availability to plants of the P_o in the rhizosphere, as pH at the root surface can differ from that in soil a few mm away by as much as 1-2 pH units[132]. For example, Hedley et al.[94] recently demonstrated that no P was supplied by hydrolysis of P_o in the rhizosphere of rape (<u>Brassica napus</u>) plants growing in a P deficient soil for 41 d, despite a marked increase in activity of soil phosphatase. Instead, the H^+ released by roots caused a sharp pH drop of 2.4 units, which was mainly responsible for dissolving non-exchangeable P_i to satisfy the plants P requirements.

Soil pH, then, provides only a rough guide to the likelihood of P being released to plants from mineralisation of organic phosphorus, because the plant itself can exploit some of the P in the rhizosphere, including recalcitrant forms, by influencing the chemistry close to the root surface.

7.7. <u>Inorganic nutrients (P, N and S)</u>

Various effects on P transformations of adding P_i to soil have been reported, including net immobilisation as P_o[29,66,93,145], net mineralisation[66] and no effect[66]. Although the reasons behind these conflicting reports are not altogether clear, field experiments involving growing plants usually show immobilisation of P as a result of P_o synthesis from stimulated plant growth. In the absence of plants on the other hand, added P_i may displace P_o from soil surfaces and favour mineralisation. Indeed, mineralisation can vary markedly depending on whether plants are present or not[41], indicating that results from laboratory incubation experiments may not be safely extrapolated to the field.

Previous fertiliser history can also strongly influence the mineralisation of P_o through its affect on soil phosphatase activity[159]. The common practice of using P_o to monitor the effects of adding P_i to soil could also mask changes in different P_o fractions[66]. Inositol polyphosphate contents, for example, were increased by P fertiliser application in some acid soils of the U.S.S.R., but no effect was found in some English arable soils[134]. No consistent trends were observed in either the total amounts of inositol phosphates, or in their proportion of the total soil P, in two groups of fertilised and unfertilised Canadian and Ghanaian soils[20]. No differences

358

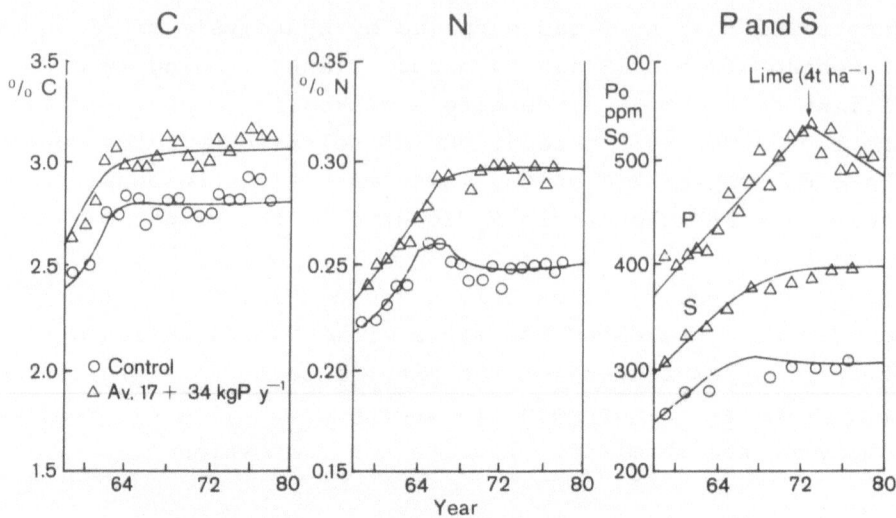

FIGURE 8.Changes in soil organic C, N, P and S in the control, and the average of the 17 and 34 kg P y^{-1} treatments, of a long-term superphosphate trial on a Udic Ustochrept. Adapted from Quin and Rickard[145].

were detected by ^{31}P-nmr in the P_o fractions of alkali extracts of soils receiving 0, 17 and 34 kg P $ha^{-1}y^{-1}$ (K.R. Tate and R.H. Newman, unpublished results) used in a long-running New Zealand superphosphate trial[145] (see Section 7.6. pH). In these soils, annual net immobilisation (Fig. 8) of P of about 4, and 10 kg P ha^{-1} (0-75 mm) was observed[145] in response to the super-phosphate additions. Although not measured, changes in the more labile P fractions could have occurred. However, no effect on the P_o fractions of a Chernozem soil of adding P_i (as KH_2PO_4, equivalent to 20 kg P ha^{-1}) over 9 months was detected in an incubation experiment[56].

Under pasture, immobilisation of fertiliser P into organic forms can be rapid[41]. Accumulation of C, N and S had ceased while accumulation of P_o continued on all treatments (Fig. 8), providing further support for McGill and Coles'[122] 'elastic' model for C, N, S and P cycling in soil. In the absence of added fertiliser and lime, soil P_i in grass-legume pastures would be expected to decrease at the expense of P_o until N became limiting, due to the failure of the P-dependent legumes[181].

Although different nutrients cycle at characteristically different rates in the same ecosystem, the amounts of biologically active N and P will be in balance in an ecosystem in a steady state condition[62]. A ratio of 8:1 for the levels of biologically active N (24 g N m^{-2}) to P (3 g P m^{-2}) was estimated for the top 20 cm of a semi-arid grassland soil[62]. This hypothesis now needs further testing in a range of different ecosystems.

The influence of N and S on P immobilisation and mineralisation has received little attention compared to the effects of P on N transformations[62]. In a non-legume cropping system, increased immobilisation of P would be expected in response to additions of N or S, based on the approximate stoichiometry of the major nutrient elements, while some P_o mineralisation should occur when these elements are withheld[66]. However, predictions based on an assumed stoichiometry between P and N can be misleading, because P_o may be relatively more stable than N and C in soils under prolonged cultivation[170]. This stability has been attributed to either an inherent property of the biological-biochemical mineralisation processes, during which P_o is mineralised on demand rather than as a consequence of C mineralisation[122], or to biocycling of P through deep rooting plants causing a relative P enrichment in the topsoil[22]. Care is also needed in using C:N:S:P ratios of plant residues as indices of their potential to mineralise[172].

The mechanisms involved in the interaction of the major nutrient elements are clearly very complex[122] and much more research is needed before the influence of P transformations of added P, N, or S can be confidently predicted.

8. AVAILABILITY OF P_o TO PLANTS

Evidence for the mineralisation in soil of P_o does not by itself indicate that the P_i released will be immediately available to plants. Much of this P_i will eventually be incorporated into plant tissue, but in the short term (of a growing season) sorption by clay colloids and to humic substances through metal cations, as well as re-immobilisation by microorganisms also occur. Despite occasional reports[101,125] that P_o may be directly absorbed by plant roots, it is generally accepted that mineralisation of P_o is a prerequisite for its availability to plants.

In undisturbed ecosystems, where the P cycle is closed, most of the P entering plants is supplied from recycling of plant residue P through microbial processes in the soil, e.g.[52,98]. Woodmansee and Duncan[194] recently found that the P mineralised from litter and dead roots in an unfertilised annual grassland in U.S.A. was sufficient to satisfy the P requirements of the plants, substantiating the importance of microorganisms in the mineralisation process. Similar conclusions were reached for two unfertilised phosphorus-deficient British ecosystems by Harrison[89], one an oak-ash woodland and the other an upland grassland.

Some of the more readily mineralised forms of P_o in a soil development sequence were recently identified by a ^{31}P-nmr technique[166]. Orthophosphate diesters, including phospholipids, and phosphonates, but not orthophosphate monoesters, appeared able to supply plant-available P in these mainly undisturbed New Zealand tussock grassland ecosystems, given soil and climatic conditions favourable to P_o mineralisation. Similarly, orthophosphate diesters were more readily mineralised than monoesters in an old grassland soil after 21 y of bare fallowing, because the diesters declined sharply during this period[143].

In agricultural ecosystems, the P cycle is more open than in undisturbed ecosystems, because some losses occur by the removal of agricultural products and in some situations by surface runoff and erosion. Applications of fertiliser are needed to compensate for these losses, as well as to raise the level of production above native levels. Where grazing animals are present, unevenly distributed returns of P through dung and, to a small extent, urine also occur. Consequently, in managed ecosystems the natural P cycle is modified by these additional factors, so that the importance of P_o mineralisation for plant nutrition is more difficult to assess. This is why

Cole et al.[63] chose a natural ecosystem for their simulation modelling studies of the soil P cycle.

Long-term monitoring of agricultural systems has nevertheless provided strong evidence of the importance of P_O in soil for crop growth[170].

Recent estimates of the average annual rates of mineralisation of P_O in three English arable soils, including two from Rothamsted Experimental Station, indicated that about one-third of the average P removal in crops could be provided from the organic matter by mineralisation[55]. The organic matter in British arable soils had earlier been estimated[78] to supply about 6 kg P ha^{-1} to crops annually, while for grassland soils, about 15 kg P ha^{-1} could be supplied from mineralised P_O. A similar value for an unmanured arable soil at Rothamsted was recently obtained by Jenkinson and Ladd[104]. They calculated a P flux through the microbial biomass of 4.6 kg P ha^{-1}y^{-1}, which was compared with about 5 kg P ha^{-1}y^{-1} removed in grain-plus straw. The calculation was based on the known flux of carbon[106] through the microbial biomass in this soil and an assumed C:P ratio for this biomass of 50. Just how much of this biomass P reaches the plant, and how much is re-immobilised in other microbial cells in soil in a growing season are as yet unanswered questions. Microorganisms, however, have a higher potential for P_i absorption than plant roots[32]. From their simulation model of a native semi-arid grassland, Cole et al.[63] predicted that the transfer of P between soil solutions and microorganisms was three times larger than the transfer between soil solution and roots. Stewart et al.[162] recently demonstrated that the ^{33}P in labelled bacteria was readily taken up by plants. In 26 d, they found that 38% of the ^{33}P was lost from the microbial P pool, 20% of the added ^{33}P had moved to the resin-extractable form (70% of this was taken up by the plant), 4% was in the more resistant P form and the remaining ^{33}P was distributed in the labile and moderately labile fractions. Although the ^{33}P labelled bacteria would have decomposed more rapidly than a native soil population, the experiment nevertheless underlines the important role of microorganisms in the P cycle, and demonstrates the availability of microbial P to plants.

Other field observations have also indicated that organic matter is an important source of P to plants. The marked seasonal variation in the response of grass-clover pastures to P fertiliser in New Zealand[152], for example, is well known. In spring, when high growth rates are sustained, the high P status of the soils is probably due to the P released by mineralisation of organic matter. Similarly, Dormaar[69] attributed the rapid decline in P_O levels from April to May in an unfertilised Canadian Chernozem to the mineralisation of P_O, following the build-up over the previous winter. Large seasonal fluctuations in the labile (NaHCO$_3$-soluble) P_O were also observed by Halm et al.[85] in a cool-temperate grassland soil, with increases coinciding with the period of maximum growth and highest microbial activity.

When laboratory incubation procedures are used to measure P_O mineralisation they usually underestimate the potential utilisation of this P by plants[67]. Recent attempts have been made[40,67] to quantify the re-utilisation by plants of the P in plant residues. Blair and Boland[40] have shown from field experiments that the net re-utilisation of P from ^{32}P-labelled white clover residues, after 48 d in a soil with an NaHCO$_3$-soluble P level of 25 ppm, was 29.3% in the presence of growing oat plants. In a soil without plants, and with an NaHCO$_3$-soluble P level of only 8 ppm, net re-utilisation was only 0.6% because of significant soil immobilisation. The plants took up the plant residue P to a greater extent than the native soil P in the low P soil, but in the high P soil, both the clover residue P and native soil P were taken up by equal amounts after 24 d. These results, and those of Dalal[67], demonstrate the necessity of incorporating plants in experiments designed to assess turnover rates of P in soils. They have also indicated that much of the P in plant residues can rapidly re-enter the labile soil P pools for re-utilisation by growing plants.

In some ecosystems, the re-utilisation of nutrients from roots may be even more important than from above ground plant

residues, as studies incorporating root mortality data from
forests have suggested[77]. The P in roots of crop residues labelled with ^{32}P is as rapidly utilised by growing plants as the P
in tops[67].

The effects on plant nutrition of discrete organic phosphorus
compounds have also been used to demonstrate the availability
to plants of soil P_O[118,125]. Inositol polyphosphates[118] are capable
of supplying P to plants in soils with low P retention, but
not in soils with high P retention properties where the formation
of insoluble salts, or sorption by clays, make them totally
resistant to mineralisation. The mineralisation of phosphatidyl
choline and DNA when added to a P-deficient Chernozem soil,
supplied P to barley plants in pot experiments. Phosphatidyl
choline was an inferior source of P compared to an equivalent
amount of P applied as KH_2PO_4, whereas DNA and KH_2PO_4 were
similarly efficient at increasing dry matter and P content per
unit of plant biomass[125]. Both forms of organic phosphorus,
chosen as models for soil phospholipid-P and nucleic acid-P,
induced unusual but dissimilar physiological effects in barley
plants. McKercher and Tollefson[125] speculated that phosphatidyl
choline may have been directly absorbed by the plant roots to
cause the unusual physiological effects. In natural soils,
however, there is no evidence to indicate that phospholipid-P
is found largely in low molecular weight forms. On the contrary,
Tate and Newman[166] observed that choline phosphate was formed
in soil extracts by alkaline hydrolysis of higher molecular
weight phospholipids. It also seems likely that in soils, the
mineralisation of phospholipid-P would be preceded by depolymerisation of these high molecular weight P compounds, before
the P and C they contain is released by soil microorganisms[112].

9. CURRENT PERSPECTIVE

The rapidly escalating cost of P fertilisers, and the problems created in inland waterways by nutrient enrichment from
agricultural land, have created an urgent need for more reliable
information on the P cycle in soil. The contribution made by
decomposing plant residues and soil organic matter to plant

364

nutrition in a growing season is now believed to be considerable and it is in this part of the P cycle that quantitative
information is most lacking.

The central role of soil microorganisms in the cycling of
P, as in nutrient cycling in general, is well established.
Quantitative description of this role is now possible by the
use of recently developed methods for measuring how much P is
held in microbial cells. However, measurement of the fluxes of
P through the microbial biomass under different environmental
and management conditions still awaits further development of
methods, such as that of Harrison[90] involving the use of radio-
isotopes.

Up until the late 1970s, lack of an integrated conceptual
framework for investigating the P cycle was a major factor
limiting progress towards a better understanding of the key
processes involved. Simulation modelling however has provided
this framework, enabling the various P pools and processes in
soil to be described, and the gaps in knowledge to be identified. A modelling approach also proved useful for examining
the interrelationship between the P cycle and C, N and S
cycles, acting as a further stimulus for more research on soil
P.

Practical expression of this integrated approach to the P
cycle has emerged in the form of a new soil P fractionation
scheme[93] that distinguishes between the different soil
inorganic and organic P fractions as well as the microbial P.

These fractions vary in the extent of their availability
to growing plants, providing for the measurement of short-
term P transformations in soil that can be related to long-
term changes in the field. Further elucidation of the
mechanisms involved will undoubtedly be aided by the use of
^{31}P-nmr, which can distinguish between the different chemical
forms of P_o in soil extracts[131,166].

The availability of plants of P_o, however, depends mainly
on mineralisation rates rather than on the amounts of P_o in
soil. A number of factors collectively influence these rates,
including the plant itself which actively competes for P with

soil organisms in the rhizosphere. Despite its complexity
more attention will need to be given in future to investigations
of P transformations in the rhizosphere of growing plants
rather than in bulk soil, if the amount of P reaching the
plant is to be distinguished from that being recycled within
rhizosphere microbial populations. The use of radioisotopes
will undoubtedly be needed to resolve such questions.

A better appreciation of the rates and pathways of P through
the organic matter in soils, and of the interaction between the
biological and physico-chemical processes that control the P
cycle, should lead to the refinement of simple predictive
models like Decide[34] and Superchoice[64] which both already
provide objective advice to farmers on the use of fertiliser P.

10. ACKNOWLEDGEMENTS

I thank my colleagues, Drs D.J. Ross and T.W. Speir for
helpful discussions, and Mrs J.E. Davin for assistance with
references.

11. REFERENCES

1. ABBOTT J.L. 1978. Importance of the organic phosphorus
   fraction in extracts of calcareous soil. Soil Science
   Society of America Journal, 42, 81-85.
2. ABBOTT J.L. and TUCKER T.C. 1973. Persistence of manure
   phosphorus availability in calcareous soil. Soil Science
   Society of America Proceedings, 37, 60-63.
3. ABBOTT L.K. and ROBSON A.D. 1982. The role of vesicular
   arbuscular mycorrhizal fungi in agriculture and the
   selection of fungi for inoculation. Australian Journal of
   Agricultural Research, 33, 389-408.
4. ADAMS S.N. and DICKSON D.A. 1973. Some short-term effects
   of lime and fertilisers on a Sitka Spruce plantation. I.
   Field studies on the forest litter and the uptake of
   nutrients by the trees. Forestry, 46, 31-37.
5. ADU J.K. and OADES J.M. 1978. Physical factors influencing
   decomposition of organic materials in soil aggregates.
   Soil Biology & Biochemistry, 10, 109-115.
6. AGBOOLA A.A. and OKO B. 1976. An attempt to evaluate
   available phosphorus in Western Nigerian soils under
   shifting cultivations. Agronomy Journal, 68, 798-801.
7. AHMED B. 1976. Mineralisation of inositol hexaphosphate in
   soil at varying static moisture levels. Plant and Soil,
   44, 253-256.
8. ALEXANDER M. 1977. Introduction to soil microbiology. 2nd.
   Ed. John Wiley & Sons Ltd.

9. ALVA A.K., LARSEN S. and BILLE S.W. 1980. The influence
   of rhizosphere in rice crop on resin-extractable P in
   flooded soils at various levels of P applications. Plant
   and Soil, 56, 17-25.
10. AMER F., BOULDIN D.R., BLACK C.A. and DUKE F.R. 1955.
    Characterisation of soil phosphorus by an ion exchange
    resin adsorption method and ^{32}P equilibration. Plant and
    Soil, 6, 391-408.
11. ANDERSON D.W., SAGGAR S., BETTANY J.R. and STEWART J.W.B.
    1981. Particle size fractions and their use in studies of
    soil organic matter. I. The nature and distribution of
    forms of carbon, nitrogen and sulphur. Soil Science
    Society of America Journal, 45, 767-772.
12. ANDERSON G. 1967. Nucleic acids, derivatives and organic
    phosphates. Chapter 3 in Soil Biochemistry, Vol. 1. Eds.
    McLaren A.D. and Peterson G.H., Marcel Dekker.
13. ANDERSON G. 1970. The isolation of nucleoside diphosphates
    from alkaline extracts of soil. Journal of Soil Science,
    21, 96-104.
14. ANDERSON G. 1975. Other phosphorus compounds. Chapter 4
    in Soil components, Vol. 1, Organic components. Ed. Giese-
    king J.E., Springer-Verlag.
15. ANDERSON G. 1980. Assessing organic phosphorus in soils.
    Chapter 15 in The Role of Phosphorus in Agriculture. Eds.
    Khasawneh F.E., Sample E.C. and Kamprath E.J. American
    Society of Agronomy, Madison, Wis., U.S.A.
16. ANDERSON J.P.E. and DOMSCH K.H. 1980. Quantities of plant
    nutrients in the microbial biomass of selected soils.
    Soil Science, 130, 211-216.
17. ANDERSON R.V., COLEMAN D.C., COLE C.V. and ELLIOTT E.T.
    1981. Effect of the nematodes Acrobeloides sp. and
    Mesodiplogaster Lheritieri on substrate utilisation and
    nitrogen and phosphorus mineralisation in soil. Ecology,
    62, 549-555.
18. ANDERSON R.V., TROFYMOW J.A., COLEMAN D.C. and REID C.P.P.
    1982. Phosphorus mineralisation by a soil pseudomonad in
    spent oil shale as affected by a rhabditid nematode. Soil
    Biology & Biochemistry, 14, 365-371.
19. APPIAH M.R. 1975. Organic phosphorus and phosphatase
    activity in cocoa soils of Ghana. Ghana Journal of Agri-
    cultural Science, 8, 45-50.
20. APPIAH M.R. and THOMAS R.L. 1982. Inositol phosphate and
    organic phosphorus contents and phosphatase activity of
    some Canadian and Ghanaian soils. Canadian Journal of
    Soil Science, 62, 31-38.
21. BAKER R.T. 1975. A new method for estimating the phospho-
    lipid content of soils. Journal of Soil Science, 26,
    432-436.
22. BARBER S.A. 1979. Soil phosphorus after 25 years of
    cropping with five rates of phosphorus application. Com-
    munications in Soil Science and Plant Analysis, 10,
    1459-1468.
23. BARBER S.A. 1980. Soil-plant interactions in the phosphorus
    nutrition of plants. Chapter 21 in The Role of Phosphorus
    in Agriculture. Eds. Khasawneh F.E., Sample E.C. and

367

Kamprath E.J., American Society of Agronomy, Madison, Wis., U.S.A.

24. BARBER S.A., WALKER J.M. and VASEY E.H. 1963. Mechanisms for the movement of plant nutrients from the soil and fertiliser to the plant root. Journal of Agriculture and Food Chemistry, 11, 204-207.

25. BARKLEY S.A., BAREL D., STONER W.A. and MILLER P.C. 1978. Controls on decomposition and mineral release in wet meadow tundra - a simulation approach. In Environmental Chemistry and Cycling Processes. Eds. Adriano D.C. and Brisbin Jr. I.L., U.S. Dept. of Energy, Washington D.C. pp.754-778.

26. BARLEY K.P. 1970. The configuration of the root system in relation to nutrient uptake. Advances in Agronomy, 22, 159-201.

27. BARROW N.J. 1960. The effects of varying the nitrogen, sulphur and phosphorus content of organic matter on its decomposition. Australian Journal of Agricultural Research, 11, 317-330.

28. BARROW N.J. 1978. Inorganic reactions of phosphorus, sulphur and molybdenum in soil. In Mineral Nutrition of Legumes in Tropical and Subtropical Soils. Eds. Andrew C.S. and Kamprath E.J., CSIRO, Melbourne, Australia. pp.189-206.

29. BATTEN G.D., BLAIR G.J. and LILL W.J. 1979. Changes in soil phosphorus and pH in a Red Earth soil during build-up and residual phases of a wheat-clover ley farming system. Australian Journal of Soil Research, 17, 163-175.

30. BATZLI G.O. 1978. The role of herbivores in mineral cycling. In Environmental Chemistry and Cycling Processes. Eds. Adriano D.C. and Brisbin Jr. I.L., U.S. Dept. of Energy, Washington D.C. pp.95-112.

31. BEEK J. and VAN RIEMSDIJK W.H. 1979. Interaction of ortho-phosphate ions with soil. Chapter 8 in Developments in Soil Science 5B. Physico-chemical models. Ed. Bolt G.H. Elsevier Scientific Publishing Co., Amsterdam.

32. BEEVER R.E. and BURNS D.J.W. 1976. Microorganisms and the phosphorus cycle : some physiological considerations. In Reviews in Rural Science III, Prospects for Improving Efficiency of Phosphorus Utilisation. Ed. Blair G.J. University of New England, N.S.W., Australia. pp.113-118.

33. BEEVER R.E. and BURNS D.J.W. 1981. Phosphorus uptake, storage and utilisation by fungi. Advances in Botanical Research. Ed Woolhouse M.W., 8, 127-218.

34. BENNETT D. and BOWDEN J.W. 1976. Decide - an aid to effi-cient use of phosphorus. In Reviews in Rural Agriculture III, Prospects for Improving Efficiency of Phosphorus Utilisation. Ed. Blair G.J. The University of New England, N.S.W., Australia. pp.77-81.

35. BERGSTROM D.W. and McGILL W.B. 1981. Controls on release and mineralisation of organic phosphorus. In Proceedings 18th Alberta Soil Science Workshop, Alberta Soil Science Society, Canada. pp.1-19.

36. BHAT K.K.S. and NYE P.H. 1974. Diffusion of phosphate to plant roots in soil. Depletion around onion roots without root hairs. Plant and Soil, 41, 383-394.

37. BIELESKI R.L. 1973. Phosphate pools, phosphate transport and phosphate availability. Annual Reviews of Plant Physiology, 24, 225-252.
38. BIRCH H.F. and FRIEND M.T. 1961. Resistance of humus to decomposition. Nature, 191, 731-732.
39. BLACK C.A. and GORING C.A.I. 1953. Organic phosphorus in soils. In Soil and Fertiliser Phosphorus in Crop Nutrition. Eds. Pierre W.H. and Norman A.G., Academic Press, pp.123-153.
40. BLAIR G.J. and BOLAND O.W. 1978. The release of phosphorus from plant material added to soil. Australian Journal of Soil Research, 16, 101-111.
41. BLAIR G.J., TILL A.R. and SMITH R.C.G. 1977. The phosphorus cycle - what are the sensitive areas? In Reviews in Rural Science III, Prospects for Improving Efficiency of Phosphorus Utilisation. Ed. Blair G.J. University of New England, N.S.W., Australia, pp.9-19.
42. BLOOM P.R. 1980. Phosphorus adsorption by an aluminium-peat complex. Soil Science Socity of America Journal, 45, 267-272.
43. BOLIN B., DEGENS E.T., DUVIGNEAUD P. and KEMPE S. 1979. The global biogeochemical carbon cycle. Chapter 1 in The Global Carbon Cycle. Eds. Bolin B., Degens E.T., Kempe S. and Ketner P. John Wiley & Sons Ltd.
44. BOWMAN R.A. and COLE C.V. 1978. Transformations of organic phosphorus substrates in soils evaluated by $NaHCO_3$ extraction. Soil Science, 125, 49-54.
45. BOWMAN R.A. and COLE C.V. 1978. An exploratory method for fractionation of organic phosphorus from grassland soils. Soil Science, 125, 95-101.
46. BROECKER W.S. 1974. Chemical Oceanography, Harcourt Brace Jovanovich Inc.
47. BROOKES P.C., POWLSON D.S. and JENKINSON D.S. 1982. Measurement of microbial biomass phosphorus in soil. Soil Biology & Biochemistry, 14, 319-329.
48. BURNS D.J.W. and BEEVER R.E. 1978. Physiology of P release from fungi-implications for the P cycle. In Microbial Ecology. Eds. Loutit M.W. and Miles J.A.R., Springer-Verlag. pp.156-160.
49. BURNS R.G. 1981. Microbial activity at soil colloid surfaces. Bulletin of the International Society of Soil Science No. 60, pp.79-84.
50. CAMPBELL C.A. and BIEDERBECK V.O. 1976. Soil bacterial changes as affected by growing season weather conditions. A field and laboratory study. Canadian Journal of Soil. Science, 56, 293-310.
51. CAMPBELL C.A., BIEDERBECK V.O., WARDER F.G. and ROBERTSON G.W. 1973. Effect of rainfall and subsequent drying on nitrogen and phosphorus changes in a dryland fallow loam. Soil Science Society of America Proceedings, 37, 909-915.
52. CHAPIN III F.S., BARSDATE R.J. and BARÈL D. 1978. Phosphorus cycling in Alaskan coastal tundra : a hypothesis for the regulation of nutrient cycling. Oikos, 31, 189-199.
53. CHARHOLM M. 1981. Protozoan grazing of bacteria in soil - impact and importance. Microbial Ecology, 7, 343-350.

54. CHARHOLM M. and ROSSWALL T. 1980. Biomass and turnover of bacteria in a forest soil and tundra peat. Soil Biology & Biochemistry, 12, 49-51.
55. CHATER M. and MATTINGLY G.E.G. 1979. Changes in organic phosphorus contents of soils from long-continued experiments at Rothamsted and Saxmundham. Report of Rothamsted Experimental Station, 2, 41.
56. CHAUHAN B.S., STEWART J.W.B. and PAUL E.A. 1979. Effect of carbon additions on soil labile inorganic, organic and microbially held phosphate. Canadian Journal of Soil Science, 59, 387-396.
57. CHAUHAN B.S., STEWART J.W.B. and PAUL E.A. 1981. Effect of labile inorganic phosphate status and organic carbon additions on the microbial uptake of phosphorus in soils. Canadian Journal of Soil Science, 61, 373-385.
58. CHESHIRE M.V. and ANDERSON G. 1975. Soil polysaccharides and carbohydrate phosphates. Soil Science, 119, 356-362.
59. CHESHIRE M.V., GREAVES M.P. and MUNDIE C.M. 1976. The effect of temperature on the microbial transformation of (^{14}C) glucose during incubation in soil. Journal of Soil Science, 27, 75-88.
60. CLARK F.E. and PAUL E.A. 1970. The microflora of grassland. Advances in Agronomy, 22, 375-435.
61. COLE C.V., ELLIOTT E.T., HUNT H.W. and COLEMAN D.C. 1978. Trophic interactions in soils as they affect energy and nutrient dynamics. V. Phosphorus transformations. Microbial Ecology, 4, 381-387.
62. COLE C.V. and HEIL R.D. 1981. Phosphorus effects on terrestrial nitrogen cycles. In Terrestrial Nitrogen Cycles. Eds. Clark F.E. and Rosswall T. Ecological Bulletin (Stockholm), 33, 363-374.
63. COLE C.V., INNIS G.I. and STEWART J.W.B. 1977. Simulation of phosphorus cycling in semi-arid grasslands. Ecology, 58, 1-15.
64. CORNFORTH I.S. and SINCLAIR A.G. 1982. Model for calculating maintenance phosphate requirements for grazed pastures. New Zealand Journal of Experimental Agriculture, 10, 53-61.
65. COSGROVE D.J. 1980. Inositol phosphates in soils and sediments. Chapter 13 in Studies in organic chemistry 4. Inositol phosphates, their chemistry, biochemistry and physiology. Elsevier Scientific Publishing Co. pp.128-138.
66. DALAL R.C. 1977. Soil organic phosphorus. Advances in Agronomy, 29, 83-113.
67. DALAL R.C. 1979. Mineralisation of carbon and phosphorus from ^{14}C and ^{32}P labelled plant material added to soil. Soil Science Society of America Journal, 43, 913-916.
68. DICK W.A. and TABATABAI M.A. 1978. Hydrolysis of organic and inorganic phosphorus compounds added to soils. Geoderma, 21, 175-182.
69. DORMAAR J.F. 1972. Seasonal patterns of soil organic phosphorus. Canadian Journal of Soil Science, 52, 107-112.
70. EDWARDS C.A. and LOFTY J.R. 1977. Biology of Earthworms. Chapman and Hall.

71. EMSLEY J. 1980. The phosphorus cycle. In The Handbook of Environmental Chemistry Vol. 1A. The Natural Environment and the Biogeochemical Cycles. Ed Hutzinger O. Springer-Verlag. pp.147-167.

72. ENWEZOR W.O. 1967. Significance of the C:organic P ratio in the mineralisation of soil organic phosphorus. Soil Science, 103, 62-66.

73. ENWEZOR W.O. 1976. The mineralisation of nitrogen and phosphorus in organic materials of varying C:N and C:P ratios. Plant and Soil, 44, 237-240.

74. FISHER J.D. 1973. Geochemistry of minerals containing phosphorus. Chapters 6 and 7 in Environmental Phosphorus Handbook. Ed.Griffith E.J., John Wiley & Sons Ltd.

75. FLOATE M.J.S. 1970. Decomposition of organic materials from hill soils and pastures. III. The effect of temperature on the mineralisation of carbon, nitrogen and phosphorus from plant materials and sheep faeces. Soil Biology & Biochemistry, 2, 187-196.

76. FLOATE M.J.S. 1970. Decomposition of organic materials from hill soils and pastures. IV. The effects of moisture content on the mineralisation of carbon, nitrogen and phosphorus from plant materials and sheep faeces. Soil Biology & Biochemistry, 2, 275-283.

77. FOGEL R. 1980. Mycorrhizae and nutrient cycling in natural forest ecosystems. New Phytologist, 86, 199-212.

78. GASSER J.K.R. 1962. Mineralisation of nitrogen, sulphur and phosphorus from soils. Welsh Soils Discussion Group, 3, 26.

79. GERRITSE R.G. 1978. Assessment of a procedure for fractionating organic phosphates in soil and organic materials using gel filtration and H.P.L.C. Journal of the Science of Food and Agriculture, 29, 577-586.

80. GHOSHAL S. and JANSSON S.L. 1975. Transformations in organic matter. Decomposition studies with special reference to the immobilisation aspect. Swedish Journal of Agricultural Research, 5, 199-208.

81. GIANINAZZI-PEARSON V. and GIANINAZZI S. 1978. Enzymatic studies on the metabolism of vesicular-arbuscular mycorrhiza. II. Soluble alkaline phosphatase specific to mycorrhizal infection in onion roots. Physiology and Plant Pathology, 12, 45-53.

82. GOULD W.D. and BOLE J.B. 1980. Phosphorus transformations at the root-soil interface. In Western Canada Phosphate Symposium Proceedings. Ed.Harapiak J.T. Alberta Society of Soil Science, Calgary. pp.323-335.

83. GREAVES M.P. and WEBLEY D.M. 1965. A study of the breakdown of organic phosphates by microorganisms from the root region of certain pasture grasses. Journal of Applied Bacteriology, 28, 454-465.

84. GRINSTED M.J., HEDLEY M.J., WHITE R.E. and NYE P.H. 1982. Plant-induced changes in the rhizosphere of rape (Brassica napus var. Emerald) seedlings. 1. pH change and the increase in P concentration in the soil solution. New Phytologist, 91, 19-29.

85. HALM B.J., STEWART J.W.B. and HALSTEAD R.L. 1972. The phosphorus cycle in a native grassland ecosystem. In Isotopes and Radiation in Soil Plant Relationships including Forestry. IAEA, Vienna. pp.571-586.
86. HALSTEAD R.L., LAPENSEE J.M. and IVARSON K.C. 1963. Mineralisation of soil organic phosphorus with particular reference to the effect of lime. Canadian Journal of Soil Science, 43, 97-106.
87. HALSTEAD R.L. and McKERCHER R.B. 1975. Biochemistry and cycling of phosphorus. Chapter 2 in Soil Biochemistry, Vol. 4. Eds Paul E.A. and McLaren A.D. Marcel Dekker.
88. HANNAPEL R.J., FULLER W.H. and FOX R.H. 1964. Phosphorus movement in a calcareous soil. II. Soil microbial activity and organic phosphorus movement. Soil Science, 97, 421-427.
89. HARRISON A.F. 1978. Phosphorus cycles of forest and upland grassland systems and some effects of land management practices. In Phosphorus in the Environment : Its Chemistry and Biochemistry, CIBA Foundation Symposium, 57. Excerpta Medica. pp.175-195.
90. HARRISON A.F. 1982. ^{32}P-method to compare rates of mineralisation of labile organic phosphorus in woodland soils. Soil Biology & Biochemistry, 14, 337-342.
91. HARRISON A.F. 1982. Labile organic phosphorus mineralisation in relationship to soil properties. Soil Biology & Biochemistry, 14, 343-352.
92. HEDLEY M.J. and STEWART J.W.B. 1982. Method to measure microbial phosphate in soils. Soil Biology & Biochemistry, 14, 377-385.
93. HEDLEY M.J., STEWART J.W.B. and CHAUHAN B.S. 1982. Changes in inorganic and organic soil phosphorus fractions induced by cultivation practices and by laboratory incubations. Soil Science Society of America Journal, 46, 970-976.
94. HEDLEY M.J., WHITE R.E. and NYE P.H. 1982. Plant induced changes in the rhizosphere of rape (Brassica napus var. Emerald) seedlings. III. Changes in L value, soil phosphate fractions and phosphatase activity. New Phytologist, 91, 45-56.
95. HIGASHI T. and SHINAGAWA A. 1981. Phosphorus sorption by Al(Fe)-humus complexes in volcanic ash soils. Memoirs of the Faculty of Agriculture, Kagoshima University, 17, 253-258.
96. HONG J.K. and YAMANE I. 1980. Proposal for a more suitable method to extract soil organic phosphorus. Soil Science and Plant Analysis (Tokyo), 26, 383-390.
97. HONG J.K. and YAMANE I. 1981. Distribution of inositol phosphate in the molecular size fractions of humic and fulvic acid fractions. Soil Science and Plant Nutrition (Tokyo), 27, 295-303.
98. HUSZ G. St. 1977. Agro-ecosystems in South America. Agro-Ecosystems, 4, 277-292.
99. HUTCHINSON K.J. and KING K.L. 1982. Invertebrates and nutrient cycling. In Proceedings 3rd Australasian Conference on Grassland Invertebrate Ecology. Ed.Lee K.E. South Australian Government Printer, Adelaide. pp.331-338.

372

100. IRVING G.C.J. and COSGROVE D.J. 1981. The use of hypo-
     bromite oxidation to evaluate two current methods for the
     estimation of inositol polyphosphates in alkaline extracts
     of soils. Communications in Soil Science and Plant
     Analysis, 12, 495-509.
101. ISLAM A., MANDAL R. and OSMAN K.T. 1979. Direct absorption
     of organic phosphate by rice and jute plants. Plant and
     Soil, 53, 49-54.
102. JENKINSON D.S. 1976. The effect of biocidal treatments on
     metabolism in soil. IV. The decomposition of fumigated
     organisms in soil. Soil Biology & Biochemistry, 8, 203-
     208.
103. JENKINSON D.S. 1981. The fate of plant and animal residues
     in soil. Chapter 9 in The Chemistry of Soil Processes.
     Eds.Greenland D.J. and Hayes M.H.B. John Wiley & Sons Ltd.
104. JENKINSON D.S. and LADD J.N. 1981. Microbial biomass in
     soil : measurement and turnover. Chapter 10 in Soil Bio-
     chemistry Vol. 5. Eds Paul E.A. and Ladd J.N. Marcel
     Dekker.
105. JENKINSON D.S. and POWLSON D.S. 1976. The effects of
     biocidal treatments on metabolism in soil. 1. Fumigation
     with chloroform. Soil Biology & Biochemistry, 8, 167-177.
106. JENKINSON D.S. and RAYNER J.H. 1977. The turnover of soil
     organic matter in some of the Rothamsted classical experi-
     ments. Soil Science, 123, 298-305.
107. JONES O.L. and BROMFIELD S.M. 1982. Macromolecular organic
     phosphorus in decomposing plants and in pasture soils.
     Soil Biology & Biochemistry, 14, 145-152.
108. JONES R. 1979. Comparative studies of plant growth and
     distribution in relation to waterlogging. VIII. The
     uptake of phosphorus by dune and dune slack plants.
     Journal of Ecology, 63, 109-116.
109. KAILA A. 1963. Organic phosphorus in Finnish soils. Soil
     Science, 95, 38-44.
110. KAWADA H., NISHADA T. and YOSHIOKA J. 1973. Chemical
     composition of soil and foliage in relation to the growth
     of Hinoki (Chamaecyparis obtusa S. et Z). Appraisal of
     carbon/organic phosphorus ratio of forest soil as an
     index of phosphorus availability. Bulletin of the Govern-
     ment Forest Experiment Station (Japan), No. 253, pp.1-37.
111. KOWALENKO C.G. 1978. Organic nitrogen phosphorus and
     sulphur in soils. Chapter 3 in Developments in Soil
     Science 8. Soil Organic Matter. Eds. Schnitzer M. and
     Khan S.U. Elsevier Scientific Publishing Co.
112. KOWALENKO C.G. and LOWE L.E. 1975. Mineralisation of
     sulphur from four soils and its relationship to soil
     carbon, nitrogen and phosphorus. Canadian Journal of Soil
     Science, 55, 9-14.
113. KOWALENKO C.G. and McKERCHER R.B. 1971. An examination of
     methods for extraction of soil phospholipids. Soil Biology
     & Biochemistry, 2, 269-273.
114. LARSEN S. 1967. Soil phosphorus. Advances in Agronomy,
     19, 151-210.
115. LEWIS D.G. and QUIRK J.P. 1967. Phosphate diffusion in
     soil and uptake by plants. Plant and Soil, 26, 445-453.

116. MACKAY A.D., SYERS, J.K., SPRINGETT J.A. and GREGG P.E.H. 1982. Plant availability of phosphorus in superphosphate and a phosphate rock as influenced by earthworms. Soil Biology & Biochemistry, 14, 281-287.

117. MANSELL G.P., SYERS J.K. and GREGG P.E.H. 1981. Plant availability of phosphorus in dead herbage ingested by surface-casting earthworms. Soil Biology & Biochemistry, 13, 163-167.

118. MARTIN J.K. and CARTWRIGHT B. 1971. The comparative plant availability of ^{32}P myo-inositol hexaphosphate and $KH_2 {}^{32}PO_4$ added to soils. Soil Science and Plant Analysis, 2, 375-381.

119. MARTIN J.K. and CUNNINGHAM R.B. 1973. Factors controlling the release of phosphorus from decomposing wheat roots. Australian Journal of Biological Science, 26, 715-727.

120. MATTINGLY G.E.G. and TALIBUDEEN O. 1967. Progress in the chemistry of fertiliser and soil phosphorus. Topics in phosphorus Chemistry, 4, 157-190.

121. MAYS D.A., WILKINSON S.R. and COLE C.V. 1980. Phosphorus nutrition of forages. Chapter 28 in The Role of Phosphorus in Agricjlture. Eds. Khasawneh F.E., Sample E.C. and Kamprath E.J. American Society of Agronomy, Madison, Wis., U.S.A.

122. McGILL W.B. and COLE C.V. 1981. Comparative aspects of cycling of organic C, N, S and P through soil organic matter. Geoderma, 26, 287-309.

123. McKERCHER R.B. and ANDERSON G. 1968. Characterisation of the inositol penta- and hexa-phosphate fractions of a number of Canadian and Scottish soils. Journal of Soil Science, 19, 302-310.

124. McKERCHER R.B. and TINSLEY J. 1982. Recovery of soil organic phosphorus with titanium. Communications in Soil Science and Plant Analysis, 13, 215-229.

125. McKERCHER R.B. and TOLLEFSON T.S. 1978. Barley response to phosphorus from phospholipids and nucleic acids. Canadian Journal of Soil Science, 58, 103-105.

126. McKERCHER R.B., TOLLEFSON T.S. and WILLARD J.R. 1979. Biomass and phosphorus contents of some soil invertebrates. Soil Biology & Biochemistry, 11, 387-391.

127. MOGHIMI A., TATE M.E. and OADES J.M. 1978. Characterisation of rhizosphere products especially 2-ketogluconic acid. Soil Biology & Biochemistry, 10, 283-287.

128. MOLLOY L.F. and BLAKEMORE L.C. 1974. Studies on a climo-sequence of soils in tussock grasslands. 1. Introduction, sites and soils. New Zealand Journal of Science, 20, 167-177.

129. NANNIPIERI P., CECCANTI B., CONTI C. and BIANCHI D. 1982. Hydrolases extracted from soil: their properties and activities. Soil Biology and Biochemistry, 14, 257-263.

130. NEWBERG D. McC. 1979. The effects of decomposing roots on the growth of grassland plants. Journal of Applied Ecology, 16, 613-622.

131. NEWMAN R.H. and TATE K.R. 1980. Soil phosphorus characterisation by ^{31}P nuclear magnetic resonance. Communications in Soil Science and Plant Analysis, 11, 835-842.

132. NYE P.H. 1981. Changes of pH across the rhizosphere induced by roots. Plant and Soil, 61, 7-26.
133. NYE P.H. and TINKER P.B. 1977. Solute movement in the root-soil system. Blackwell Scientific Publications.
134. ONIANI O.G., CHATER M. and MATTINGLY G.E.G. 1973. Some effects of fertilisers and farmyard manure on the organic phosphorus in soils. Journal of Soil Science, 24, 1-9.
135. PARFITT R.L. 1978. Anion Adsorption by soils and soil materials. Advances in Agronomy, 30, 1-50.
136. PATTEN D.K., BREMNER J.M. and BLACKMER A.M. 1980. Effects of drying and air-dry storage of soils on their capacity for denitrification of nitrate. Soil Science Society of America Journal, 44, 67-70.
137. PAUL E.A. and VAN VEEN J.A. 1978. The use of tracers to determine the dynamic nature of organic matter. Transactions, 11th International Congress of Soil Science, Vol. 3, Edmonton, Canada. pp.61-102.
138. PAUL E.A. and VORONEY R.P. 1980. Nutrient and energy flows through soil microbial biomass. In Contemporary Microbial Ecology. Ed. Ellwood D.C. Academic Press. pp.216-227.
139. PETERSON G.W. and COREY R.B. 1966. A modified Chang and Jackson procedure for routine fractionation of inorganic soil phosphate. Soil Science Society of America Proceedings, 30, 563-565.
140. PERROTT, K.W. 1982. Personal communication.
141. PIERROU U. 1976. Nitrogen, phosphorus and sulphur global cycles. SCOPE Report 7, Eds. Svensson B.H. and Söderlund R. Ecological Bulletin (Stockholm), 22, 75.
142. PONNAMPERUMA F.N. 1972. The chemistry of submerged soils. Advances in Agronomy, 24, 29-96.
143. POWLSON D.S., TATE K.R. and RANDALL E.W. 1982. Characterisation of soil phosphorus by ^{31}P nuclear magnetic resonance spectroscopy. Report of Rothamsted Experimental Station, 1, 256.
144. PRATT P.F. and LAAG A.E. 1981. Effect of manure and irrigation on sodium bicarbonate-extractable phosphorus. Soil Science Society of America Journal, 45, 887-889.
145. QUIN B.F. and RICKARD D.S. 1983. Pasture production and changes in soil fertility on a long-term irrigated superphosphate trial at Winchmore, New Zealand. In Proceedings XIV International Grassland Congress, Lexington, Kt., U.S.A. Eds. Smith A.J. and Hays V.W. Westview Press, Colorado. pp.323-326.
146. RACZ G.J. 1979. Release of P in organic soils under aerobic and anaerobic conditions. Canadian Journal of Soil Science, 59, 337-339.
147. RAMIREZ-MARTINEZ J.R. and McLAREN A.D. 1966. Some factors influencing the determination of phosphatase activity in native soils and in soils sterilised by irradiation. Enzymologia, 31, 23-38.
148. READ D.W.L. and CAMPBELL C.A. 1981. Biocycling of phosphorus in soil by plant roots. Canadian Journal of Soil Science, 61, 587-589.
149. RUSSELL E.W. 1973. Soil Conditions and Plant Growth, 10th Edition. Longmans.

150. RYDEN J.C. and SYERS J.K. 1977. Origin of the labile phosphate pool in soils. Soil Science, $\underline{123}$, 353-361.
151. SADLER J.M. and STEWART J.W.B. 1975. Changes with time in form and availability of residual fertiliser phosphorus in a catenary sequence of chernozemic soils. Canadian Journal of Soil Science, $\underline{55}$, 149-159.
152. SAUNDERS W.M.H. and METSON A.J. 1971. Seasonal variation in phosphorus in soil and pasture. New Zealand Journal of Agricultural Research, $\underline{14}$, 307-328.
153. SAXENA S.N. 1979. Biochemistry of soil phosphorus. Bulletin of the Indian Society of Soil Science, $\underline{12}$, 42-57.
154. SHARPLEY A.N. and SYERS J.K. 1977. Seasonal variation in casting activity and in the amounts and release to solution of phosphorus forms in earthworm casts. Soil Biology & Biochemistry, $\underline{9}$, 227-231.
155. SIBBESEN E. 1977. A simple ion-exchange procedure for extracting plant-available elements from soil. Plant and Soil, $\underline{46}$, 665-669.
156. SMECK N.E. 1973. Phosphorus : an indicator of pedogenetic weathering processes. Soil Science, $\underline{115}$, 199-206.
157. SMITH A.M. 1977. Anaerobic microsites in the rhizosphere of plants as mechanisms for increasing phosphate availability. In Reviews in Rural Science III. Prospects for Improving Efficiency of Phosphorus Utilization. Ed Blair G.J. University of New England, N.S.W., Australia. pp. 119-121.
158. SPEIR T.W. and ROSS D.J. 1978. Soil phosphatase and sulphatase. Chapter 6 in Soil Enzymes. Ed. Burns R.G. Academic Press.
159. SPIERS G.A. and McGILL W.B. 1979. Effects of phosphorus addition and energy supply on acid phosphatase production and activity in soils. Soil Biology & Biochemistry, $\underline{11}$, 3-8.
160. STANIER R.Y., DOUDEROFF M. and ADELBERG E.A. 1970. General Microbiology. 3rd edition. Macmillan.
161. STEWARD J.H. and OADES J.M. 1972. The determination of organic phosphorus in soils. Journal of Soil Science, $\underline{23}$, 38-49.
162. STEWART J.W.B. and HEDLEY M.J. 1980. Phosphorus immobilisation, mineralisation and redistribution in soils. 72nd Annual Meeting, American Society of Agronomy, Detroit. Agronomy Abstracts, p.176.
163. STEWART J.W.B. and McKERCHER R.B. 1982. Phosphorus cycle. Chapter 14 in Experimental Microbial Ecology. Eds. Burns R.G. and Slater J.H. Blackwell scientific Publications.
164. STOUT J.D. and LEE K. 1980. Ecology of soil micro- and macroorganisms. Chapter 17 in Soils with Variable Charge. Ed Theng B.K.G., New Zealand Society of Soil Science.
165. TATE K.R. 1979. Fractionation of soil organic phosphorus in two New Zealand soils by use of sodium borate. New Zealand Journal of Science, $\underline{22}$, 137-142.
166. TATE K.R. and NEWMAN R.H. 1982. Phosphorus fractions of a climosequence of soils in New Zealand tussock grassland. Soil Biology & Biochemistry, $\underline{14}$, 191-196.

167. TATE K.R. and THENG B.K.G. 1980. Organic matter and its interactions with inorganic soil constituents. Chapter 12 in Soils with Variable Charge. Ed. Theng B.K.G., New Zealand Society of Soil Science.

168. THENG B.K.G. 1979. Formation and Properties of Clay-Polymer Complexes. Developments in Soil Science 9, Elsevier.

169. THOMPSON L.M., BLACK C.A. and ZOELLNER J.A. 1954. Occurrence and mineralisation of organic phosphorus in soils with particular reference to associations with nitrogen, carbon and pH. Soil Science, 77, 185-196.

170. TIESSEN H., STEWART J.W.B. and BETTANY J.R. 1982. Cultivation effects on the amounts and concentrations of carbon, nitrogen and phosphorus in grassland soils. Agronomy Journal, 74, 831-835.

171. TILL A.R. and BLAIR G.J. 1978. The utilisation by grass of sulphur and phosphorus from clover litter. Australian Journal of Agricultural Research, 29, 235-242.

172. TILL A.R., BLAIR G.J. and DALAL R.C. 1982. Isotopic studies of the recycling of carbon, nitrogen, sulphur and phosphorus from plant material. In Cycling of Carbon, Nitrogen, Sulphur and Phosphorus in Terrestrial and Aquatic Ecosystems. Eds Freney J.R. and Galbally I.E. Springer-Verlag. pp.51-59.

173. TISDALL J.M. and OADES J.M. 1982. Organic matter and water-stable aggregates in soils. Journal of Soil Science, 33, 141-163.

174. TORSVIK V.L. 1980. Isolation of bacterial DNA from soil. Soil Biology & Biochemistry, 12, 15-21.

175. TORSVICK V.L. and GOKSOYR J. 1978. Determination of bacterial DNA in soil. Soil Biology & Biochemistry, 10, 7-12.

176. VAN DIEST A. and BLACK C.A. 1959. Soil organic phosphorus and plant growth. II. Organic phosphorus mineralised during incubation. Soil Science, 87, 145-154.

177. VAN VEEN J. and PAUL E.A. 1979. Conversion of biovolume measurements of soil organisms grown under various moisture tensions, to biomass and their nutrient content. Applied Environmental Microbiology, 37, 686-692.

178. VEINOT R.L. and THOMAS R.L. 1972. High molecular weight organic phosphorus complexes in soil organic matter : Inositol and metal content of various fractions. Soil Science Society of America Proceedings, 36, 71-73.

179. WADA K. and GUNJIGAKI N. 1979. Active aluminium and iron and phosphate adsorption in Ando soils. Soil Science, 128, 331-336.

180. WADA K. and HARWARD M.E. 1974. Amorphous clay constituents of soils. Advances in Agronomy, 26, 211-260.

181. WALKER T.W. and ADAMS A.F.R. 1958. Studies on soil organic matter. 1. Influence of phosphorus content of parent materials on accumulations of carbon, nitrogen, sulphur and organic phosphorus in grassland soils. Soil Science, 85, 307-318.

182. WALKER T.W. and SYERS J.K. 1976. The fate of phosphorus during pedogenesis. Geoderma, 15, 1-19.

183. Westin F.C. 1978. Organic phosphorus changes over a growing season in some Borolls and associated Aquolls of South Dakota. Soil Science Society of America Journal, 42, 472-477.
184. WHITE R.E. 1980. Retention and release of phosphate by soil and soil constituents. In Soils and Agriculture. Ed Tinker P.B. Blackwell Scientific Publications. pp. 71-114.
185. WHITE R.E. 1981. Pathways of phosphorus in soil. In Proceedings of a Symposium on Phosphorus in Sewage Sludge and Animal Waste Slurries. Eds.Hucker T.W.G. and Catroux G. Reidel, pp.21-46.
186. WILLETT I.R. 1979. The effects of flooding for rice culture on soil chemical properties and subsequent maize growth. Plant and Soil, 52, 373-383.
187. WILLETT I.R. and HIGGINS M.L. 1978. Phosphate sorption by reduced and re-oxidized rice soils. Australian Journal of Soil Research, 16, 319-326.
188. WILLIAMS C.H. and ANDERSON G. 1968. Inositol phosphates in some Australian soils. Australian Journal of Soil Research, 6, 121-130.
189. WILLIAMS J.D.H. and WALKER T.W. 1969. Fractionation of phosphate in a maturity sequence of New Zealand basaltic soil profiles : 2. Soil Science, 107, 213-219.
190. WILLIAMS R.J.P. 1978. Phosphorus biochemistry. In Phosphorus in the Environment : Its Chemistry and Biochemistry. CIBA Foundation Symposium 57. Excerpta Medica. pp.95-108.
191. WILLIAMS R.J.P., GILES R.G.F. and POSNER A.M. 1981. Solid state phosphorus n.m.r. spectroscopy of minerals and soils. Journal of the Chemical Society, Chemical Communications, 20, 1051-1052.
192. WILSON J.M. and GRIFFIN D.M. 1975. Water potential and the respiration of microorganisms in the soil. Soil Biology & Biochemistry, 7, 199-204.
193. WILSON M.A. 1981. Applications of nuclear magnetic resonance spectroscopy to the study of the structure of soil organic matter. Journal of Soil Science, 32, 167-186.
194. WOODMANSEE R.G. and DUNCAN D.A. 1980. Nitrogen and phosphorus dynamics and budgets in annual grasslands. Ecology, 61, 893-904.
195. YOSHIDA T. 1975. Microbial metabolism of flooded soils. Chapter 3 in Soil Biochemistry. Vol. 3. Eds. Paul E.A. and McLaren A.D. Marcel Dekker.
196. ZUNINO H., BORIE F., AGUILERA S., MARTIN J.P. and HAIDER K. 1982. Decomposition of ^{14}C-labelled glucose, plant and microbial products and phenols in volcanic ash-derived soils of Chile. Soil Biology & Biochemistry, 14, 37-43.

CHAPTER 10

SULPHUR IN SOILS AND PLANTS

N.M. SCOTT

The Macaulay Institute for Soil Research, Aberdeen, Scotland.

CONTENTS

380

1.   INTRODUCTION

For many years sulphur was a neglected element in soil science[12], although it has a vital role in plants in the production of protein, vitamins, chlorophyll, glucoside oils, and in structurally and physiologically important sulphide linkages in cell walls and sulphydryl groups of enzymes[47]. In recent years, however, sulphur has received more attention due mainly to reports of sulphur deficiency in crops, to concern over the effects of acid rain, and sulphur dioxide pollution, and to work on ochre deposition in field drainage systems.

The amount of sulphur required by plants depends on the species but is substantial (Table 1). Furthermore, with the introduction of higher yielding plant varieties and more intensive farming practices, greater amounts of sulphur will be required in the foreseeable future.

Table 1. Sulphur and primary nutrients contained in various crops.

Crop	Yield Tonnes ha^{-1}	Sulphur	Nitrogen kg.ha^{-1}	Phosphorus	Potassium
Alfalfa	25	63	570	57	456
Clover	12.6	34	285	25	188
Grasses	15	34	205	31	171
Rapeseed	5	24	120	23	80
Turnips	75	57	148	29	217
Tobacco	4	25	108	11	182
Onions	63	40	148	29	103
Cabbage	63	57	160	23	143

Adapted from The Fourth Major Nutrient
The Sulphur Institute, 1982.

To ensure that adequate sulphur is available, it is becoming more essential to monitor sulphur inputs and outputs and to construct sulphur balances and cycling models that will simulate the cycle of this nutrient

381

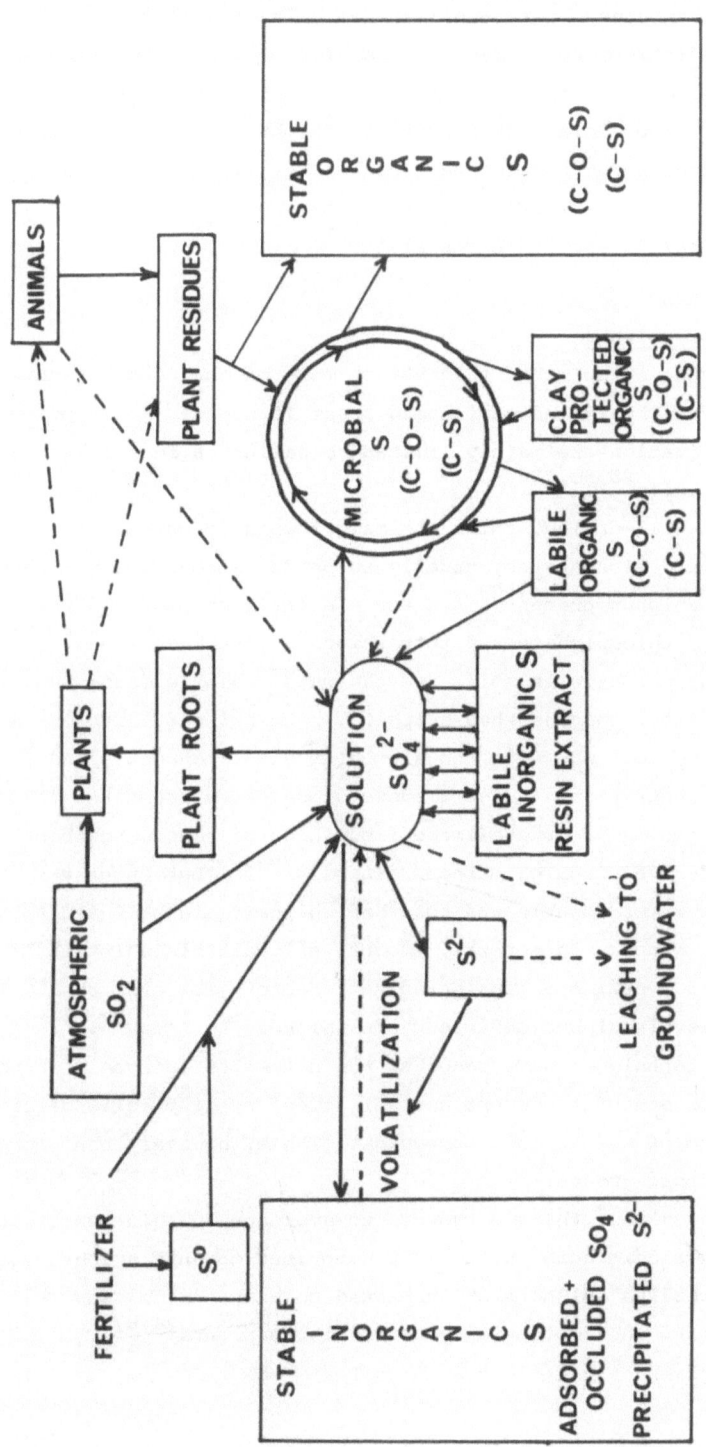

FIGURE 1. A conceptual flow diagram of the main forms and transformations of sulphur in the soil-plant system. (adapted from Bettany, J.R. and Stewart, J.W.B.⁴ International Conference "Sulphur 82" Proceedings: British Sulphur Corporation).

through the atmosphere-plant-soil system. These will enable more
accurate predictions to be made of the effects of atmospheric inputs and
of the large removal of sulphur by crops, so that remedial action may be
taken before crop yields are reduced by sulphur deficiency. Recent work
in Canada is directed towards this goal (Fig. 1). However, before such
models can provide realistic answers, a basic understanding of sulphur
transformations in soil is necessary.

2. SOIL SULPHUR

2.1. Forms of sulphur

It is well established that 95% or more of the total sulphur in
most soils from humid and semi-humid areas is organic. Exceptions to
this generalization are mainly limited to saline, acid sulphate and
gypsiferous soils[25,50,78].

In most well-drained soils the main inorganic component is sulphate,
with lower oxidation states, usually sulphide, accounting for less than
1% of total sulphur present[23]. Where elemental sulphur has been used as
a fertiliser, thiosulphate and tetrathionate have been found[60,61]. The
main forms of sulphate in soils are the water-soluble magnesium, calcium
and sodium salts, and adsorbed sulphate. The principal factors affecting
adsorption are soil pH, iron and aluminium oxide contents, and the
sulphate concentration. Large amounts of sulphate are found as
co-precipitated or co-crystallised impurities of calcium carbonate in
some soils derived from calcareous minerals[78,96], but these are insoluble,
and only the water-soluble and adsorbed sulphate are believed to be
available to crops. In general, amounts of available sulphate in excess
of 15 mg S kg^{-1} soil are considered satisfactory for crop production and
above this level sulphur deficiency is unlikely to occur[76].

Modern techniques have resolved the nature of inorganic sulphur in
soils, but present methods are not suitable for the precise determination
of the nature of the organic compounds. These, however, can be resolved
into three broad groups:-

(a) Organic sulphur that is reduced to hydrogen sulphide on treatment
    with hydriodic acid (HI). This comprises organic sulphate, believed
    to be choline sulphate or sulphated polysaccharide, and is considered
    labile and thus potentially available to plants[24,87].

383

(b) Organic sulphur that is reduced to inorganic sulphide by Raney-
     nickel. This seems to consist mainly of sulphur-containing amino
     acids such as cysteine and methionine[30,51,76].

(c) Organic sulphur that is not reduced by either HI or Raney-nickel
     and is thought to be in the form of highly resistant carbon-bonded
     sulphur compounds.

It has been shown that for Iowa soils from 0.29 to 0.45% of the
total soil sulphur is associated with the lipid fraction, with an average
of 29% of the lipid sulphur reduced by HI, 41% reduced by Raney-nickel,
and 30% remaining unidentified[43]. Similar results have been reported
for Canadian soils[11].

Much effort has been directed to finding extractants that would
remove sulphur fractions from soils for a more precise analysis, but the
production of artifacts has made this difficult. However, an extraction
procedure using 0.01M acetylacetone with ultrasonic dispersion looks
promising, extracting from 60-90% of the total sulphur in relatively
unaltered form. Further separation of the acetylacetone extract with
gel permeation chromatography provides freeze-dried samples high in
HI-reducible sulphur and carbon-bonded sulphur compounds[77].

Two humic acids extracted from contrasting soils were fractionated
with respect to molecular weight by gel chromatography[86]. It was found
that nitrogen and phosphorus contents were greatest in the high molecular
weight fractions and decreased with decreasing humic acid molecular
weight. The change in nitrogen content was accounted for by the loss of
amino acid nitrogen. In contrast, sulphur contents remained constant
throughout the molecular weight range. It was considered that these
changes may reflect part of the process whereby nutrient elements,
combined organically within humic acid are made available for plant
growth.

2.2. Microbial biomass

No sulphur model would be complete without a knowledge of the
sulphur associated with the soil biomass (see also Chapter 6). Simple
sulphur compounds are rapidly assimilated by microbial cells, and it has
been shown that 2-3% of the soil sulphur is contained in the microbial
biomass. The major proportion of sulphur in bacteria appears to be in
amino acid form, with only about 10% present as HI-reducible sulphur[74].

Reports of sulphur forms other than amino acids are few, but small amounts of choline sulphate have been found in bacterial cell walls[18]. In fungi the major sulphur portion is in the amino acid form, similar to bacteria, but fungi contain larger amounts of organic sulphates such as choline sulphate[18]. Furthermore, the percentage of total sulphur present as HI-reducible sulphur in the biomass increases significantly with an increase of sulphate in the growing medium[74]. Significant amounts of various arylsulphates may be present in fungi, and even elemental sulphur is contained in some species[9,63]. Although, at any one time, the soil microbial biomass contains only a small amount of the total organic sulphur this fraction is extremely labile and is undoubtedly the mainspring for sulphur turnover in soils[5].

3. SULPHUR TRANSFORMATIONS

Sulphur transformations in soils are considered to result primarily from microbial activity[27], although chemical processes such as the oxidation of iron sulphide are also possible[46,81]. The major microbial processes are:

(a) Mineralisation - the breakdown of large organic sulphur molecules to smaller units and finally to inorganic sulphate.

(b) Immobilisation - the conversion of simple inorganic molecules to organic compounds.

(c) Oxidation - the conversion of inorganic sulphur compounds of lower oxidation state to sulphate.

(d) Reduction - the reduction of sulphate and intermediate compounds to sulphide.

3.1. Mineralisation

Despite many studies the mechanisms of sulphur mineralisation are still largely unknown. Originally it was thought that N and S were closely associated in soil organic matter, and that the two elements would mineralise in related amounts. However, numerous experiments have shown that this is not so. On incubation, sulphur mineralisation in some soils occurred concurrently with N at ratios wider than that of the soil organic matter[34,44,94], while in other soils mineralisation occurred at ratios narrower than the N:S ratio in soil organic matter[59,88] (Table 2). It has even been reported that N mineralisation can occur simultaneously with sulphur immobilisation[74].

Table 2. Effect of incubation temperature on the mineralisation rate of N and S, in organic matter.

Soils	N:S ratio of original soil	Ratio of mineralized N:S after incubation		Reference number
		20°C	30°C	
Australia	7.4	14.4	22.3	94
	8.9	21.1	21.1	
	9.3	17.0	14.3	
Iowa mean	6.8	3.3	5.1	88

Part of the reason for the variability of the results may lie in the conditions used. For example, with eastern Australian soils it was shown that all soils released sulphate when dried, but the amounts released depended upon the manner of drying, and that S mineralisation was sharply reduced at moisture contents appreciably above or below field capacity (Fig. 2).

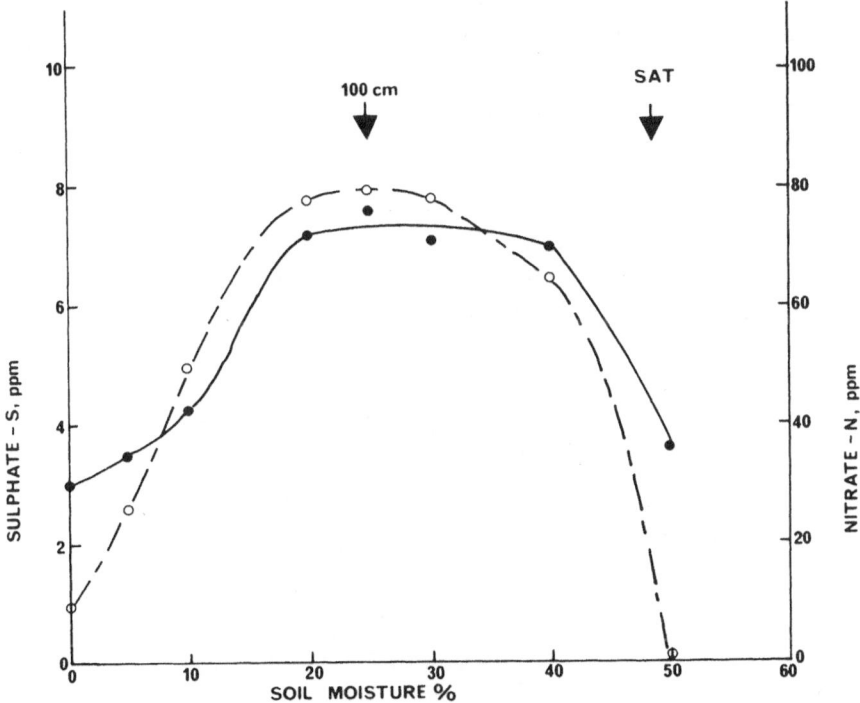

FIGURE 2. Effect of moisture content on sulfate and nitrate production in a Yellow podzolic soil during two weeks of incubation at 30°C. (Adapted from Williams, 1967) 94.

For the same soils it was shown that the amounts released could not be related to any single soil property, such as total amounts of C, N, S, pH, or the C:S, N:S and C:N ratios[94].

A reason suggested for the variable results for N and S mineralisation is that cycling patterns of C, N and S within soil humus can best be interpreted within the framework of the dichotomous system whereby elements stabilised by direct association with C (e.g. N and S) can be mineralised as a result of C oxidation to provide energy, in contrast to elements in ester form, which are stabilised through reactions of esters with soil components and may be considered to be mobilised by extra-cellular or periplasmic hydrolases, controlled by end product supply[52]. The former is the classical biological mineralisation process whereas the latter is termed biochemical mineralisation because it operates largely outside the cell membrane and is controlled by supply of the end product. Such a model (Fig. 3) implies that separate controls may regulate individual elements or types of organics although they are eventually interrelated and integrated within organisms. These considerations may well explain the variable behaviour associated with sulphur mineralisation.

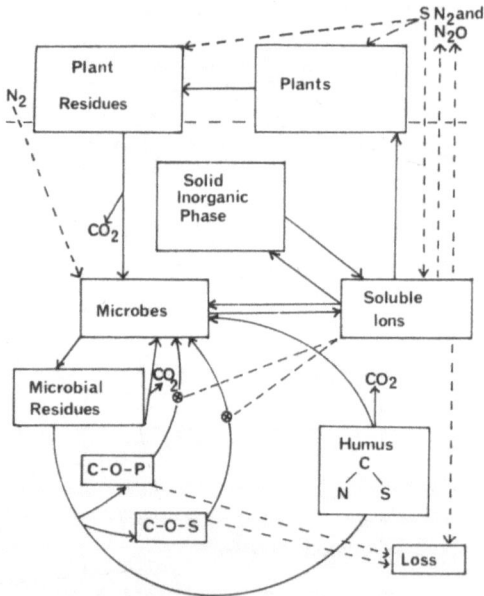

FIGURE 3. Interrelation of C, N, S and P cycling within soil-plant systems. A schematic illustration from McGill and Cole[52].

3.2. Immobilisation

This is the uptake of sulphur into soil microorganisms and eventual incorporation into humus, and is an important factor in the sulphur cycle. Under conditions when organic matter is rapidly accumulating, considerable amounts may be transferred from inorganic sulphur forms to organic forms. A pronounced example was recorded in South Australia where vigorous legume growth was associated with immobilisation of sulphur applied as ordinary superphosphate[25]. Details of the immobilisation process have been well documented[24]. Briefly it depends on the activation of the sulphate ion by a two step process leading to the production of energy rich sulphate nucleotides APS (adenosine 5'-phosphosulphate) and PAPS (3'-phosphoadenosine-5'-phosphosulphate). The nucleotides are then used in the synthesis of the sulphur amino acids. The pathway is still uncertain, but it is thought that PAPS is reduced to sulphide via sulphite. The amino acid serine is then combined with sulphide to produce cysteine. Thiosulphate is readily used by microbes in place of sulphate and is thought to be on the direct reduction pathway. The overall reaction may be represented as follows –

$$SO_4^{2-}\text{-S} \rightarrow APS \rightarrow PAPS \rightarrow \text{(active sulphite)} \rightarrow \begin{array}{c} \text{sulphide} \\ + \\ \text{serine} \end{array} \longrightarrow \text{cysteine}$$

Cysteine is then used as the building block for other S-amino acids, which are then combined into proteins.

Little is known about S transformations during the breakdown of plant, animal and microbial compounds into soil organic matter, but studies in which rye plants were grown with labelled $^{35}SO_4^{2-}$-S and the residues then incorporated into soil, showed that appreciable amounts of the labelled sulphur is found in the soil humic acids[75]. Similarly another study showed that after 168 days incubation, 50% of added labelled sulphate was incorporated into the soil organic fraction. The radioactive label in this case was found in both the HI-reducible and the carbon-bonded S fractions, the former showing greater specific activity; this study further showed that the fulvic acid fraction contained about 75% of the immobilised sulphur of which approximately 90% was in the HI-reducible form[27].

3.3 Oxidation

The oxidation process is of considerable importance within the sulphur cycle, both because the ultimate source of sulphur in soil is that in sulphur bearing minerals, and because of the increasing use of elemental sulphur as a fertiliser. Numerous studies have shown that the rate of oxidation is inversely proportional to the particle size[3,20,48], and such information has been used to select particle sizes which can either accelerate oxidation or minimize leaching losses, and hence give considerable residual effects.

Other factors which affect the oxidation process are temperature and moisture. Optimal conditions are 25°-40°C, with oxidation proceeding only slowly below 10°C [92] and soil moisture levels near to field capacity[42,57].

The most thoroughly studied group of sulphur oxidising organisms belong to the family Thiobacteriaceae. These autotrophic organisms appear to be ubiquitous and capable of very rapid oxidation rates in vitro. However, in most agricultural soils the heterotrophic sulphur organisms greatly outnumber the autotrophs[53,91].

The various members of the Thiobacilli can oxidise hydrogen sulphide (see Chapter 4), elemental sulphur, thiosulphate and poly-thionates, the end product being sulphate[85], while oxidation of reduced sulphur compounds in flooded soil can be brought about by the photo-synthetic bacteria such as Chromatium and Chlorobium.

As already stated, oxidation rates can be extremely variable. In a study of 273 Australian soils, 19% showed no appreciable oxidation of added elemental sulphur, while 39% showed very rapid oxidation[91]. When four particle size fractions of elemental sulphur were applied to pots in which Canola plants were grown, the total sulphur uptake increased significantly with decreasing particle size, and was highly correlated with the surface area of the applied sulphur. However, even at the lowest particle size, the elemental sulphur was not oxidised quickly enough to prevent sulphur deficiency[37]. In contrast oats grown in pots filled with a sulphur deficient Scottish soil responded quickly to a micro particulate elemental sulphur foliar spray at growth stage four (5 unfolded leaves + 3 tillers) and gave a grain yield comparable to that with sulphur added as potassium sulphate. A field experiment with swedes confirmed the rapid oxidation and availability of sulphur applied as a foliar spray, and the mean fresh weight values were 27 kg and 37 kg per metre square for the So and S+ treatments respectively.

The oxidation of sulphur may under certain circumstances give rise to ochre deposition, a major problem in agriculture. The problem arises when iron sulphide, formed as a result of anaerobic conditions, is oxidized by Thiobacilli in newly drained soils to form soluble ferrous sulphate[46]. This is further oxidized either chemically, particularly at high pH values, or by metal oxidizing and precipitating bacteria such as Leptothrix and Gallionella. Eventually, ferric oxide is precipitated in the drains as a brownish-red to yellow coloured, gelatinous sludge called ochre. The ferric oxide content of ochre varies from 3 to 66% and other components such as Al, Mn, Ca, Mg, S and Si may also be present[46]. Severe ochre deposition may occur as a result of draining peaty soils (dystric histosols), or acid soils close to peat deposits. In the worst cases, the problem may persist for more than 20 years although generally the problem has ended within the first 7 years after installation of the drainage system.

Several palliative measures have been attempted to control the problem, such as flushing out the drainage system with hypochlorite solutions, but the most successful method to date relies on removing the ferrous salts from solution before oxidation to the ferrous form.

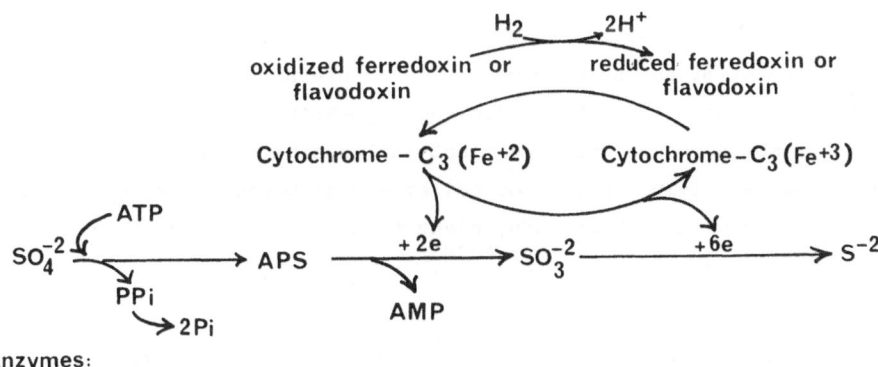

FIGURE 4. Pathway of dissimilatory sulphate reduction in Desulfovibrio. (adapted from Roy & Trudinger[73]).

3.4. Reduction

Dissimilatory sulphate reduction is brought about exclusively by
certain bacteria that use sulphate as the terminal electron acceptor in
their respiratory processes. The predominant microorganisms in this
reaction are obligate anaerobes belonging to the Desulphotomaculum and
Desulphovibrio, usually under anaerobic conditions. In soils subject
to periodic flooding, large amounts of sulphides can be produced. The
suggested pathway is shown in figure 4[73]. The sulphate reducing bacteria
all contain cytochrome respiratory pigments, which act as electron donors
in the reduction process. Sulphite, thiosulphate and tetrathionate can
all be reduced by Desulphuricans but the pathways for the process have
yet to be defined.

The sulphuretta system is ecologically important and occurs when an
anaerobic zone is produced by gross pollution with organic residues and
there are fluctuating or cyclic aerobic conditions. The reduced sulphur
compounds produced in anaerobic zone are reconverted to oxidised sulphur
forms and precipitated as sulphuretta in a cyclic manner. Sulphuretta
are ubiquitous and vary widely in size from particulate to vast areas
such as in the Dead Sea[64]. It was suggested that the operation of
sulphuretta in soils may explain why little loss of volatile sulphides
is observed in soils that are sometimes anaerobic in the subsoil layers.
The presence of an oxidised surface soil layer effectively traps the
volatile S by both chemical and microbiological reactions[4].

4. SULPHUR DEFICIENCY IN PLANTS AND SOILS

An aspect of the sulphur cycle now receiving a great deal of
attention because of sulphur deficiency appearing in plants, is the sulphur
status of soils. Both soil and plant analyses have been used to measure
the sulphur supplying ability of soils, and although moderate success has
been achieved with both methods, plant techniques are generally
considered to be more reliable[1,6,29].

4.1. Soil analysis

Soil extractants can be conveniently placed in three groups on the
basis of the soil sulphur form removed.
(a) Readily soluble sulphate usually extracted with cold water[22], 0.1M
     lithium chloride solution[72] or 0.15% calcium chloride solution[95].

(b) Readily soluble sulphate plus adsorbed sulphate extracted with phosphate solutions, either 0.016M potassium dihydrogen phosphate[16] or 0.01M calcium dihydrogen phosphate[20]. The latter is usually preferred since it prevents deflocculation of the soil, and the suspension is then easier to filter.

(c) Readily soluble and adsorbed sulphate plus some organic sulphur. The usual extractants are 0.5M NaHCO$_3$[41] and sodium chloride after heat treatment (heat soluble sulphur)[95].

No single method has received universal application and the method chosen depends mainly on the soil type. Group (a) extractants are used with soils which contain no adsorbed sulphate[20,72,93], while those in group (b) are preferred for weathered soils which contain iron and aluminium oxides, allophane and kaolinitic clays, in which adsorbed sulphate is an important component[64,76,87,90]. The extractants used in group (c) are more useful for soils containing higher amounts of organic matter, and good relationships have been obtained with both plant uptake and yield responses, indicating that labile organic sulphates do contribute to plant supply[2,56,65,76].

The main factors to be considered with soil analysis are the temperature of extraction, drying of the soil, and sampling depth. It is sometimes difficult to achieve an optimum sampling depth. Several workers were unable to demonstrate any advantage in sampling below 30 cm[35,93], while others maintain that the whole profile should be examined before accurate predictions can be made[32,66]. This is probably a result of differences in root penetration between species, and variations in the distribution of extractable sulphate throughout the profile[66]. It is essential that extraction time, and particularly extraction temperature, be standardised since errors can arise from these sources, and that the temperature at which the soils are dried is carefully defined. Increases of 1-12 ppm sulphate have been obtained by increasing the temperature at which the soils were dried from 17°C to 40°C[17].

4.2. Plant analysis

Analysis of the plant tissues can give a more accurate assessment of sulphur status than soil analysis, but it still is not without problems concerning the techniques employed, the diagnostic criteria to be applied, and the effect of plant part, plant age and plant species[6,7,56,89]. Plant analysis must give results early in the growing season if effective remedial action is to be taken. The practical usefulness of plant analysis depends on the concept of a critical value. This is often defined "as the concentration or ratio present in the plant which is just sufficient for unrestricted plant growth"[16]. The critical value to be used depends on the function the element performs within the plant, and since this is sometimes not fully known, the value is usually arrived at by fitting mathematical equations to yield and composition data derived from pot culture and field experiments for different plant species[81]. The critical value chosen is usually the concentration or ratio associated with 90% of the optimal yield of the plant[55].

The plant indices most commonly used for assessing the sulphur status are total sulphur, total sulphate and the total N:total S ratio, (N:S)t.

4.2.1. Total sulphur.

Total sulphur was popular as a plant index because of the strong correlations obtained between it and plant yield, and because of the belief that plant sulphur content was directly related to the sulphur supply[10,38,71]. However, total sulphur contents of plants depend on the part of the plant being sampled, contents in leaves and stems usually being greater than in the roots[54], but reports of a precise detailed examination of sulphur distribution within different plant species are scarce. It is customary to analyse the whole of the above-ground plant, but even then the sulphur concentration varies with stage of growth. For example, a general decline in total sulphur concentration with age was found for cereals[69], and for other species[39,84], this decline being independent of the amount of sulphur fertiliser added. Sulphur concentration can also be reduced in the early stages of growth by rapid dry matter production stimulated by sulphur addition, the Piper-Steenbjerg effect[39]. Thus the total sulphur content of plants is of limited use as a diagnostic criterion for sulphur status, unless the tissue and age of the plant are carefully standardised.

4.2.2. Total sulphate.

Total sulphate has become increasingly used as a plant index, and

the critical value is based on the premise that sulphate only appears
in the plant tissue when protein synthesis has been satisfied. The
excess sulphate accumulated within the plant seems to have no
physiological purpose, other than to act as a reserve should the sulphur
supply from the soil become limiting[56]. Total sulphate is strongly
correlated with the yield[84], and appears to be a more sensitive indicator
of sulphur status than total sulphur[31,38,84]. As with total sulphur
however, the index is affected by plant part and age, and these need to be
defined and standardised[29,31,39,82,83].

A more recent innovation is to express the total sulphate as a
percentage of the total sulphur. Good correlations have been found with
the yield of wheat in a pot culture experiment[29], and with rapeseed[54],
while a field experiment with ryegrass in Scotland showed that adequate
sulphur was present when the total sulphate exceeded 30% of the total
sulphur[79]. The index also has also the advantage that it does not seem
to be affected by age or the plant part examined[82].

4.2.3. N:S ratio.

The total N to total S ratio (N:S)t, has been used extensively as
a plant index, and its usefulness is·based on the close relationship
between N and S in their principal function in plants – the synthesis
of protein. In greenhouse trials with subterranean clover (N:S)t was
shown to be less variable with plant part, age, and nitrogen supply than
total sulphur and total sulphate[28], but more recent evidence with
rapeseed showed that the (N:S)t ratio of rapeseed tops sampled at the
rosette stage, was too insensitive, and changes due to different sites,
years and seed varieties were sometimes greater than differences between
sulphur sufficient and sulphur deficient rapeseed[54]. However, numerous
reports have demonstrated the usefulness of values between 8 and 14 for
the (N:S)t ratio as a critical index, and strong correlations between·
(N:S)t ratios and yield with different plants have been obtained[55,69,82].

4.2.4 Amide nitrogen.

A further approach to the diagnosis of sulphur deficient plants is
the use of amide nitrogen content as an indicator of sulphur deficiency,
based on the fact that insufficient sulphur for protein synthesis
increased the non-protein nitrogen content of the plant. The method

has not found much support since amide accumulation can be caused by factors other than low sulphur. A further disadvantage is that while amide nitrogen content can distinguish between sulphur deficient and sulphur sufficient plants, it does not provide an index of the degree of deficiency[71].

4.2.5. <u>Seed tissue and grain test.</u>

A more recent approach in plant diagnostic techniques is the study of relationships between seed yield as affected by sulphur supply, and the total sulphur concentration in seed[19,62,67,68,69]. Although the range of sulphur concentrations found in seeds is often narrower than in leaves, significant correlations were obtained between total sulphur concentrations in seed and yield, and critical values have been derived for upland rice[62], cowpea[19], and wheat[67]. Piper-Steenbjerg effects have been noticed in some pot experiments, and it·has been observed that nitrogen deficiency depressed grain sulphur concentrations to values which may be below the critical value determined in the presence of adequate nitrogen[67]. The latter problem may be resolved by determining the (N:S)t ratio to distinguish low N supply from low S supply as the cause of reduced grain sulphur[28].

Recently it has been shown that low sulphur grain soaked in glutaraldehyde solution turns a deep brown or purplish brown, while high sulphur grain retains normal straw colour. The difference in colour obtained has been proposed as the basis for a simple test for sulphur deficiency[58].

5. ANTHROPOGENIC SULPHUR INPUTS

The burning of fossil fuels is injecting an increasingly large amount of sulphur into the atmosphere. Global estimates of anthropogenic sulphur emissions to the atmosphere are currently about 90 Tg S per year, compared with a total natural atmospheric component (from volcanoes, sea spray and biogenic sources) of about 140 Tg S per year[13]. Most of this anthropogenic sulphur enters the atmosphere as sulphur dioxide, in contrast to ionic sulphate from sea spray which accounts for about a third of the natural component. About half of the sulphur dioxide is oxidised and deposited by rainfall as sulphuric acid, and the remainder is deposited dry onto vegetation, soil or water at the earth's surface, either by direct adsorption of the gas or in particulate form, and is presumably oxidised subsequently.

Although atmospheric sulphur from anthropogenic emissions is not yet as great as from natural sources, it can have serious environmental consequences because it is localised and therefore in higher concentration[40]. There are many recorded instances of either direct toxic effects of high SO_2 concentrations on vegetation[49], or the indirect effects of acid rain on the biota living on land or, more particularly, in rivers and lakes[15]. Clean air legislation has done much to reduce the damage, and in particular the advent of high stacks on smelters and power stations has spread the emissions over a much wider geographical area. In many cases this dilution of both SO_2 concentrations and the acidity in rainfall has usually rendered them relatively harmless, because most soils have a large buffering capacity for neutralizing the relatively small acid inputs to the land surface. There are, however, areas of the world where soils are shallow and poorly-buffered, and are already strongly acid due to the natural processes of soil formation[45]. In these areas, lakes and streams are naturally poor in nutrients and have a low acid-base buffering capacity. Populations of fish and other aquatic life in such environments are already stressed, and the extra acid input has reduced, and in some cases eliminated, a number of species from these waters. Salmonoid fish species are particularly at risk because of the sensitivity of their eggs and fry to the extra acidity in the water[36]. The areas affected in this way include the Adirondack mountains of the USA, parts of the Canadian Shield, and certain areas in Scandinavia[15].

Most of the concern for the environment seems to be over the acidity arising from the extra sulphur input, although nitrogen oxides also contribute up to 30% of the acidity in rainfall. The sulphur itself, however, may be beneficial, and were it not for the atmospheric input a considerable amount of sulphur would need to be added to crops grown under improved agricultural conditions, in order to obtain optimum growth. Sulphur deficiency in crops is widespread in the world, but generally speaking it is uncommon in industrialised societies such as Western Europe because of the sulphur supplied from the atmosphere that is derived from anthropogenic sources.

6. REFERENCES

1. ANDREW C.S. 1975. Evaluation of plant and soil sulphur tests in Australia. pp. 196-200. In Sulphur in Australasian Agriculture. Ed. K. MacLachlan, Sydney University Press.

2. ARORA B.R. and SEKHON G.S. 1977. Evaluation of soil tests for the estimation of available sulphur. Journal of Agricultural Science, 88, 203-206.

3. BARROW N.J. 1971. Slowly available sulphur fertilizers in S.W. Australia. 1. Elemental sulphur. Australian Journal of Experimental Agriculture and Animal Husbandry, 11, 211-216.

4. BETTANY J.R. and STEWART J.W.B. 1982. Sulphur cycling in soils. "Sulphur 82". Conference Proceedings Vol 1 British Sulphur Corporation, London.

5. BIEDERBECK V.O. 1978. Soil Organic Sulphur and Fertility. In Soil Organic Matter. Eds. Schnitzer M. and Khan S.U. Elsevier Press, New York.

6. BLAIR G.J. 1979. Sulphur in the Tropics. International Fertiliser Development Centre, Muscle Shoals, Alabama.

7. BLAIR G.J. 1978. NOMARIL C.P. and MONUAT E. Sulphur nutrition of wet land rice. International Rice Research Institute, paper series 21, 29-33.

8. BLOOMFIELD C. and COULTER J.K. 1973. Genesis and management of acid sulphate soils. Advances in Agronomy, 25, 265-326.

9. BURNS G.R.J. and WYNN C.H. 1975. Studies on arylsulphatase and phenol sulphotransferase activities of Aspergillus oryzae. Biochemical Journal, 149, 697-705.

10. CAIRNS R.R. and CARSON R.B. 1961. Effect of sulphur treatment on yield and nitrogen and sulphur content of alfalfa grown on sulphur deficient and sulphur sufficient grey wooded soils. Canadian Journal of Plant Science, 40, 709-715.

11. CHAE Y.M. and LOWE L.E. 1980. Distribution of lipid sulphur and total lipids in soils of British Columbia. Canadian Journal of Soil Science, 60, 633-640.

12. COLEMAN R. 1966. The importance of sulphur as a plant nutrient in world crop production. Soil Science, 101, 230-239.

13. CULLIS C.F. and HIRSCHLER N.M. 1979. Emissions of sulphur into the atmosphere. Sulphur Emissions and the Environment. Society of Chemical Industry, London, 1-21.

14. DIJKSHOORN W. and VAN WYK A.L. 1967. The sulphur requirements of plants as evidenced by the sulphur-nitrogen ratio in organic matter. A review of published data. Plant and Soil, 26, 129-157.

15. DRABLOS D. and TOLLAN A. 1980. Ecological Impact of Acid Precipitation. SNSF Project, Ås, Norway.

16. ENSMINGER L.E. 1954. Some factors affecting the adsorption of sulphate by Alabama soils. Soil Science Society America Proceedings, 18, 259-264.

17. ENSMINGER L.E. and FRENEY J.R. 1966. Diagnostic techniques for determining sulphur deficiencies in crops and soils. Soil Science, 101, 283-290.

18. FITZGERALD J.W. and LUSCHINSKI P.C. 1977. Further studies on the formation of choline sulphate. Canadian Journal of Microbiology, 23, 483-490.

19. FOX R.L., KANG B.T. and NANGJU D. 1977. Sulphur requirements of cowpea and implications for production in the tropics. Agronomy Journal, 69, 201-205.
20. FOX R.L., OLSON R.A. and RHOADES H.F. 1964. Evaluating the sulphur status of soils by plant and soil tests. Soil Science Society America Proceedings, 28, 243-246.
21. FOX R.L., ATESALP H.M., KAMPBELL D.H. and RHOADES H.F. 1964. Factors influencing the availability of sulphur fertilisers to alfalfa and corn. Soil Science Society America Proceedings, 28, 406-408.
22. FRENEY J.R. 1958. Determination of water soluble sulphate in soils. Soil Science, 86, 241-244.
23. FRENEY J.R. 1961. Some observations on the nature of organic sulphur compounds in soils. Australian Journal of Agricultural Research, 12, 424-432.
24. FRENEY J.R. 1967. Sulphur containing organics. pp. 25. Soil Biochemistry. Eds. McLaren A.D. and Peterson G.H. Marcel Dekker, New York.
25. FRENEY J.R. and SWABY R.J. 1975. Sulphur Transformations in Soils. Sulphur in Australasian Agriculture. Ed. McLachlan K. Sydney University Press.
26. FRENEY J.R., BARROW N.J. and SPENCER K. 1962. A review of certain aspects of sulphur as a soil constituent and plant nutrient. Plant and Soil, 17, 295-308.
27. FRENEY J.R., MELVILLE G.E. and WILLIAMS C.H. 1971. Organic sulphur fractions labelled by addition of ^{35}S-sulphate to soil. Soil Biology and Biochemistry, 3, 133-141.
28. FRENEY J.R., SPENCER K. and JONES M.B. 1977. On the constancy of the ratio nitrogen to sulphur in the protein of subterranean clover tops. Communications in Soil Science and Plant Analysis, 8, 241-249.
29. FRENEY J.R., SPENCER K. and JONES M.B. 1978. The diagnosis of sulphur deficiency in wheat. Australian Journal of Agricultural Research, 29, 727-738.
30. FRENEY J.R., STEVENSON F.J. and BEAVERS A.H. 1972. Sulphur containing amino acids in some Scottish soils. Journal of Science Food and Agriculture, 32, 21-24.
31. GRANT P.M. and ROWELL A.W.G. 1978. The distribution of sulphate and total sulphur in maize plants in relation to the diagnosis of deficiency. Rhodesian Journal of Agricultural Research, 16, 43-69.
32. GREGG P.E.E., GOH K.M. and BRASH D.W. 1977. Isotopic studies on the uptake of sulphur in pasture plants. New Zealand Journal Agricultural Research, 20, 229-233.
33. HANG C.H., LIEM T.H. and MIKKELSON D.S. 1976. Sulphur deficiency a limiting factor in rice production in the lower Amazon basin. Bulletin 48, International Rice Research Institute, New York.
34. HAQUE I. and WALMSLEY D. 1972. Incubation studies on mineralisation of organic S and N. Plant and Soil, 37, 255-264.
35. HOEFT R.G., WALSH L.M. and KEENEY D.R. 1973. Evaluation of various extractants for available soil sulphur. Soil Science Society America Proceedings, 37, 401-404.
36. HOWELLS G. and HOLDEN A.V. 1979. Effects of acid waters on fish. Sulphur Emissions and the Environment, Society of Chemical Industry, London, 401-415.
37. JANZEN H.H., BETTANY J.R. and STEWART J.W.B. 1982. Sulphur oxidation and fertiliser sources. Proceedings Alberta Soil Science Workshop, Alberta, Canada.

38. JONES M.B. 1962. Total sulphur and sulphate-sulphur content in subterranean clover as related to sulphur responses. Soil Science Society America Proceedings, 26, 482-484.

39. JONES M.B., RUCKMAN J.E., WILLIAMS W.A. and KOENIGS R.L. 1980. Sulphur diagnostic criteria as affected by age and defoliation of sub-clover. Agronomy Journal, 72, 1043-1046.

40. KEDDIE A.W.S. 1979. The spatial and temporal distributions of sulphur dioxide and of sulphates in urban and rural atmospheres of the U.K. Sulphur Emissions and the Environment. Society of Chemical Industry, London, 419-427.

41. KILMER V.J., NEARPASS D.C. 1960. The determination of available sulphur in soils. Soil Science Society America Proceedings, 24, 337-340.

42. KITTAMS H.A. and ATTOE O.J. 1965. Availability of phosphorus in rock phosphate-sulphur fusions. Agronomy Journal, 57, 331-334.

43. KOWALENKO C.G. 1978. Organic nitrogen, phosphorus and sulphur in soils. In Soil Organic Matter. Eds. Schnitzer M. and Khan S.U. Elsevier, New York.

44. KOWALENKO C.G. and LOWE L.E. 1975. Mineralisation of sulphur from four soils and its relationship to C, N and P. Canadian Journal Soil Science, 55, 9-14.

45. KUBIENA W.L. 1953. The soils of Europe, Thomas Murby, London.

46. KUNTZE H. 1982. Iron clogging in soils and pipes. Analysis and treatment. Bulletin of the German Association for water resources and land improvement. (DVWK) Bonn. Published by Verlag Paul Parey, Hamburg.

47. LAMPORT D.T.A. (1965). The protein component of primary cell walls. Advances in Botanical Research, 2, 151-218.

48. LI P. and CALDWELL A.C. 1966. The oxidation of elemental sulphur in soils. Soil Science Society America Proceedings, 30, 370-372.

49. LINES R. 1979. Airborne pollutant damage to vegetation. Sulphur Emissions and the Environment. Society of Chemical Industry, London, 234-241.

50. LOWE L.E. 1964. An approach to the study of the sulphur status of soils and its application to selected Quebec soils. Canadian Journal Soil Science, 44, 176-179.

51. LOWE L.E. and DeLONG W.A. 1963. Carbon-bonded sulphur in selected Quebec soils. Canadian Journal Soil Science, 43, 151-155.

52. McGILL W.B. and COLE C.V. 1981. Comparative aspects of cycling of organic C, N, S and P through soil organic matter. Geoderma, 26, 267-286.

53. MAHMOUD S.A.Z., ZAKI M.M. and ABD-EL-HAFEX A.E., 1977. A survey of sulphur oxidising and sulphate reducing bacteria in Egyptian soils. Journal of Microbiology, 12, 15-22.

54. MAYNARD D.C., STEWART J.W.B. and BETTANY J.R. 1982. Diagnosis of sulphur deficiency in plants and soils. Alberta Soil Science Workshop.

55. METSON A.J. 1973. Sulphur in forage crops. Technical Bulletin No. 20. The Sulphur Institute, London.

56. METSON A.J. 1979. Sulphur in New Zealand soils. A review of sulphur in soils with particular reference to adsorbed sulphate-sulphur. New Zealand Journal Agricultural Research, 22, 95-114.

57. MOSER U.S. and OLSEN R.V. 1953. Sulphur oxidation in four soils as influenced by soil moisture tension and sulphur bacteria. Soil Science, 76, 251-257.

58. MOSS R., RANDALL P.J. and WRIGLEY C.W. 1982. A simple test to detect sulphur deficiency in wheat. Australian Journal Agricultural Research, 33, 443-452.

59. NELSON L.E. 1964. Status and transformations of sulphur in Mississipi soils. Soil Science, 97, 300-306.

60. NOR Y.M. and TABATABAI M.A. 1976. Extraction and colorimetric determination of thiosulphate and tetrathionate in soils. Soil Science, 122, 171-178.

61. NOR Y.M. and TABATABAI M.A. 1977. Oxidation of elemental sulphur in soils. Soil Science Society America Proceedings, 41, 736-741.

62. OSINAME O.A. and KANG B.T. 1975. Response of rice to sulphur applications under upland conditions. Communications in Soil Science and Plant Analysis, 6, 585-598.

63. PEZET R. and PLANT V. 1977. Elemental sulphur accumulation in different species of fungi. Science, 196, 428-429.

64. POSTGATE J.R. 1968. The sulphur cycle. In Inorganic Sulphur Chemistry. Ed. Nickless G., Elsevier Press, New York.

65. PROBERT M.E. 1976. Studies on available and isotopically exchangeable sulphur in some N. Queensland soils. Plant and Soil, 45, 461-475.

66. PROBERT M.E. and JONES R.K. 1977. The use of soil analysis for predicting the response to sulphur of pasture legumes in the Australian tropics. Australian Journal of Soil Research, 15, 137-146.

67. RANDALL P.J., SPENCER K. and FRENEY J.R. 1981. Sulphur and nitrogen fertiliser effects on wheat. I. Concentrations of sulphur and nitrogen and the sulphur to nitrogen ratio in grain in relation to yield response. Australian Journal of Agricultural Research, 32, 203-212.

68. RANDALL P.J., THOMSON J.A. and SCHROEDER H.E. 1979. Cotyledonary storage proteins in Pisum sativum IV. Effects of S, P, K and Mg deficiencies. Australian Journal of Plant Physiology, 6, 11-24.

69. RASMUSSEN P.E., RAMIG R.E., ALLMARAS R.R. and SMITH C.A. 1976. N:S sulphur relations in soft white winter wheat. II. Initial and residual effects of S application on nutrient concentration uptake and N:S ratio. Agronomy Journal, 67, 224-228.

70. RASMUSSEN P.E., RAMIG R.E., EKIN L.G. and ROHDE C.R. 1976. Tissue analysis guidelines for diagnosing sulphur deficiency in white wheat. Plant and Soil, 46, 153-156.

71. RENDIG V.V. 1956. Sulphur and nitrogen composition of fertilised and unfertilised alfalfa grown in a sulphur deficient soil. Soil Science of America Proceedings, 20, 237-240.

72. ROBERTS S. and KOEHLER F.E. 1968. Extractable and plant available sulphur in representative soils of Washington. Soil Science, 106, 53-59.

73. ROY A.B. and TRUDINGER P.A. 1970. The Biochemistry of inorganic compounds of sulphur. Cambridge University Press, Cambridge.

74. SAGGAR S., BETTANY J.R. and STEWART J.W.B. 1981. Measurement of microbial sulphur in soils. Soil Biology and Biochemistry, 13, 493-498.

75. SCHARPENSEEL H.W. and KRAUSE R. 1963. (Radiochromatographic investigation on the cycle of sulphate and other S-amino acids. Cystine and Methionine in soils and humic acids). Zeitschrift für Pflanzenernährung Düngung Bodenkunde. 101, 11-23.

76. SCOTT N.M. 1981. Evaluation of sulphate status of soils by plant and soil tests. Journal of Science of Food and Agriculture, 32, 193-199.
77. SCOTT N.M. and ANDERSON G. 1976. Sulphur, Carbon and Nitrogen contents of organic fractions from acetylacetone extracts of soils. Journal of Soil Science, 27, 324-330.
78. SCOTT N.M. and ANDERSON G. 1976. Organic Sulphur Fractions in Scottish Soils. Journal of Science of Food and Agriculture, 27, 358-366.
79. SCOTT N.M. and WATSON M.E. 1982. Agricultural Sulphur Research and Responses to Sulphur in north Scotland. "Sulphur 82", Conference Proceedings Vol. I, pp. 579-586, British Sulphur Corporation, London.
80. SCOTT N.M., BICK W. and ANDERSON H.A. 1981. The measurement of sulphur-containing amino acids in some Scottish soils. Journal of Science of Food and Agriculture, 32, 21-24.
81. SMITH F.W. and DOLBY G.R. 1977. Derivation of diagnostic indices for assessing the sulphur status of Panicum maximum var. trichoglune. Communications in Soil Science and Plant Analysis, 8, 221-240.
82. SPENCER K. and FRENEY J.R. 1980. Assessing the sulphur status of field grain wheat by plant analysis. Agronomy Journal, 72, 469-472.
83. SPENCER K., FRENEY J.R. and JONES M.B. 1978. Diagnosis of sulphur deficiency in plants. In Proceedings of 8th International Colloquium in Plant Analysis and Fertilizer problems, Auckland Government Printer, 507-513.
84. SPENCER K., JONES M.B. and FRENEY J.R. 1977. Diagnostic indices for sulphur status of subterranean clover. Australian Journal of Agricultural Research, 28, 401-412.
85. SUZUKI I. 1974. Mechanisms of inorganic oxidation and energy coupling. Annual Review of Microbiology, 28, 85-101.
86. SWIFT R.S. and POSNER A.M. 1972. Nitrogen, phosphorus and sulphur contents of humic acids fractionated with respect to molecular weight. Journal of Soil Science, 23, 50-57.
87. TABATABI M.A. and BREMNER J. 1972. Forms of sulphur and carbon nitrogen and sulphur relationships in Iowa soils. Soil Science 114, 380-386.
88. TABATABAI M.A. and AL-KHAFAJI A.A. 1980. Comparisons of nitrogen and sulphur mineralisation in soils. Soil Science Society America Proceedings, 44, 1000-1006.
89. TISDALE S.L. 1971. Why market fertiliser sulphur? Joint symposium, T.V.A., Alabama.
90. TSUJI T. and GOH K.M. 1979. Evaluation of soil sulphur fractions as sources of plant available sulphur using radioactive sulphur. New Zealand Journal Agricultural Research, 22, 595-602.
91. VITOLINS M.I. and SWABY R.J. 1969. Activity of sulphur-oxidising microorganisms in some Australian soils. Australian Journal of Soil Research, 7, 171-183.
92. WEIR R.G. 1975. The oxidation of elemental sulphur and sulphides in soil. In Sulphur in Australian Agriculture. Ed. McLachlan K. Sydney University Press.
93. WESTERMANN D.T. 1974. Indexes of sulphur deficiency in alfalfa. 1. Extractable soil sulphate. Agronomy Journal, 66, 578-581.

94. WILLIAMS C.H. 1967. Some factors affecting the mineralisation of organic sulphur in soils. Plant and Soil <u>26</u>, 205-273.
95. WILLIAMS C.H. and STEINBERGS A. 1959. Soil and sulphur fractions as classical indices of available sulphur in some Australian soils. Australian Journal of Agricultural Research, <u>10</u>, 340-352.
96. WILLIAMS C.H. and STEINBERGS A. 1962. The evaluation of plant available sulphur in soils. I. The Chemical nature of sulphate in some Australian soils. Plant and Soil, <u>17</u>, 279-294.
97. WILLIAMS C.H. and STEINBERGS A. 1964. The evaluation of plant available sulphur in soils. II. The availability of adsorbed and insoluble sulphates. Plant and Soil <u>21</u>, 50-62.

CHAPTER 11

ORGANIC MATTER AND TRACE METALS IN SOILS

D.J. LINEHAN

The Macaulay Institute for Soil Research, Aberdeen, Scotland.

CONTENTS

1. INTRODUCTION

In the soil the availability to plants of any nutrient element is
conventionally described in terms of two parameters. These are the
intensity factor, which is the concentration of the element in solution,
and the capacity factor which is the ability of the solid phases in the
soil to replenish the nutrient element as it is depleted from the solution.
The fundamental relationship between intensity and capacity factors depends
on the solubility relationships between soil minerals[32]. However, with
respect to transition metal trace elements, soil organic matter, in its
various forms, has important effects on both intensity and capacity factors.

The stability in solution of transition elements is much increased by
their sequestration by soluble organic ligands. In the absence of such
natural ligands these elements would be rapidly, and in some cases almost
completely, removed from the soil solution by hydrolysis to insoluble forms
or by adsorption onto mineral or organic surfaces in the soil[10]. Despite
the formation of such soluble organic complexes a large proportion of
transition metal trace elements, after release from primary minerals, are
removed from solution[38]. Adsorption onto insoluble organic matter plays
a variable but generally important part in their disappearance from
solution[66]. Although these metals are removed from solution they remain
in equilibrium with the soluble species and thus constitute a significant
part of the capacity factor.

2. SOLUTION CHEMISTRY OF TRACE METALS

2.1. Hydrolysis of transition metals

The solution chemistry of transition metal trace elements is dominated
by their tendency to undergo hydrolysis which, in the context of metal ion
chemistry, can be viewed as the stepwise removal of protons from hydrate

water molecules, e.g. $M(H_2O)_6^{n+} \rightarrow M(H_2O)_5OH^{(n-1)+} \rightarrow M(H_2O)_4(OH)_2^{(n-2)+}$ .....
Hydrolysis may proceed from the mononuclear species through the formation
of defined high polymers and finally to insoluble precipitates[2].   The
nutrient metals manganese, iron, cobalt, copper and zinc all undergo
hydrolysis to a greater or lesser extent under the conditions occurring in
soils.   Manganese exists in a number of valency states of which the +2 is
the most stable in solution[2] and thus most important in soil solutions.
It does not hydrolyse very appreciably at pH values less than 8, at which
pH precipitation of $Mn(OH)_4^{2-}$ occurs[2].   Iron exists in the +2 and +3
oxidation states.   Ferrous iron readily hydrolyses to form various mononuclear species at pH values above 7.0.   However, the ease of oxidation
of ferrous to ferric iron makes it of doubtful importance in the soil/plant
system except under reducing conditions where insoluble ferrous hydroxide,
$Fe(OH)_2$, is important because of its relatively high solubility and
reactivity compared with ferric hydroxides.   Ferric iron begins to
hydrolyse at about pH 1.0[2] with the formation of $Fe(H_2O)_5OH^{2+}$.   With
increased pH, hydrolysis continues to produce an array of mononuclear and
polynuclear oxide and hydroxide species[2].   Cobalt has only one stable
valency state in aqueous solution; Co^{2+}.   The cobaltic ion Co^{3+} is a
powerful oxidising agent which decomposes water.   The cobaltous ion
hydrolyses in a fashion similar to manganese ions except that hydrolysis
may begin at a slightly lower pH.   Thus hydrolysis of cobalt to insoluble
forms is only important in neutral or alkaline soils[2,32].   The stable
valency state of copper under soil conditions is cupric Cu^{2+}.   This begins
to hydrolyse at around pH 5.0 to $Cu_2(OH)_2^{2+}$.   Hydrolysis rapidly proceeds
to the colloidal or insoluble cupric hydroxide $Cu(OH)_2$, so that under most
soil conditions concentrations of Cu^{2+} are extremely low[32,48].   The
hydrolysis of Zn^{2+}, the only stable valency state, starts at near neutrality
and rapidly proceeds to the insoluble $Zn(OH)_2$.   However, in air-saturated
water $Zn_5(OH)_6(CO_3)_2$ is the stable solid phase[32].

Thus the hydrolysis of Fe and Cu, to insoluble forms which are
unavailable to plants, is likely to be important in most soils.   Hydrolysis
to insoluble forms of Mn, Co and Zn is only likely to be important in
neutral or alkaline soils[31].   However, hydrolysed species are important
also because they are involved in the adsorption of metal ions upon solid
surfaces.   Hydrolysis of metal ions at the interface between an aqueous

solution and a solid surface is thought to occur at lower pH than is the case in bulk solution[24].

2.2. Chelation of metals

The hydrolysis of metals can be prevented, or at least impeded, by the pre-emptive replacement of the water molecules surrounding a metal ion by other molecules or ions with the formation of a coordination compound. The molecule which combines with the metal is referred to as the ligand. The complex formed by metal and ligand is a chelate when two or more coordinate positions about the metal ion are occupied by donor groups of a single ligand molecule to form an internal ring structure. If the ligand forms two bonds with the metal it is described as bidentate, if three, four or five then terdentate, tetradentate or pentadentate, respectively. The formation of additional bonds generally increases the stability of the chelate. Nevertheless, different organic groups have different affinities for metal ions. The order of increasing affinity is generally:-

$\diagdown$C=O	-O-	-COO$^-$	$\diagdown$N$\diagup$	-N=N-	-NH$_2$	-O$^-$
carbonyl,	ether,	carboxylate,	ring N,	azo,	amine,	enolate

although some metals may have different relative affinities for the organic groups. Two biochemical molecules involving chelation through some of these groups are shown in figure 1.

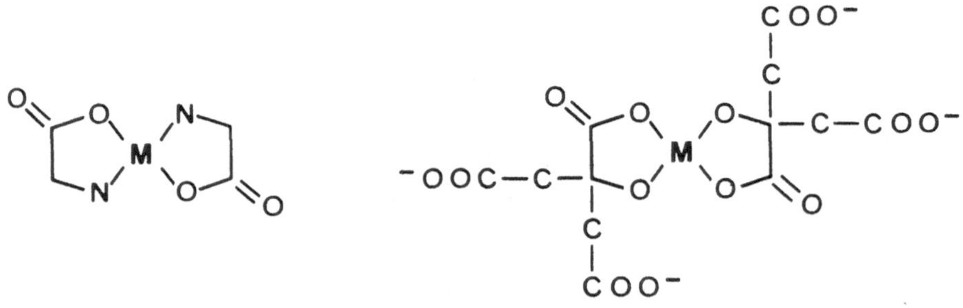

(a)   (b)

FIGURE 1. Metal complexes of (a) glycine, (b) citrate.

The stability of a metal chelate is determined by a number of factors other than the nature of the organic groups involved. These are the number of bonds and the number of rings formed, the pH of the system and the nature of the metal ion. The stability for divalent trace metals is often presented as the Irving Williams[22] series which is:-

$$Cu^{2+} > Ni^{2+} > Co^{2+} > Zn^{2+} > Fe^{2+} > Mn^{2+}$$

The stabilities of complexes of trivalent metal ions such as Fe^{3+} and Cr^{3+} are generally higher than those of divalent ions.

The tendency of a metal ion and ligand to form a complex is defined in terms of the formation constant for the reaction:-

$$M + A \rightleftharpoons MA$$

where M is the metal ion and A the ligand. The overall formation constant, K, is defined as :-

$$K = \frac{(MA)}{(M) \ (A)}$$

The term formation constant is often replaced in discussions of metal chelation by the term stability constant. The two terms are used interchangeably.

3.   HUMIC SUBSTANCES

3.1. Chemical Structures

Fully characterised organic molecules such as organic acids, amino acids and other biochemical products of the life processes of soils are without doubt of importance in the chelation of trace metals in soils, but they must constitute only a tiny proportion of the total organic matter content of any soil.   Of much greater significance quantitatively are the humic substances.   The chemical structure of humic substances are, despite very considerable investigation using many chemical and physical techniques, rather poorly understood (see Introduction).   From the point of view of their effects on trace metals in soils, a most important physical distinction is between the very complex organic colloids, the insoluble humic acids and humins, and the relatively low molecular weight soluble molecules generally described as fulvic acids. Molecular size apart, there are similarities between humic (HA) and fulvic (FA) acids[55].   The major elements in both are carbon and oxygen. Carbon in HA's range from 50-60% but in FA's it is generally in the range 40-50%.   Oxygen in HA's varies from 30 to 35% and in FA's from 45 to 50%. Thus FA's tend to be more oxygen rich and thus have more oxygen containing

functional groups capable of being involved in metal chelation.   These
groups include carboxyl, phenolic hydroxyl, enolic hydroxyl, hydroxy-
quinone, lactone, ether and alcoholic hydroxyl.

Humic substances, when isolated from soils, invariably contain a
small proportion of nitrogen.   This may amount to <1% for some highly
purified FA's to >5% for some HA's.   Thus as well as the oxygen-
containing functional groups, there are also possibly nitrogen-containing
groups including amino, amide, imino and peptide together with heterocyclic compounds such as porphyrins which might be involved in metal
chelation.   There are considerable uncertainties regarding the amounts
of various groups present in humic substances because many of the methods
used in their determination, whilst suitable for the determination of such
groups in fully characterised molecules, are less appropriate to their
determination in imperfectly characterised natural substances.   Stevenson[55]
has recently undertaken a critical review of the methods used in the
characterisation of humic substances and of the problems associated with
the interpretation of such data.   While considerable reservation must
attach to any particular data, some generalisations are possible,
particularly with regard to oxygen-containing functional groups.   Total
acidities of FA's are generally higher than those of HA's, the former
falling in the range 6.0-15.0 meq/g and the latter more likely to be in
the range 5-8 meq/g[54].   Both carboxyl and acid hydroxyl groups contribute
to total acidity, with carboxyl predominating especially in FA, where acid
hydroxyls might contribute only 2-5% of the total acidity.

3.2. Metal complexing by humic substances

The metal-complexing ability of HA and FA arises largely from their
carboxyl and acid hydroxyl groups.   Perhaps the most extensive evidence
for this comes from infrared spectroscopy but, as with so much of the
chemistry of humic substances much of the evidence is open to more than
one plausible interpretation[43,57,62].   The technique of electron spin
resonance (ESR) spectroscopy is capable of supplying very precise
information on the environment of a paramagnetic metal such as copper
which should provide definitive evidence of the nature of copper humic
complexes.   However, even this technique has produced controversial
results.   The technique has apparently shown that only carboxyls are
involved in the complexing of copper by HA[33].   The same technique has

also provided evidence that nitrogen groups as well as carboxyls are involved[25].

Perhaps of greater importance than the precise chemistry of the bonding of trace metals by humic substances is the extent of complexation, the stability of the complexes and the effect of complex formation on properties, such as solubility, of the ligand molecules. Total acidity is generally thought to provide an adequate measure of the ability of humic substances to bond metals despite the possibility that non-oxygen groups might be involved. This conclusion has been reached from studies involving methods as varied as proton release[54,65], ESR[13], voltammetric analysis[52,64], competition between humic complexes and cation exchange resins[60,68], equilibrium dialysis[67] and, for metals such as copper, free metal determination by means of ion selective electrodes[5,65]. These methods have been used to determine the maximum binding capacity of humic substances isolated from a wide variety of soils. Calculations based on these determinations indicate that, for example, one copper atom might be bound per 20 to 60 carbon atoms of an isolated HA. Values of maximum binding capacity are of considerable value for the insoluble components of soil organic matter because they may indicate the potential size of the reservoir of trace metals immobilised on the organic matter of the soil. With respect to the soluble components of the soil organic matter they can be of little value because it is unlikely that FA type molecules fully saturated by trace metals would remain in solution[50]. Polyvalent cations such as iron, copper and other trace metals are likely to form interchain bonds between individual fulvic molecules until their molecular size is such as to produce insolubility.

More important than the total metal binding capacity of soluble soil organic matter are the formation constants of particular metal organic complexes, because these provide a quantitative measure of the affinities of a particular metal for that organic matter. The most extensive range of measurement of stability constants (log K values) for FA were made by Schnitzer[49] and his various co-workers. Log K values were determined at two pH values and were in all cases greater at the higher pH. This probably resulted merely from the increased ionisation of functional groups at higher pH. For reasons which are not clear the order of increasing stability constant for a range of metals was different at the two pH values :-

at pH 3.5

Cu > Ni > Co > Pb > Ca > Zn > Mn > Mg

Log K 3.3 3.2 2.8 2.7 2.7 2.2 2.2 1.9

at pH 5.0

Ni > Co > Cu > Pb > Mn > Zn > Ca > Mg

Log K 4.2 4.1 4.0 4.0 3.7 3.6 3.3 2.1

The order at pH 3.5 closely follows the Irving-Williams stability series referred to earlier, but at pH 5.0 the order deviates very substantially from it.   That the FA copper stability constant should be lower than that of cobalt, if only marginally, is particularly surprising. Other investigations of copper FA complexes using different methods have produced overall stability constants similar, or slightly higher[5,7,12,] [52,54,56,60] than those of Schnitzer.   Generally the stability constants for metal FA complexes are somewhat lower than those of biochemicals produced by plants and soil micro-organisms[37].

4.   BIOCHEMICAL SUBSTANCES IN THE RHIZOSPHERE

4.1. Root exudation

A possible major source of organic substances having metal chelating properties are plant root exudates[18].   These are substances either lost passively from the roots or possibly, in some cases, actively secreted. It has been suggested that just about any soluble compound found within the plant can also be found in root exudates.   The extent of exudation and the composition of the exudate depends on the plant species and on the physiological condition of the plant as well as on the physical condition of the soil or other growth medium[17].   The appearance of exudates in the region of the root must alter the chemistry and biological processes of the soil in this region (see Chapter 6).   Hiltner, in 1904, was one of the first to observe that there were differences in the micro- biological population close to a root compared with populations more distant from it and coined the term rhizosphere to designate this region[46,47] As a consequence of the increased microbial activity in the rhizosphere there are increases in the amounts of biochemicals of microbial origin, some of which will be identical to those of plant origin and some of which will be entirely different[39,40,47].   Many of these substances will have metal complexing capabilities.

4.2. Amino acids

As far as the chelation of trace metals is concerned, amino acids and peptides, whether of plant or microbial origin, are probably of greater importance than any other group of biochemicals. It has been shown that over 20 different free amino acids may be present in plant root exudates. These include: leucine, isoleucine, valine, aminobutyric acid, glutamine, asparagine, alanine, serine, glutamic acid, aspartic acid, glycine, phenylalanine, threonine, tyrosine, lysine, proline and methionine. In addition to amino acids, substantial amounts of peptides are also generally present in root exudates. The exudation of amino acids has been shown to be influenced by plant species[59] and by environmental conditions, including nutrient status[4,44,45], root temperatures[21,45,58] and light intensity[45]. It seems likely that any environmental factor which alters the nitrogen metabolism of the plant or the leakiness of its membranes will influence the exudation of amino acids.

The amounts of free amino nitrogen present in soil can be as high as 16 µg N/g soil[42]. Few data have yet appeared relating levels of soluble nitrogen in the soil to plant species growing in the soil, but it has been shown that the amino acid content of rhizosphere soil, which can be up to twenty nine times higher than in adjacent bulk soil[23], is substantially higher for meadow fescue than for either timothy or sorghum. However these differences might have been due to differences in the age of the plants or moisture conditions at sampling rather than dependent on the plant species.

4.3. Organic acids

Organic acids of the citric acid cycle are effective trace metal chelating agents[37]. Their exudation by plants is less well investigated than that of amino acids. Nevertheless their presence has been observed in the exudates of several plant species although not all[59]. Exudates of barley, bean and wheat have been found to contain malic and fumaric acids, whereas cucumber exudates do not[59].

Organic acids, whether of plant of microbial origin, can be isolated from soils. Recent work[61] has shown that lactic, oxalic, malic and citric acids occur at 10^{-6} - 10^{-5} M concentrations in soil capillary water. Similar work elsewhere[19] indicates comparable concentrations of citric acid but not the other acids in capillary water isolated from permanent pasture soils. Because of the high rooting density of pastures, it is reasonable to assume that capillary water from this source represents rhizosphere soil solution. This work[19] has also demonstrated that much of the soluble copper in these soils is complexed by citric acid.

4.4. Relationships between biochemicals, humic substances and trace metals

Where biochemicals such as Krebs cycle acids and amino acids occur at appreciable concentrations in a soil solution, they are likely to complex a proportion of metals such as copper because their stability constants[36] are generally higher than those for FA's. Mixed ligand complexes involving both FA and biochemicals such as citric acid and aspartic acid have been demonstrated in model systems[35]. Because of microbial activity, the existence of these biochemicals in the soil may be rather transient so that their involvement in maintaining a pool of soluble trace metals in the soil solution may be less than might be assumed from their stability constants alone. Nevertheless, they may play an important role at the soil-root interface where their concentrations are likely to be highest. Some evidence for this suggestion comes from an ESR study of copper uptake by wheat seedlings conducted in the author's laboratory[26]. Wheat plants growing in nutrient solutions, free of micro-organisms, were provided with copper complexed by FA as their sole copper source. This was confirmed by an ESR examination of the uptake solution. Re-examination of the nutrient solutions after 48h, showed that much of the copper was complexed by amino acids and small peptides released by the wheat seedling roots. It can be speculated that copper, and by analogy other trace metals, is retained in soluble forms and transported through the soil complexed by FA type organic molecules. However, at the soil root interface the metals may be transferred to small biochemical molecules of either plant or microbial origin.

5. ADSORPTION OF TRACE ELEMENTS ON SOLID SURFACES

After release from primary minerals, a major proportion of trace element cations are immobilised by adsorption upon inorganic and organic surfaces in the soil[10]. Adsorption may be defined as the adhesion, in an extremely thin layer, of gas molecules, liquids or dissolved substances, to the surfaces of solids. Micronutrient cations Mn, Fe, Co, Cu and Zn may be held by cation exchange reactions and are, therefore, held near the soil surfaces by electrostatic or coulombic forces[34]. In addition to cation exchange, many of the micronutrient cations may enter into specific adsorption processes through covalent bonding to functional groups on the surfaces[11,24,51]. Details of the adsorption mechanism are not relevant here, suffice to say that the principal requirement of an adsorbing surface is that it possess a net negative charge. It has been suggested that the nature of the surface will only influence adsorption in a small, although possibly important, way[24].

For each metal there is a critical pH range, usually one pH unit or less, over which adsorption increases with increasing pH from negligible to almost complete adsorption. Metal cations can thus be arranged in an order related to the pH of the onset of adsorption upon solid surfaces. For trace metals the order is: Fe(III), Cu(II), Cr(III), Zn(II), Co(II), Mn(II)[16]. Adsorption of Fe(III) begins to occur at about pH 2.0 whilst adsorption of Mn(II) occurs only at above pH 6.5[16]. The other listed trace metals are adsorbed at intermediate pH values. It is probable that micronutrient cations adsorbed upon surfaces in the soil are not directly available for uptake by plants and must pass, even if only transiently, into solution[53]. Thus factors which alter the extent of adsorption must be important in regulating plant nutrition. It is likely that chelation of a trace metal by a soluble organic ligand will alter the extent of adsorption of trace metals at the solution/surface interface[8,9,63]. Addition of FA to a system containing goethite, which is active in adsorbing both iron and copper from solution, decreases adsorption from >90% in the absence of FA to <50% in the presence of 100 mg/l FA. Adsorption of Mn(II), Co(II) and Zn(II) only occurs at rather higher pH than is the case for Fe(III) and Cu(II) and to a somewhat smaller extent. Nevertheless, it is possible to show that the complexation of these metals by FA depresses their adsorption by goethite[26].

The effectiveness of FA in retaining trace elements in solution in the presence of an adsorptive surface is in conformity with the extent of chelation of these metals in soil solution. Fe(III) and Cu(II) are generally regarded as being present in soil solution at >99% complexed by organic ligands[3,14,20,48]. The proportion of Mn, Co and Zn thought to be complexed varies from >99% down to <25% depending on factors such as pH[14,15,20].

The effectiveness of non-humic organic ligands such as amino acids and organic acids, in influencing adsorption of trace elements upon soil

surfaces is a little-investigated area.   It appears that the effect of a
particular ligand depends on its interaction with the adsorptive surface.
Complexing ligands that are not adsorbed upon the surface, e.g. citrate[8,63],
decrease metal adsorption.   Where the ligand is adsorbed, e.g. glutamate[9],
then adsorption of metal ions tends to be increased.   The effects are
complicated and depend on the relative concentrations of ligands and metal
ions and on pH.

6.   DISTRIBUTION OF SOIL ORGANIC MATTER IN DIFFERENT SOILS

6.1. Total organic matter

The total organic matter content of a soil may be important with
regard to trace element nutrition of plants because it defines the size of
an important part of the reservoir of bound and immobilised trace elements.
A large reservoir of immobilised trace elements must be of advantage to the
long term fertility of the soil because as the organic matter is mineralised
trace elements will be released into forms available to plants.   However,
a very high organic matter content may cause difficulties in the short term
because the release of trace elements may not be sufficiently rapid to
sustain the high productivity of modern intensive agricultural practice.
This is known to be a problem for copper where on highly organic soils,
having adequate total copper content, copper levels in plants may be
inadequate[1].

A major factor in determining the total organic content of a mineral
soil under any particular set of conditions is the nature of the mineral
components of the soil.   A soil of high clay content will tend to have a
higher total organic matter content than a sandy soil having a rather low
clay content.   The nature of the clay is also a factor of some importance.
Montmorillonite, and other three-layer clay minerals, appear to be
associated with higher organic matter contents than are two-layer minerals
such as kaolinite[1].

The agricultural practice having the greatest influence on the total
organic matter content of a mineral soil is the cropping regime to which
the soil is exposed.   It is well established that soils maintained under
permanent pasture tend to develop much higher levels of total organic
matter than those subjected to regular cultivation.   Cultivation of soil
previously under permanent pasture results in an initial rapid mineralisation
of the organic matter with the release of immobilised trace elements into

the more labile pools. Whether they will be available for uptake by
plants then depends on other soil factors including the amounts of soluble
organic matter capable of complexing these elements in forms available for
uptake by plants.

6.2. Soluble organic matter

It is generally agreed that the major coloured component of the
soluble organic matter of soil solutions is chemically similar[27] to FA
(see Chapter 2). Little is known of its distribution in soils or how its
concentrations are related to the total organic matter of the soil or to
the amounts of chemically extracted HA and FA which can be isolated from a
soil. From the very limited data available it does not appear that any
direct relationships exist, so that it is impossible to predict from the
relative wealth of information on total soil organic matter how much soluble
organic matter might be present in a soil. Similarly data on the amounts
of chemically extractable organic matter prove of little worth in predicting
the amounts of organic matter present in soil solutions. For example,
soils of the same association, but having different drainage characteristics,
can have much higher soluble organic matter where drainage is most impeded
whilst the amounts of FA, chemically indistinguishable from the soluble
organic matter[27], are much lower in the poorly drained soils[28]. It can
be argued that the chemically similar FA and soluble organic matter have
the same origin in the soil and that different proportions are immobilised
in the soil under different drainage conditions[28]. Thus the factors
controlling the solubility of the FA-type organic matter are more important
in controlling the levels of soluble chelated trace metals in the soil
solution than are the total amounts of this type of organic molecule present
in the soil. The importance of the amounts of soluble organic matter in
this system is indicated by the fact that poorly-drained soils of a
particular soil association have higher levels of available trace metals
than well-drained examples of the association.

7. TRACE ELEMENT UPTAKE BY PLANTS

It is clear from the chemistry of nutrient trace metals and of soil
organic matter that the latter must be of considerable importance in
influencing the availability to plants of these nutrients. It is less
easy to demonstrate precisely the nature and magnitude of such effects.

It was demonstrated more than half a century ago that iron nutrition was of central importance in explaining the plant growth stimulatory effects of HA[6,41] (see also Chapter 2). It was recognised then that Fe was rapidly hydrolysed in nutrient solutions to form insoluble ferric hydroxides so that plants growing in these solutions rapidly suffered from Fe deficiency. It was suggested that soluble humates provided a more stable soluble source of Fe, thus preventing Fe deficiency[6,41]. Since these pioneering investigations many others have examined the effects of soluble humic substances on the Fe nutrition of plants. The results of these investigations illustrate some of the problems of defining the effects of soil organic matter on plant nutrition even in relatively well-defined nutrient solutions. It has been variously concluded that Fe uptake by plants is either increased or decreased by humic substances[29,30]. It is likely that these contradictory conclusions result from the use of different experimental procedures. If the effect of a humic substance is compared with an inorganic Fe source then it is likely to show significantly enhanced iron uptake. If, however, it is compared with Fe complexed by an organic chelating agent it may show depressed or enhanced uptake depending on the particular chelating agent and the pH and ionic composition of the nutrient solution[30]. Furthermore, consideration of Fe uptake into plant roots may indicate a different result from that of the whole plant[30]. A major hydrolysis product of Fe(III) is a colloid having a net positive charge. Such colloidal material is bound to the negatively charged root surface giving a spurious indication of high Fe uptake. The amounts of Fe transported to the shoots and leaves of the plant may be quite small. Humic substances complexed with Fe will probably retain a net negative charge and will not be bound substantially to root surfaces, but, because of their solubility, be absorbed into plant roots and transported through the plant. Thus one can have the apparent paradox of increasing amounts of a humic substance appearing to depress the Fe content of plant roots, but to increase the amounts present in the shoots. Humic metal complexes may vary substantially in their lability so that some complexes may release their metal component easily to the plant whilst others may retain the metal and thus prevent its uptake by plants. It appears that FA complexes of Fe and similar complexes present in soil solutions, are rather labile and thus able to transfer Fe easily to plants[30]. HA, in contrast, may produce complexes of both high and low lability so that at high metal ligand ratios labile complexes dominate whilst at low metal ligand ratios complexes of low lability predominate[30].

Experimental evidence for the importance of soil organic matter in influencing the uptake by plants of other trace metals is less plentiful than that for Fe. Cu, Zn, Mn and Co are much less prone to hydrolysis than is Fe(III), so that deficiency resulting from hydrolysis to insoluble forms is less likely except under rather alkaline conditions. It seems to be generally the case in nutrient solution experiments that plants obtain these metals more easily from soluble ionic species than from soluble metal organic ligand complexes. However, this is not always the situation which occurs in soils where ionic species tend to be largely adsorbed onto organic or inorganic surfaces in the soil. This can be simulated in the laboratory by adding an adsorbing substance such as goethite to the nutrient solution[26]. As was discussed earlier, goethite adsorbs appreciable amounts of Cu, Zn, Mn and Co from solution, an effect which can to some extent be prevented by addition of a soluble FA-type ligand. Under these conditions uptake by plants of Zn and Co can be shown to be enhanced by the organic ligand. Only Mn uptake is unaffected.

8. A PERSPECTIVE

Soil organic matter in its various forms is important in controlling trace metal availability to plants because of the chemistry of the trace metals and because of the chemistry of the soil. In the absence of soluble organic substances Fe and Cu would, in most soils, be almost completely unavailable to plants, and Zn and Co might be only marginally available. Changes in agricultural practice may cause more problems with trace metal availability in the future. More intensive use of fertilizers and the continuing development of new high yielding varieties will increase the absolute requirements for trace elements. Replacement of traditional rotation of crops by monoculture may cause decreases in the amounts of organic matter or, of more importance for trace element availability, might depress the levels of the soluble organic ligands present in soil solutions. New varieties of crop plants may not release into the rhizosphere sufficient quantities of organic ligands capable of complexing trace metals.

In order that we might be able to respond to such possible changes a number of questions require answers. What are the amounts and nature of metal chelating ligands in soil solutions and how do these respond to changes in agricultural practice? We require to isolate and quantify the amounts of these substances and determine their metal chelating properties. These priorities would require a move away from the traditional approach of chemical extraction of organic matter from soil and determination of the gross chemical structure and biological activity of humic substances. These might be found to be of relatively trivial importance in influencing trace element uptake by plants when compared with biochemicals, of plant and microbial origin, present in the rhizosphere and thus not warrant the very considerable efforts required to further elucidate their structures.

9. REFERENCES

1. ALLISON F.E. 1973. Soil organic matter and its role in crop production. Elsevier, Amsterdam.
2. BAES C.F. and MESMER R.E. 1976. The Hydrolysis of Cations. Wiley-Interscience, New York.
3. BENIANS G., SCULLION P. and FITZHUGH G.R. 1977. Concentrations and activities of ions in solutions displaced from basaltic soils. Journal of Soil Science, 28, 454-461.
4. BOWEN G.D. 1969. Nutrient status effects on loss of amides and amino acids from pine roots. Plant and Soil, 30, 139-142.
5. BRESNAHAN W.T., GRANT C.L. and WEBER J.H. 1978. Stability constants for the complexation of copper (II) ions with water and soil fulvic acids measured by an ion selective electrode. Analytical Chemistry, 50, 1675-1679.
6. BURK D., LINEWEAVER H. and HORNER C.K. 1932. Iron in relation to the stimulation of growth by humic acid. Soil Science, 33, 413-435.
7. CHEAM V. 1973. Chelation studies of copper (II): fulvic acid system. Canadian Journal of Soil Science, 53, 377-382.
8. DAVIS J.A. and LECKIE J.O. 1978. The effect of complexing ligands on trace metal adsorption at the sediment/water interface. In Environmental Biogeochemistry and Geomicrobiology. Vol. 3. Ed. Krumbein W.E., Ann Arbor Science, Ann Arbor, Michigan.
9. DAVID J.A. and LECKIE J.O. 1978. Effect of adsorbed complexing ligands on trace metal uptake by hydrous oxides. Environmental Science and Technology, 12, 1309-1315.
10. ELLIS B.G. and KNEZEK B.D. 1972. Adsorption reactions of micronutrients in soils. In Micronutrients in Agriculture. Soil Science Society of America. Madison. Eds. Mortvedt J.J., Giordano G.M. and Lindsay W.L.
11. GADDE R.R. and LAITINEN H.A. 1974. Studies of heavy metal adsorption by hydrous iron and manganese oxides. Analytical Chemistry, 46, 2022-2026.
12. GAMBLE D.S., SCHNITZER M. and HOFFMAN I. 1970. Cu^{2+} - fulvic acid chelation equilibrium in 0.1 m KCl at 25.0°C. Canadian Journal of Chemistry, 48, 3197-3204.

13. GAMBLE D.S., SCHNITZER M. and SKINNER D.S. 1977. Mn(II) fulvic acid complexing equilibrium measurements by electron spin resonance spectrometry. Canadian Journal of Soil Science, 57, 47-53.

14. GEERING H.R. and HODGSON J.F. 1969. Micronutrient cation complexes in soil solution: III Characterisation of soil solution ligands and their complexes with Zn^{2+} and Cu^{2+}. Soil Science Society of America Proceedings, 33, 54-59.

15. GEERING H.R., HODGSON J.F. and SDANO C. 1969. Micronutrient cation complexes in soil solution IV. The chemical state of manganese in soil solution. Soil Science Society of America Proceedings, 33, 81-85.

16. GRIMME H. 1968. Die Adsorption von Mn, Co, Cu und Zn durch Goethit aus verdünnten Lösungen. Zeitschrift für Pflanzenernährung und Bodenkunde, 121, 58-65.

17. HALE M.G., FOY C.L. and SHAY F.J. 1971. Factors affecting root exudation. Advances in Agronomy, 23, 89-109.

18. HALE M.G., MOORE L.D. and GRIFFIN G.J. 1978. Root exudates and exudation. In Interactions between non-pathogenic soil micro-organisms and plants. Eds. Dommergues Y.R. and Krupa S.V. Elsevier, Amsterdam.

19. HASWELL S.J. Personal communication.

20. HODGSON J.F., LINDSAY W.L. and TRIERWEILER J.F. 1966. Micronutrient cation complexing in soil solution: II complexing of zinc and copper in displaced solution from calcareous soils. Soil Science Society of America Proceedings, 30, 723-726.

21. HUSSAIN S.S. and McKEEN W.E. 1963. Interactions between strawberry roots and Rhizoctonia fragariae. Phytopathology, 53, 541-545.

22. IRVING H. and WILLIAMS R.J.P. 1948. Order of stability of metal complexes. Nature 162, 746-747.

23. IVARSON K.C. and SOWDEN F.J. 1969. Free amino acid composition of the plant root environment under field conditions. Canadian Journal of Soil Science, 49, 121-127.

24. JAMES R.O. and HEALY T.W. 1972. Adsorption of hydrolyzable metal ions at the oxide-water interface I Co(II) adsorption on SiO_2 and TiO_2 model systems. Journal of Colloid and Interface Science, 40, 42-52.

25. LAKATOS B., TIBAI T. and MEISEL J. 1977. ESR spectra of humic acids and their metal complexes. Geoderma, 19, 319-338.

26. LINEHAN D.J., GOODMAN B.A. and McPHAIL D.B. Unpublished observations.

27. LINEHAN D.J. 1977. A comparison of the polycarboxylic acids extracted by water from an agricultural top soil with those extracted by alkali. Journal of Soil Science, 28, 369-378.

28. LINEHAN D.J. 1978. Polycarboxylic acids extracted by water and by alkali from agricultural top soils of different drainage status. Journal of Soil Science, 29, 373-377.

29. LINEHAN D.J. 1978. Humic acid and iron uptake by plants. Plant and Soil, 50, 663-670.

30. LINEHAN D.J. and SHEPHERD H. 1979. A comparative study of the effects of natural and synthetic ligands on iron uptake by plants. Plant and Soil, 52, 281-289.

31. LINDSAY W.L. 1972. Inorganic phase equilibria of micronutrients in soils. Micronutrients in Agriculture. Eds. Mortvedt J.J., Giordano G.M. and Lindsay W.L. Soil Science Society of America, Madison.

32. LINDSAY W.L. 1979. Chemical Equilibria in Soils. Wiley-Interscience. New York.

33. McBRIDE M.B. 1978.   Transition metal bonding in humic acid: An ESR study.   Soil Science, 126, 200-209.

34. McBRIDE M.B. 1981.   Forms and distribution of copper in solid and solution phases of soil.   In Copper in soils and plants.   Eds. Loneragan J.F., Robson A.D. and Graham R.D.   Academic Press, New York.

35. MANNING P.G. and RAMAMOORTHY S. 1973.   Equilibrium studies of metal-ion complexes of interest to natural waters VII.   Mixed-ligand complexes of Cu(II) involving fulvic acid as primary ligand.   Journal of Inorganic and Nuclear Chemistry, 35, 1577-81.

36. MARTELL A.E. and SMITH R.M. 1974.   Critical Stability Constants. Plenum Press, New York.

37. MARTELL A.E. 1975.   The influence of natural and synthetic ligands on the transport and function of metal ions in the environment. Pure and Applied Chemistry, 44, 81-113.

38. MATTIGOD S.V., SPOSITO G. and PAGE A.L. 1981.   Factors affecting the solubilities of trace metals in soils.   In Chemistry in the Soil Environment.   ASA special publication No. 40.   Ed. Stelly M. American Society of Agronomy and Soil Science Society of America. Madison.

39. NANNIPIERI P., PEDRAZZINI F, ARCARA P.G. and PIOVANELLI C. 1979. Changes in amino acids, enzyme activities and biomasses during soil microbial growth.   Soil Science, 127, 26-34.

40. NEWMAN E.I. 1978.   Root microorganisms their significance in the ecosystem.   Biological Reviews, 53, 511-554.

41. OLSEN C. 1930.   On the influence of humus substances on the growth of green plants in water culture.   Comptes rendus des travaux du Laboratoire Carlsberg, 18, 1-16.

42. PAUL E.A. and SCHMIDT E.L. 1960.   Extraction of free amino acids from soil.   Soil Science Society of America Proceedings, 24, 195-198.

43. PICCOLO A. and STEVENSON F.J. 1982.   Infrared Spectra of Cu^{2+}, Pb^{2+} and Ca^{2+} complexes of soil humic substances.   Geoderma, 27, 195-208.

44. RATNAYAKE M., LEONARD R.T. and MENGE J.A. 1978.   Root exudation in relation to supply of phosphorus and its possible relevance to mycorrhizal formation.   New Phytologist, 81, 543-552.

45. ROVIRA A.D. 1959.   Root excretions in relation to the rhizosphere effect IV.   Influence of plant species, age of plant, light temperature and calcium nutrition on exudation.   Plant and Soil, 11, 53-64.

46. ROVIRA A.D. and DAVEY C.B. 1974.   Biology of the rhizosphere.   In The Plant Root and its Environment.   Ed. Carson E.W.   University Press, Virginia.

47. ROVIRA A.D. and McDOUGALL B.M. 1967.   Microbiological and Bio-chemical aspects of the Rhizosphere.   In Soil Biochemistry.   Eds. McLaren A.D. and Peterson G.H.   Marcel Dekker, New York.

48. SANDERS J.R. 1982.   The effect of pH upon copper and cupric ion concentrations in soil solutions.   Journal of Soil Science, 33, 679-690.

49. SCHNITZER M. and HANSEN E.H. 1970.   Organo-metallic interactions in soils: 8.   An evaluation of methods for the determination of stability constants of metal-fulvic acid complexes.   Soil Science, 109, 333-340.

50. SEQUI P., GUIDI G. and PETRAZZELLI G. 1975.   Influence of metals on solubility of soil organic matter.   Geoderma, 13, 153-161.

51. SHUMAN L.M. 1977.   Adsorption of Zn by Fe and Al hydrous oxides as influenced by aging and pH.   Soil Science Society of America Journal, 41, 703-706.

52. SHUMAN M.S. and CROMER J.L. 1979. Copper association with aquatic fulvic and humic acids. Estimation of conditional formation constants with a titrimetric anodic stripping voltammetry procedure. Environmental Science and Technology, 13, 543-545.

53. STANTON D.A. and BURGER R. Du T. 1967. Availability to plants of zinc sorbed by soil and hydrous iron oxides. Geoderma, 1, 13-17.

54. STEVENSON F.J. 1976. Stability constants of Cu^{2+}, Pb^{2+} and Cd^{2+} complexes with humic acids. Soil Science Society of America Proceedings, 40, 665-672.

55. STEVENSON F.J. 1982. Humus Chemistry, Genesis, Composition, Reactions. Wiley-Interscience. New York.

56. STEVENSON F.J., KRASTANOV S.A. and ARDAKANI M.S. 1973. Formation constants of Cu^{2+} complexes with humic and fulvic acids. Geoderma, 9, 129-141.

57. TAN K.H. 1978. Formation of metal-fulvic acid complexes by titration and their characterisation by differential thermal analysis and infrared spectroscopy. Soil Biology and Biochemistry, 10, 123-129.

58. VANCURA V. 1967. Root exudates of plants III. Effect of temperature and "cold shock" on the exudation of various compounds from seeds and seedlings of maize and cucumber. Plant and Soil, 27, 319-328.

59. VANCURA V. and HOVADIK A. 1965. Root exudates of plants II. Composition of root exudates of some vegetables. Plant Soil, 22, 21-32.

60. VAN den BERG C.M.G. and KRAMER J.R. 1979. Determination of complexing capacities of ligands in natural waters and conditional stability constants of the copper complexes by means of manganese dioxide. Analytica Chimica Acta, 106, 113-120.

61. VEDY J.C. and BRUCKERT S. 1979. Soil solution: Composition and pedogenic significance. Chapter 8 in Constituents and properties of soils. Ed. Bonneau M. and Souchier B. Academic Press. London.

62. VINKLER P., LAKATOS B. and MEISEL J. 1976. Infrared spectroscopic investigations of humic substances and their metal complexes. Geoderma, 15, 231-242.

63. VUCETA J. and MORGAN J.J. 1978. Chemical modelling of trace metals in fresh waters: role of complexation and adsorption. Environmental Science and Technology, 12, 1302-1308.

64. WILSON S.A., HUTH T.C., ARNDT R.E. and SKOGERBOE R.K. 1980. Voltammetric methods for determination of metal binding by fulvic acid. Analytical Chemistry, 52, 1515-1518.

65. YOUNG S.D., BACHE B.W. and LINEHAN D.J. 1982. The potentiometric measurement of stability constants of soil polycarboxylate Cu^{2+} chelates. Journal of Soil Science, 33, 467-475.

66. ZUNINO H. and MARTIN J.P. 1977. Metal binding organic macromolecules in soil. 1. Hypothesis interpreting the role of soil organic matter in the translocation of metal ions from rocks to biological systems. Soil Science, 123, 65-76.

67. ZUNINO H. and MARTIN J.P. 1977. Metal binding organic macromolecules in soil. 2. Characterization of the maximum binding ability of the macromolecules. Soil Science, 123, 188-202.

68. ZUNINO H., PEIRANO M., AGUILERA M. and ESCOBAR I. 1972. Determination of the maximum complexing ability of water-soluble complexants. Soil Science, 114, 414-416.

CHAPTER 12

ORGANIC FARMING

J.W. PARSONS

Department of Soil Science, University of Aberdeen, Scotland.

CONTENTS

"You will have to work hard and sweat to make soil produce anything until
you go back to the soil from which you were formed. You were made from
the soil and you will become soil again" - Genesis 3:19.

1. INTRODUCTION

Agriculture by its very definition represents a disturbance of the
natural ecosystem through cultivations, the husbandry of single species
and the removal of nutrients from the soil, either in crops or animal
carcasses. The degree of disturbance is related to the intensity of the
system of husbandry practised. Traditional agricultural systems have
generally been based on mixed farming where animal wastes were returned
to the soil, and nutrients, exported from the farm, were replaced by
biological nitrogen fixation, or by importation of feeding materials for
stock, and of lime, phosphate and potassium for application to soils.
In such a system soil organic matter plays a crucial role and its
important characteristics, which create an environment in soil conducive
to healthy plant growth, have been reviewed in preceding chapters.
Modern intensive systems appear to pay less attention to the importance
of organic matter in soils, sometimes to their peril, and the increasing
use of chemicals, in the form of fertilizers and pesticides, has been
criticised from the standpoint of the increasing energy demands of the
system and the potential damage which may be inflicted on the environment.

2. HISTORICAL PERSPECTIVE

The practice of adding plant nutrients to soils goes back a long time
even though the reasons for doing so were not understood, but the
tradition no doubt developed from careful observations by enlightened

farmers. The Chinese[20] recognised the value of green manuring, probably
with legumes, on the growth of subsequent crops and the Romans applied
dung, marl and chalk to their crops. The four course rotation, the basis
of British agriculture for a very long period, included a legume and the
return of animal wastes to the soil. In the Middle Ages many English
land tenancy agreements included a clause requiring the tenant to herd
his flocks on the landlord's ground overnight.

An understanding of plant nutrition had to await the development of
the science of chemistry (see Introductory Chapter). It was Liebig, in
his writings and lectures in the middle of last century, who destroyed
the Humus Theory and convinced people that plants obtain their inorganic
constituents from the soil. He considered that nitrogen was absorbed from
the atmosphere as ammonia but this idea was quickly dispelled by Lawes
and Gilbert in their first systematic field experiments, which established
the importance of applying combined nitrogen to crops in the form of
ammonium sulphate. They obtained a direct relationship between dry matter
yield and nitrogen applied. Ammonium sulphate was a bi-product of coal
gas production and as supplies became insufficient to satisfy farmers'
requirements sodium nitrate was imported from Chile. At the time of
maximum production of Chilean nitrate in 1899, the UK was importing
112,000 t and in addition a further 65,000 t of ammonium sulphate was
used as fertilizer. This profligate use of nitrogen by farmers in the
production of cereals caused Sir William Crookes, in a lecture[10] in 1898,
to warn farmers that it would lead 'not only to a catastrophe little
short of starvation for the wheat eaters but indirectly to scarcity for
those who exist on inferior grains, together with a lower standard of
living for meat-eaters'. His predictions fortunately were never realized
because of the development in 1913 of the Haber-Bosch process for the
chemical fixation of atmospheric nitrogen.

The application of phosphate to soils in the form of bones is a very
old practice, but the first reliable accounts of their use date from the
middle of the 17th Century. They were applied whole or broken but later
grinding was observed to improve their fertilizer value and eventually
steamed bone flour was used. In the 19th Century the demand for bones
increased to the point where Liebig accused the British of ransacking the
battlefields of Europe to satisfy the demands of her farmers but it led
eventually to the development of the vast mineral deposits of phosphate

in North Africa as an alternative source. Rock phosphate is water
insoluble and not readily available to plants except in acid soils.
Experiments in the early 19th Century demonstrated the increased
solubility of phosphate resulting from sulphuric acid treatment and in
1842 Lawes patented the process and went on to prove its value in field
experiments. Production of superphosphate in the UK increased from
150,000 t in 1862 to 800,000 t in 1907 and, as demand for home produced
food in World War I grew, it was matched by increased use of nitrogen
and phosphorus fertilizers. At about this time criticisms of the use of
water-soluble, readily available forms of fertilizers were expressed and
the development of an uncoordinated 'organic' movement was born. Sir
Albert Howard who, as a comparatively new graduate, was appointed
Imperial Chemical Botanist to the Government of India in 1907 where he
developed the theory that healthy crops, resistant to infection, can
only be grown in soils where fertility is maintained by adequate supplies
of freshly prepared humus in the complete absence of inorganic fertilizers.
Over the next twenty years he developed the Indore process of making
compost from animal and vegetable waste products and preached the theory
of feeding the soil with humus rather than the plants with inorganic
salts.

At about the same time, Rudolph Steiner, the founder of the spiritual-
istic doctrine of Anthroposophy, gave a series of lectures to a group of
German farmers worried by the new developments in agriculture and from
this developed the biodynamic movement. Koepf[22] summarised the basic
concepts of the movement as:

(i) Sound farming and gardening techniques.

(ii) Diversification, recycling, excluding objectionable chemicals,
decentralised production and distribution.

(iii) Those which evolve from Steiner's spiritual teaching which
mould the method into a consistent whole.

The dynamic movement has spread from its origins in Germany to the
rest of Europe and the USA.

In the UK, a number of individuals were farming without the use of
inorganic chemicals in the 1930's and their faith in the importance of
soil organic matter was confirmed by the catastrophic loss of soils
through erosion in the wheatlands of America, later known as the dust
bowl. The most influential member of this group was Lady Eve Balfour who

published her book, The Living Soil[3], in 1943. She was a founder member
of the Soil Association organised to bring together all those interested
in the vital interrelationships between soil, plant, animal and man and
to assist in research and dissemination of information on this topic.
Similar associations developed in the USA, mainly on a regional basis, of
which the best known, partly through its publishing activities, is the
Rodale Press in Pennsylvania.

A large number of fragmented groups adhere to the precepts of organic
farming and although there are differences between them they agree on an
holistic approach to food production and maintain the view that a healthy
society must be based on a 'healthy' and fertile soil.

In recent years the term organic farming has been replaced by
biological agriculture which can be defined[17] as 'a system that attempts
to provide a balanced environment in which maintenance of soil fertility
and the control of pests and diseases are achieved by the enhancement of
natural processes and cycles, with only moderate inputs of energy and
resources while maintaining an optimum productivity'. The emphasis is
on the conversion to a low energy input system of agriculture rather than
complete rejection of the use of modern agrochemicals. Inevitably a
reduction in the energy input will reduce the quantities of fertilizer
and pesticide used and thereby minimise the threat to the environment.

3. THE PRESENT SITUATION

Agriculture, like most other industries, passed through a severe
depression in the 1930's, but increasing demand for home produced food
in the UK caused a rapid expansion in production in the 1940's and here
we can discern the beginning of the development towards our present
intensive agriculture, which is a feature of most developed countries'
economies today. Taking a simplified view, four developments can be
identified which promoted the intensification of agriculture:

(i) The availability of cheap energy and the development of substantial
and effective traction power have drastically reduced the time required
for both harvesting and cultivation operations so that continuous
production of high yielding winter-sown cereals is now a common feature,
even on heavy clay soils.

(ii) The availability of comparatively cheap, concentrated, water-soluble
fertilizers.

(iii) Breeding of new, high yielding crop varieties and particularly short strawed cereal cultivars which respond to large applications of nitrogen fertilizer.

(iv) The development of chemical herbicides, fungicides and pesticides.

Farming has become more specialised in recent years and even though there are still many mixed farms in the UK, individually producing a narrower range of products, there are also vast areas of arable land with no livestock associated with the farming enterprises. The argument is now much wider than the question as to whether the use of inorganic as opposed to organic manures is harmful to soils; equally important are the problems of efficiency of energy use and possible damage which may be done to the environment. In energy terms, use of pesticides may not be significant but they do represent a serious hazard to the environment if misused.

Much of the very impressive increase in crop yields, witnessed over the past thirty years, can be attributed to·the work of plant breeders. Jensen[19] partitioned the gain in productivity of wheat varieties over forty years into 49% arising from genetic improvement and 51% from improved technology.

I wish to restrict discussion here to the questions of energy use, soil fertility and soil structure, and the criticism, made by organic farmers, that conventional agriculture is mining the soil and thereby reducing its inherent fertility.

In a natural ecosystem an equilibrium is established in the cycling of nutrients between soil and plants but cultivations and removal of crops and/or animal products induce a deficit in the system. Failure to correct this deficit will diminish the fertility of the soil. In Table 1, the amounts of phosphorus and potassium removed from a mixed farm in the south of England (Bridget's Experimental Husbandry Farm)[13] are presented.

The amounts removed, particularly of potassium, are quite substantial but a proportion will be returned through animal wastes derived from the conserved grass and silage. Plants obtain phosphorus through mineralisation of soil organic phosphorus, recycling of animal and crop residues and from weathering of mineral phosphates. Any removal from soil must be made good by importing animal wastes on to the farm, importing animal feeding stuffs or applying phosphate fertilizers. Recycling organic wastes is very important and must be encouraged wherever possible but a

Table 1. Amounts of phosphorus and potassium removed in crops.

Product	Yield ha^{-1}	P kg ha^{-1}	K
Grass - grazed	11 200 l milk	10	20
- conserved	7.5 t	24	208
Wheat - grain + straw	4.6 t	6	18
Barley - grain + straw	4.3 t	5.5	17
- whole silage	8 t	10	104
Potatoes	30 t	23.5	150
Peas	3.1	12	26

serious geographical imbalance in their distribution means that some soils receive too much and others too little. I will return to this later but inevitably some deficit must be made good by additions of mineral phosphates. Organic farmers generally accept ground mineral phosphate as a fertilizer because it is water insoluble and therefore not immediately available to plants. Phosphorus added in soluble forms, such as triple superphosphate or ammonium phosphates, rapidly becomes 'fixed' in soils in forms of variable availability to plants so that leaching of excess is not a problem, although small amounts may be lost through surface run-off.

World reserves of phosphorus have been estimated to last another thousand years[27] and, as little is lost from soils and a comparatively small amount of energy goes into the production of phosphorus fertilizers, the only strong argument for reducing amounts applied to soils is to reduce farmers' production costs. Cooke[7] has calculated a small positive balance for the phosphorus budget in UK agriculture as a whole but this has been achieved through a substantial input of fertilizer phosphorus. Little data is available for the comparison of nutrient budgets in conventional and organic systems but Lockeretz et al.[23] in a survey of fourteen paired farms reported a positive phosphorus balance on the conventional farms but a deficit of about 5 kg P ha^{-1}y^{-1} averaged over the organic farms.

The story for potassium is similar except that the soil supply originates from weathering of minerals only and potassium, recycled through animals, is found almost exclusively in animal urine, rather than in the faeces, and is therefore water soluble. This makes the return

to grazed grassland patchy. Almost the sole form of potassium fertilizer
is potassium chloride although the more expensive potassium sulphate is
sometimes used. Both forms dissolve in water and for this reason many
organic farmers prefer to use a slow release potassium mica, glauconite.

Cooke[7] suggests there is a massive imbalance between input and output
of potassium in UK agriculture, made good by the weathering of soil
reserves, which can be quite substantial in some soils, and the recycling
of animal and crop wastes. In the survey by Lockeretz et al.[23] the
conventional farms were in balance but the organic farms were removing
34 kg ha^{-1} y^{-1} more potassium from the soil than they were returning.
The organic farmers were therefore relying on the soil's capacity to
supply this deficit. Leaching losses of potassium are small and generally
there is no evidence to indicate a serious wastage of potassium fertilizers
on conventional farms.

Nitrogen warrants special consideration because it is the element
which, for many reasons, forms a major bone of contention between
conventional farmers on the one hand and organic farmers and environ-
mentalists on the other. Conventional agriculture is described as a high
input/high output system but, in terms of overall energy use, output to
the farm gate accounts for only 3.9% of total UK energy consumption;[34] food
processing, distribution, preparation and storage accounts for a further
11.9%. Similar figures are reported for the USA.[28] 29.7% of the energy
consumed in agricultural production, to the farm gate, is used in the
manufacture of fertilizers[34] and nitrogen fertilizers account for 91% of
that value. Further energy will of course be consumed in the application
of fertilizers to crops. Consequently one area where savings of energy
might be looked for is by reduction of nitrogen fertilizer use. The saving
may not be massive in UK terms but any decrease will diminish farmers'
production costs; nitrogen fertilizer accounts for about 6% of farm
expenditure on inputs. The reverse of this situation has in fact been the
trend over forty years as illustrated in Table 2. There has been a twenty
two-fold increase in the amount of nitrogen fertilizer used over that
period. Much of this increase has been applied to grass but in recent
years there has been a significant increase in the amount applied to
arable crops.

Table 2.Use of fertilizers in the UK from 1939 to 1980[18].

| | N | P | K |
		kt	
1939–40	61	75	63
1979–80	1268	192	369

Crops obtain their nitrogen from rainfall and dry deposition; through mineralization of soil organic reserves; from the recycling of animal and crop residues; from fertilizer applications and through biological nitrogen fixation. Of all the plant nutrients, nitrogen produces the greatest response in terms of growth but it is also the element most liable to loss from the soil system by a variety of pathways. Ammonia is lost by volatilization at soil pH values greater than 7, particularly when urea is used as a surface application; nitrate is leached from well-drained, aerated soils at times of high rainfall and gaseous losses occur by denitrification under anoxic conditions. All forms represent a serious economic loss to the farmer but nitrate leaching in particular causes a threat to the environment through eutrophication of rivers and lakes and as a possible danger in drinking water supplies. A concentrated effort has been made in recent years to obtain quantitative estimates of nitrogen losses under different farming systems and gradually the picture is becoming clearer. For winter-sown cereal crops the system is most vulnerable to leaching losses in the early stages when there is insufficient plant cover to hold nitrate. Nitrate, formed in the warm soil following harvesting of the previous crop, is flushed out by the autumn and early winter rainfall. The second period is during heavy spring rainfall following fertilizer application; this is also the danger period for spring-sown crops. The amount of nitrate lost by leaching is extremely variable and depends on the amount and timing of rainfall at these two critical periods. During crop growth, evapotranspiration exceeds precipitation so that the drains run only in periods of excessive summer rainfall. Nitrate from soil finds its way via the drains to river systems and, in the predominantly cereal growing areas of south-east England, water authorities have monitored a progressive increase in nitrate content of river water over the last twenty years[24].

Denitrification losses are much more difficult to monitor but they appear to be at a maximum in spring during rainfall following fertilizer application.

Loss of nitrogen from soils raises two important questions:

(i) Can losses be reduced under a conventional system of agriculture?

(ii) Does the use of organic manures reduce losses of nitrogen?

Farmers and research scientists are very much aware of the need to conserve nitrogen and, as we begin to identify the pathways of loss, methods for plugging the gaps are being sought. The first essential is to develop a satisfactory method for estimating the crop's requirements for nitrogen. Mathematical models, derived from data collected over a period of years, and which can integrate information on the crop, soil and climatic conditions to predict the crop's requirements, look the most promising means of achieving this end. Unfortunately, even the most sophisticated models will be invalidated by the vagaries of the weather. In the case of winter-sown cereals, there are a number of practical steps which can be taken to reduce losses. Early establishment of the crop will provide sufficient root growth to reduce nitrate leaching and this may be achieved by more careful timing and placement of fertilizer and possibly by the use of an ammonium based fertilizer. Unfortunately, the farmer is faced with the problem of having to work over a very short time period when weather conditions may be unsuitable and so prevent him from optimising all the operations required.

There is no straightforward answer to the second question because two very different systems of production are being compared, one which attempts to maximise yields by tight control of the variables of the system, the other one dependent on lower energy inputs which necessarily accepts lower yields. Many people would argue that, at a time when the developed countries are overproducing in a number of agricultural commodities, we should accept lower yields. In the context of the developed countries this may be an acceptable argument but on a world basis it is not. On the assumption that we are willing to accept a lower level of production can we meet the nitrogen requirements of crops without the use of nitrogen fertilizers? Animal wastes, as discussed earlier, are produced in areas of higher rainfall and on the heavier soils which are not conducive to cereal production. The Royal Commission on Environmental Pollution[30] recommended the desirability of examining the benefits to be derived from a 'manure bank' system which operates in the Netherlands where farmers are encouraged to import animal waste on to their farms

with the help of government subsidised transport. No action has been taken
in this country since the Committee reported in 1979, possibly because of
the longer distances and high transport costs involved. Gostick[14] has
calculated the total nitrogen produced in animal excreta in the UK in
1980 to be 975 kt. This figure compares with 1268 kt of nitrogen
fertilizer used in the same year and represents 44% of the total nitrogen
consumed, excluding that contributed by mineralisation of soil organic
nitrogen. As an estimate of the amount of nitrogen available to crops it
is a gross overestimate because losses from storage and spreading of
animal wastes can be as high as 50%, mainly in the form of ammonia, and
much of the nitrogen in animal excreta is in organic forms and is not
immediately available to plants. Table 3 summarises the composition of
various forms of animal excreta and provides a comparison of the availability of the nitrogen with that of an inorganic source.

Table 3. Mean dry matter and nitrogen content of animal excreta and its
efficiency compared to inorganic nitrogen[14].

	% Dry matter	%N	% Efficiency of inorganic-N
Cattle, farmyard manure	25	0.6	25
Pig, farmyard manure	25	0.6	25
Poultry, broiler litter	70	2.4	60
Battery, hen droppings	70	4.2	60
Cattle slurry	10	0.5	50
Pig slurry	10	0.6	65

A high proportion of the animal waste will have been recycled on to
grassland by the grazing animal and we must therefore conclude that, even
if an equitable system of distribution of animal wastes was available in
this country, there would certainly be insufficient to maintain present
yields of arable crop production.

Yields of wheat grain from the continuous wheat plots on Broadbalk,
Rothamsted, averaged 5.85 t ha^{-1}, over the period 1970–78, on the plot
supplied with 35 t FYM ha^{-1} y^{-1}. The yield was 0.5 t ha^{-1} more than the
maximum achieved with fertilizers, estimated from a response curve[12].
The difference between yields in the presence and absence of FYM is

difficult to explain but a very substantial amount of FYM has been added to those plots over a period of one hundred years and the soil organic matter content has almost doubled in that time. This may confer improved soil physical characteristics on the soil which in turn enhance plant growth, particularly in dry years. A yield of 5.85 t ha^{-1} is very respectable and compares very favourably with the UK average winter wheat yields of 5.23 t ha^{-1} in 1979 and 5.88 t ha^{-1} in 1980. We know very little about loss of nitrogen from soil where large annual additions of FYM have been made, but over the period 1852 to 1982, in the Broadbalk experiment, precentage recovery of added fertilizer nitrogen was twice the value of that recovered from FYM. Over that period, 15% of the nitrogen applied as FYM remained in the soil so that a simple calculation shows that more nitrogen was lost from the FYM plots, through leaching and denitrification, than was actually added to the plots receiving fertilizer[21]. This represents a very substantial loss of nitrogen.

An alternative source of nitrogen is biological fixation of atmospheric nitrogen. Legumes constitute the largest contributor but free living bacteria and blue-green algae may be responsible for smaller inputs. Forage legumes are the most important but they are cycled through the animal to produce animal protein, which is not only inefficient but of little value in an all-arable system. We do not at present have a suitable pulse crop available which can be introduced into an arable system, both to increase nitrogen input and to provide a break between cereals. Even if one were available, the input of nitrogen from one year's growth would be small.

Green manuring, the growth of a crop for the specific purpose of incorporating it into the soil to improve soil structure and nitrogen supplying power, is an ancient agricultural practice but it is rarely used in modern intensive systems. There are many advantages associated with the practice[26] but it is difficult to see how a break of this kind can be fitted into a modern system where timing of cultivations is critically important.

In grassland systems white clover is rarely a major constitutent in swards nowadays because of the large increase in nitrogen fertilizer applications. Cowling[9] has estimated that the amount of nitrogen fixed annually in grassland is about 11 kg N ha^{-1} compared to a potential value of 150 - 200 kg N ha^{-1} in the UK. The great advantage to farmers of fertilizer nitrogen over legume nitrogen is the flexibility in management provided by the former in providing increased production early and late in the season. Cooper[8] denounced the conflict between fertilizer and legume nitrogen in grassland production and argues that the two can be used in a complementary way. He suggested that grass production on one part of the farm should be maintained with fertilizers and no legumes and on the remaining area with legumes but no fertilizer.

Nitrogen fertilizer applications to grassland have increased dramatically over the last thirty years. For many years nitrogen losses from grassland were considered to be minimal. In some of the earliest lysimeter experiments[16], measured loss of nitrate by leaching was almost negligible, even where 100 kg N ha^{-1}, as ammonium sulphate, was applied, an extremely heavy dressing in 1931. Estimates of leaching losses of 2 to 11% of the nitrogen added have been made in more recent ^{15}N lysimeter studies[11] and in field experiments leaching losses were small for nitrogen applications of 250 kg N ha^{-1} but increased with additions above that value[6]. In all these experiments the grass crop was cut and removed but, in more recent work on systems where the grazing animal was included, this happy situation no longer holds apparently. Approximately 75 to 95% of the nitrogen consumed by the grazing animal may be returned directly to grassland in the form of faeces and urine. Urine contains urea and Whitehead[35] has estimated that 14% of the area grazed by cattle will receive urine in one year, possibly receiving as much as 576 kg N ha^{-1} in a localised area. Measurements of ammonia volatilization, denitrification and leaching from urine affected patches in New Zealand[4] and the UK[5] indicate substantial losses from the system. In addition much of the animal excreta produced during winter, when animals are housed indoors, is returned to grassland but the losses during storage and spreading by ammonia volatilization and winter surface run-off are high. Quantitative data for losses of this kind are difficult to obtain.

4. SOIL STRUCTURE

In addition to its role as a reservoir of plant nutrients, soil organic matter makes a substantial contribution to the formation of a good soil structure and to the stability of aggregates. Organic materials, mainly of microbial origin, bind primary particles together to form aggregates sufficiently stable to withstand the impact of falling rain

drops and the movement of animals and machinery over the soil surface.
Cultivation encourages the oxidation of organic components and in a
system where no organic material, other than roots and stubble of the
previous crop, are returned to the soil, oxidation may exceed addition
and organic content will decline. The amounts of roots and stubble
returned are influenced by the addition of nutrients to the crop and
where heavy applications of nitrogen are made the actual reduction in
soil organic content may be small. Obviously, the practice of straw
burning now prevalent in areas of intensive winter cereal production,
because of the very short period of time available for cultivations
between harvest and sowing the following crop, will seriously reduce the
amount of organic matter returned and in the long term may lead to a
reduction in soil organic content. In the climatic conditions of UK, soil
organic contents have remained relatively constant even under intensive
cultivation. Strutt[31] in his report on 'Modern Farming and the Soil'
suggested that organic matter contents of 3% or less may be satisfactory
for a long period of cereal production on stable soils but they are
inadequate for crops on unstable soils, particularly those high in sand
or silt. Greenland et al.[15] report that non-calcareous soils with organic
contents below 3.4% are liable to structural deterioration whereas soils
containing 4.3% or more may be considered to be very stable.

In mixed farming systems in this country there is no problem of
declining organic contents as the period under grass restores the organic
material, but in all-arable systems the problem becomes much more acute
and there are increasing signs of structural damage and erosion problems
on some soil types, although many of these problems have resulted from
poor management rather than through a reduction of organic matter per se.

5. COMPARISON OF LEVELS OF PRODUCTION

Increasing concern over problems introduced through intensification of
agriculture has stimulated interest in alternative systems of production
and has encouraged a number of comparative surveys to be made of organic
and conventional farming. Comparisons of this type are particularly
difficult to make, not only because of the diversity of farm products and
the fact that organic farms in general tend to be small family units, but
organic systems usually evolve as an integral part of the farmer's

philosophy of life and they are constricted by his relationship with his land.

Lockeretz et al.[23] compared the economic performance and energy consumption of fourteen organic farms, paired as closely as possible with an equal number of conventional farms, in the corn belt of the USA. Overall production per unit area of land was highest on conventional farms but, because of lower production costs on the organic farms, the profitability of the two systems was similar. Energy consumption on the organic enterprises was just over a third of that on conventional farms. However, as mentioned earlier, the authors expressed concern over the depletion of soil reserves of phosphorus and potassium on the organic farm.

An extensive survey of organic farms in England and Wales, assessing the future role of organic farming, has been published by Vine and Bateman[33]. They also report a lower production per unit area of land on average from organic farms but they are careful to stress that a number of enterprises maintained levels of production comparable to conventional farms by substantial imports of nutrients in the form of farmyard manure and animal food. Incomes of organic farmers are generally lower than can be earned through conventional agriculture but, when the former can benefit from premium prices on their products, the economic returns may be high. Vine and Bateman[33] confirm the finding that energy consumption per unit area is lower on organic farms but performance, in terms of output per unit of energy used, was similar in the two systems. They conclude that, contrary to general belief, organic farms do not employ more labour than their conventional counterparts.

A report produced by a task force of twenty-four scientists for the Council for Agricultural Science and Technology[25] in the USA suggests that a shift from conventional to organic farming will reduce crop yields by 15 to 25% and to offset this reduction, an extra 18 to 33% of the same class of land would need to be brought into production. Overall, the reduction in yield would be even greater because of the transfer of land from grain to legume production to supply nitrogen to the system.

Large sums of government money invested in research in conventional farming systems and the dissemination of the results through the advisory services have produced spectacular increases in agricultural production. However, research in organic farming methods has been underfinanced and

left to a small number of private institutions where the standard of
scientific experimentation has not always been as rigorous as one might
wish. A shift in attitudes towards alternative systems can now be detected
and in 1979 the USDA commissioned a team of scientists to make a study of
organic farming in the USA and Europe[29]. They produced recommendations on
new lines of research, the establishment of educational programmes and
the encouragement of the Extension Service to provide information and
advice on organic farming. In 1982, a bill promoting research into
organic-based systems was approved by a committee of the House of
Representatives in the USA but it was never enacted. The USDA indicated
sympathetic support for the bill but it was unwilling to provide the
finance required out of its over-stretched budget.

6. THE FUTURE

Agricultural production in the developed countries has reached a
watershed. Yields of agricultural products have continued to rise, as a
result of increasing inputs of chemicals and energy, to a point where
we are now producing more than can be consumed, but at a price which
developing countries cannot afford. Farmers have been continually
encouraged by governments and forced by economics to strive for maximum
production per unit area of land and, however one may view their achieve-
ments (see Table 4), the results are remarkable. Having reached self-
sufficiency in a number of agricultural products farmers are now being
criticised for over-production and their retort is 'don't criticise
farmers with a full mouth'.

Table 4. Changes in areas and yields of wheat and barley over the period
         1975 to 1983[2] in the UK.

	1975	1983	Diff.	% increase
Area 000's ha				
Wheat	1034	1691	651	63.5%
Barley	2354	2151	-194	-8.3%
Yield 000's t				
Wheat	4488	10677	6189	138.0%
Barley	8513	9957	1146	13.5%

Unfortunately we do not understand how to cut back production except by the crude controls of price reduction and land quotas. The argument is advanced that now is the opportune time to change from a high to a low input system, but if we achieved that overnight we would face a traumatic drop in food production. Any change of this type on a large scale must be carefully planned and researched. The change from a conventional to an organic system requires a transition of three to five years during which weeds, pests and diseases become a serious problem and yields are seriously reduced.

The comparison is always made between conventional and organic farming systems, the two extremes of the spectrum, when we should perhaps be thinking in terms of a change to an alternative lower input system. But how do we make this change and which system do we choose? This is an area where we need much more information and a lot more research, not just of the type where we test scientific hypothesis, but research which includes measurements of the economic and energy efficiencies of the system under test. Farmers will not make the change of their own accord, they will either be forced into it by reduced prices of agricultural products or by tax incentives or disincentives.

There are two ways of looking at the problem, one is in terms of saving energy, the other is in terms of a reduction in nitrate pollution of rivers. Swanson[32] has reviewed some of the linear programming models, developed in the USA on a national and regional basis, to examine the affects of reduced nitrogen fertilizer use on nitrate leaching. There is a large geographical variation, but in almost all cases the decrease in fertilizer application required caused a substantial increase in food prices. The same models can be used to predict changes in fertilizer application to reduce crop yields. Control of pollution might be achieved by placing an excise tax on nitrogen fertilizers, or by providing a market rights issue to use fertilizer, for which farmers would bid or by a pollution tax on the producer[32]. No attempt has been made to use these fiscal controls in practice.

To be realistic, the difference between our present over-production and self-sufficiency is not very large and the drop in production required to achieve self-sufficiency and no more, is small. Even if we could solve the problems of weeds, pests and diseases by cultivations, rotations and the use of immune varieties, a substantial quantity of nitrogen will still be required to provide the food we need and considerably more nitrogen than can be supplied as animal wastes, even if the geographical problems of supply were solved. Nevertheless, we must make every effort to recycle animal wastes, including sewage sludge, on to our soils as efficiently as possible. To provide the nitrogen required, by biological fixation, means taking land out of cash crop production or bringing new land into production. Much more research is required to produce a legume for a temperate climate and which is suitable as a source of protein for human consumption. The equivalent of soya bean is required which will grow in a cool temperate climate. Excellent as this goal may appear, it is most unlikely that high yields can be maintained through nitrogen fixation alone, unless the problem of introducing the nitrogenase system into cereal roots is solved. In one step we would then achieve a reduction in yield, due to diversion of photosynthate to nitrogen fixation, and a reduction in nitrogen requirement of the crop.

Two other criticisms made by the organic farming community are that modern agriculture is mining and destroying our primary resource, the soil, and that the quality and flavour of agricultural products have declined and are inferior to those produced by organic methods. In answer to the first, the long term fertility experiments in temperate countries have failed to indicate any deterioration in soil structure resulting from fertilizer use. Problems do occur as a result of the tightness of a farmer's timetable and the necessity of taking heavy equipment on to land under adverse climatic conditions. The question of flavour of food is very subjective but there is little evidence to suggest a decline due to fertilizer use. Inevitably changes in flavour, possibly adverse, will result from breeding new varieties for maximum production. On the question of nutritional quality we lack information and much more research is required on this topic and in particular on the trace element content and balance of human food. Allaway[1] sums up the present situation in his statement that 'the use of fertilizers has been just one of a complex series of agricultural practices which have permitted people in the developing countries to exchange the nutritional problems of the hungry for the nutritional problems of the overfed'.

In the developed countries, with static population size, there is an argument for improving the efficiency of agricultural production but, until we can solve the economic problems of distributing excess food to third world countries, there is no argument for increasing production. Our problems pale into insignificance by comparison with those of the developing countries. World population almost tripled between 1900 and 1980 to over 4300 millions and it is expected to reach over 6000 millions by the year 2000. A significant section of the world's population is under-nourished at the present time, but to provide food for so many extra mouths requires an increase in productivity in the developing countries similar to that experienced here over the last twenty to thirty years. They are faced with the problems of expensive energy supplies, the dangers of erosion and climatic limitations to production. The introduction of new rice and cereal varieties, which respond to nitrogen fertilizers, have already provided impressive increased productivity but organic methods as such cannot supply the nutrients needed to sustain that level of production and fertilizers must be used, even on a limited basis.

The alternative to increasing productivity on established farm land is to bring new land into production. We have already witnessed the clearance of tropical rainforest in some parts of the world to establish new agricultural land but this has long term implications, not least of which are the problems of destruction of soil structure and loss of soils through erosion.

We need to exploit the tropical legumes as an alternative source of nitrogen and to encourage the work in progress on agro-forestry. Agro-forestry is an alternative system of food production which will appeal to the supporters of biological agriculture, although it does not preclude the use of fertilizers. Food crops are grown in between rows of trees and nutrients are recycled via the deep tree roots to the leaves which fall on to the soil surface to give cover and provide nutrients. The wood can eventually be used as fuel.

In developed countries the proportion of farmers using organic methods of production is very small and in the present economic climate and on a world basis it is likely to remain so. A shift towards lower input systems in the developed countries is a real possibility over the next five to ten years and the development of new systems based on low energy inputs in third world countries is almost a necessity. To make these changes we need well planned and carefully executed research programmes to provide the information on which changes of this type and magnitude can be based.

7. REFERENCES

1. ALLAWAY W.H. 1975. The effects of soils and fertilizers on human and animal nutrition. United States Department of Agriculture. Information Bulletin No. 378. Washington: U.S. Government Printing Office.
2. ANNUAL REVIEW OF AGRICULTURE 1980 and 1984. COMMAND PAPERS 7812 and 9137. Ministry of Agriculture, Food and Fisheries, HMSO, London.
3. BALFOUR E.B. 1943. The living soil. Faber and Faber, London.
4. BALL R., KEENEY D.R., THEOBOLD P.W. and NES R. 1979. Nitrogen balance in urine-affected areas of New Zealand pastures. Agronomy Journal, 71, 309-314.
5. BALL R. and RYDEN J.C. 1984. Nitrogen relationships in intensively managed temperate grassland. Plant and Soil, 76, 23-33.
6. BARRACLOUGH D., HYDEN M.J. and DAVIES G.P. 1983. Fate of fertilizer nitrogen applied to grassland I. Field leaching results. Journal of Soil Science, 34, 483-497.
7. COOKE G.W. 1977. Waste of fertilizers. Philosophical Transactions of the Royal Society of London. Series B, 281, 231-241.
8. COOPER M.McG. 1970. Group Report, 6th Symposium of British Grassland Society, Belfast, 273-275.
9. COWLING D.W. 1982. Biological nitrogen fixation and grassland production in the United Kingdom. Philosophical Transactions of the Royal Society of London, Series B, 296, 397-404.
10. CROOKES W. 1905. The wheat problem. The Chemical News, London.
11. DOWDELL R.C. and WEBSTER C.P. 1980. A lysimeter study using nitrogen-15 on the uptake of fertilizer nitrogen by perennial ryegrass swards and losses by leaching. Journal of Soil Science, 31, 65-75.
12. DYKE G.V., GEORGE B.J., JOHNSTON A.E., POULTON P.R. and TODD A.D. 1982. The Broadbalk wheat experiment 1968-78: Yields and plant nutrients in crops grown continuously and in rotation. Rothamsted Annual Report, Pt 2, 5-44.
13. FRANCIS A.L. 1969. Fertilizing Bridget's in the 70's. Bridget's Experimental Husbandry Farm, Annual Report No. 10, Ministry of Agriculture, Fisheries and Food.
14. GOSTICK K.G. 1982. Agricultural Development and Advisory Services (ADAS) recommendations to farmers on manure disposal and recycling. Philosophical Transactions of the Royal Society of London, Series B, 296, 329-332.
15. GREENLAND D.J., RIMMER D. and PAYNE D. 1975. Determination of the structural stability class of English and Welsh soils, using a water coherence test. Journal of Soil Science, 26, 294-303.
16. HENDRICK J. and WELSH H.D. 1932. Intensive manuring of grassland. Transactions of the Highland and Agricultural Society of Scotland, 44, 86-96.
17. HODGES R.D. 1981. An agriculture for the future. Biological Husbandry. Ed. Stonehouse B. 1-14, Butterworths, London.
18. HOOD A.E.M. 1982. Fertilizer trends in relation to biological productivity within the U.K. Philosophical Transactions of the Royal Society of London, Series B, 296, 315-328.

19. JENSEN N.F. 1978. Limits to growth in world food production. Science, 201, 317-320.
20. JIAO B. 1983. Utilization of green manure in raising soil fertility in China. Soil Science, 135, 65-69.
21. JOHNSTON A.E. 1984. Private Communication.
22. KOEPF H.H. 1984. The principles and practice of biodynamic agriculture. Biological Husbandry. Ed. Stonehouse B. 237-250, Butterworths, London.
23. LOCKERETZ W., KLEPPER R., COMMONER B., GERTLER M., FAST S. and O'LEARY D. 1976. Organic and conventional crop production in the corn belt: A comparison of economic performance and energy use on selected farms. Center for the Biology of Natural Systems, ST. Louis, Missouri, U.S.A.
24. MARSH J.J. 1980. Towards a nitrate balance for England and Wales. Water Services, 84, 601-606.
25. ORGANIC AND CONVENTIONAL FARMING COMPARED. 1980. Council for Agricultural Science and Technology, Report No 82, Ames, Iowa, U.S.A.
26. PARSONS J.W. 1984. Green manuring. Outlook on Agriculture, 13, 20-23.
27. PHOSPHORUS: A RESOURCE FOR UK AGRICULTURE. 1978. Centre for Agricultural Strategy, Report No 2, Reading, England.
28. PIMENTAL D., BERARDI G. and FAST S. 1983. Energy efficiency of farming systems: Organic and conventional agriculture. Agriculture, Ecosystems and Environment, 9, 359-372.
29. REPORT AND RECOMMENDATIONS ON ORGANIC FARMING 1980. United States Department of Agriculture, Washington, U.S.A.
30. ROYAL COMMISSION ON ENVIRONMENTAL POLLUTION 1979. Agriculture and pollution. Seventh Report, HMSO, London
31. STRUTT N. 1970. Modern farming and the soil. Agricultural Advisory Council, HMSO, London.
32. SWANSON E.R. 1982. Economic implications of controls on nitrogen fertilizer use. Nitrogen in agricultural soils. Ed. Stevenson F.J. American Society of Agronomy No 22, Madison, Wisconsin, USA.
33. VINE A. and BATEMAN D. 1981. Organic farming systems in England and Wales: Practice, performance and implications. University College of Wales, Aberystwyth.
34. WHITE D.J. 1980. Agriculture - energy user and producer. Proceedings of a Conference on Energy from farm waste - towards reality. National Agricultural Centre, Stoneleigh, Warwick.
35. WHITEHEAD D.C. 1970. The role of nitrogen in grassland productivity. Commonwealth Bureau of Pastures and Field Crops, Bulletin No 48, Hurley, Berkshire.

LIST OF ABBREVIATIONS

ABA	Abscisic acid
ADP	Adenosine diphosphate
ADP:O	Adenosine diphosphate : Oxygen ratio
APS	Adenosine 5'phosphosulphate
ATP	Adenosine triphosphate
CEC	Cation Exchange Capacity
Cyt	Cytochrome
2,4–D	2,4 –Dichlorophenoxyacetic acid
DNA	Desoxyribose nucleic acid
DNP	2,4–Dinitrophenol
DOPA	3,4–dihydroxyphenylalanine
EDTA	Ethylenediamine tetraacetic acid
FA	Fulvic acid
FYM	Farmyard manure
GA_3	Gibberellic acid
HA	Humic acid
HS	Humic substances
IAA	Indole–3–acetic acid
IAA oxidase	Indole–3–acetic acid oxidase
Km	Michaelis constant
MRT	Mean Residence Time
MW	Molecular Weight
NAD	Nicotinamide adenine dinucleotide
NMR	Nuclear Magnetic Resonance
PAPS	Adenosine 3'–phosphoadenosine–5'–phosphosulphate
Po	Organic phosphate
Pi	Inorganic phosphate
PMA	Polymaleic acid
PVP	Polyvinyl pyrrolidone
RNA	Ribose nucleic acid
R.Q.	Respiratory Quotient
SOM	Soil organic matter
V (max)	Maximum velocity of reaction

INDEX

Abbreviations used: FA – fulvic acid, HA – humic acid, HS – humic substances, FYM – farm yard manure, SOM – soil organic matter.

448

452

458

Manganese 61, 127, 165, 240, 389, 405,
407, 410, 412–413, 417
Mannosamine 294
Mannose 206, 268, 278, 294
Manures, (see Farm Yard Manure and
Fertilizers)
–   , green 301, 311–312, 425, 434
Mass spectrometry 296
Meadow fescue (*Festuca elatior* L) 411
Mean Residence Time (MRT) 5, 21–22
Medic (*Medicago* spp.) 237
Membranes 41, 54, 61–62, 92, 97, 134,
386, 411
Mercaptans 193
Metabolic inhibitors 60–61, 83, 87, 89,
167, 227
Metals, complexing 46–47, 407–416
–   , hydrolysis 406
–   , ligands (see chapter 11)
–   , precipitation 47–48, 405
–   , toxicity 26, 47
–   , uptake into plants 58–61, 413,
415–417
Methacrylate gels 302
Methane 126
Methanol in soils 167
Methionine 156, 158, 293, 296, 383, 411
Methionine sulphone 293
Methylcytosine 294
Mevalonate 159–160, 164
Mexacarbate 269
Microbial, biomass (see chapters 6 & 9)
128, 138–139, 188, 197, 199,
213, 294, 312, 355
–   , degradation 153
–   , growth 52–55, 80, 197
–   , inoculants 164
–   , numbers 236
Microcalorimetry 231
Mildew 168
Millet (*Setaria* spp.) 344
Mineral, phosphates 428–429
–   , *Theory* 3, 40–41
Mineralization, biological 346, 386
–   , C 242, 244, 246, 350, 356
–   , N 26, 229, 242, 244, 246,
291, 305–308, 312–313,
316, 346, 350, 356,
384–386, 431, 433
–   , P 26, 230, 245–246,
341–363, 428

Mineralization, S 26, 190, 230, 346, 352,
384–386
Mitochondria 56, 82–83
Mnium affine 62
Model, enzyme complexes 200, 213
–   , HS 14, 51, 84, 94, 96–97
Modelling, C 2, 20–23
–   , for aggregation 352–353
–   , N 291, 314–317, 432
–   , P 352, 360, 364
–   , S 380–382, 386
Moisture content of soil 25, 176, 349–350,
385, 389
Molecular weights 15, 17, 22, 56–57, 195,
300, 335–336, 383
Monoamines in soils 295
Mononucleotides 294
Montmorillonite 25, 61, 127, 267, 273,
300
Mor humus 5
Moraxella 126
Mucigel 269, 272, 275, 279, 311, 344
Mucopeptides 293
Mucopolysaccharides 293
Mucor hiemalis 158
Mucoraceous fungi 164
Mull, coal 7
–   , humus 5
Mungbeans (*Vigna radiata*, formerly
Phaseolus aureus) 84, 136
Muramic acid 231–232, 277, 294
Mushroom compost 233
Mustard (*Sinapsis alba*) 43, 58
Mycorrhiza 162–163, 240–241, 346
Myricetin 139

Nematodes 224, 246, 342–343
Nickel 193, 407
Nicotiana tabacum 21, 43–44, 51, 58–60,
380
Nicotinamide adenine dinucleotide phos-
phate (NAD) 82, 92–93, 185
Ninhydrin 292
Nitella gracilis 62
Nitrate, 41, 246, 277, 291, 296–297,
306–308, 310, 341, 351, 385,
391, 431–432, 439
–   , respiration 126
Nitrification 139, 240, 306–307, 310,
316, 351